1st International Conference on Efficient Building Design 2014

Materials and HVAC Equipment Technologies

Beirut, Lebanon
2-3 October 2014

ISBN: 978-1-5108-3462-0

Printed from e-media with permission by:

Curran Associates, Inc.
57 Morehouse Lane
Red Hook, NY 12571

CURRAN ASSOCIATES INC. proceedings.com

Some format issues inherent in the e-media version may also appear in this print version.

Printed by Curran Associates, Inc. (2017)

ISBN: 978-1-5108-3462-0
ASHRAE ISBN: 978-1-936504-84-8 (USB Flash Drive)

American Society of Heating, Refrigerating and Air-Conditioning Engineers (ASHRAE)
1791 Tullie Circle
N.E. Atlanta, GA 30329
USA

Phone: (404) 636-8400
Fax: (404) 321-5478

ashrae@ashrae.org

Additional copies of this publication are available from:

Curran Associates, Inc.
57 Morehouse Lane
Red Hook, NY 12571 USA
Phone: 845-758-0400
Fax: 845-758-2633
Email: curran@proceedings.com
Web: www.proceedings.com

Thursday, October 2

11:15 AM-12:45 PM

3:00 PM-4:45 PM

Friday, October 3

10:15 AM-12:15 PM

2:30 PM-4:45 PM

Keynote Addresses

Thursday, October 2, 10:15 AM-11:15 AM

Application of Simulations for the Development and Evaluation of ASHRAE Standard 90.1 N/A
Ronald Jarnagin, Presidential Fellow Life Member ASHRAE, RLK Technical Consultants, LLC, Richland, WA

Thursday, October 2, 2:00 PM-3:00 PM

Phasing out HCFCs without Increasing Greenhouse Gas Emissions: The Challenges N/A
Didier Coulomb, International Institute of Refrigeration, Paris, France

Friday, October 3, 9:00 AM-10:00 AM

Personalized and Localized Cooling and Ventilation Devices and Systems N/A
Kamel Abou Ghali, Member ASHRAE, American University of Beirut, Beirut, Lebanon

Friday, October 3, 1:30 PM-2:30 PM

UNEP-UNIDO High-Ambient Project (PRAHA) N/A
Ayman Eltalouny, PEng, Member ASHRAE, UNEP, Bahrain, Bahrain

Steering Committee

Nesreen Ghaddar, American University of Beirut, Conference Chair
Samir Traboulsi, American University of Beirut (CEC Representative)
Bassam Elassaad, ASHRAE Sub Regional Chair B
Walid Chakroun, Kuwait University

2014 International Conference on Efficient Building Design— Materials and HVAC Equipment Technologies

October 2–3, Beirut, Lebanon
American University of Beirut

ISBN 978-1-936504-84-8

This conference presents the latest research and developments to improve building designs and state-of-the-art technologies in building materials and systems. Reducing energy consumption and carbon dioxide emissions, implementing energy efficiency standards, and introducing new equipment and technologies are areas needing improvement in the Arab region. The conference serves as a platform for presenting research and finding direction on technology to use when phasing out HCFC refrigerants in developing countries.

ASHRAE
ASHRAE Lebanese Chapter
1866 AUB American University of Beirut
Munib and Angela Masri Institute of Energy and Natural Resources

Sponsors

Gold Level

AWAL GULF
MANUFACTURING Co. B.S.C. (C)

dar al-handasah
shair and partners

United for Industry & Contracting S.A.L. (UNIC)

Silver Level

CLIMAVENETA/Majdalani

Engineering and Trading sal

Al Salem Johnson Controls (KSA)

AWAL GULF
MANUFACTURING Co. B.S.C. (C)

Bronze Level

Khatib & Alami

N. Hermes & Co.

For more information, visit the ASHRAE Bookstore at www.ashrae.org/bookstore.

1791 Tullie Circle, NE, Atlanta, GA 30329-2305 USA
Telephone: 404-636-8400
www.ashrae.org

A Simplified Mathematical Model for Predicting Cross-Contamination in Office Building air Conditioned by Displacement Ventilation

Carine Habchi

Kamel Abou Ghali
Member ASHRAE

Nesreen Ghaddar
Member ASHRAE

ABSTRACT

In indoor environments, the occupants constitute one of the main sources of particles from the ultrafine to the coarse mode. As a result, diseases can be transmitted through exhaled droplets nuclei produced by the different respiratory activities. The behavior of the particles depends largely on their weight and the flow field determined by the air conditioning system adopted. To insure good indoor air quality (IAQ) it is advisable to design a ventilation system covering the inhalable range of particle diameters lower than 15μm (4.92x10⁻⁵ft). Recently, displacement ventilation system (DV) is widely used in offices and buildings for its effectiveness in providing high indoor air quality and its ability to carry contaminants from the lower occupied zone. The purpose of this work is to study cross-infection between occupants in typical internal offices by the development of a simplified model that simulates active particle behavior in spaces ventilated by displacement ventilation system. The developed model incorporates the deposition of particles on walls and the gravitational settling effect on particles distribution within the space. The model is able of capturing the physics of the problem with significant reduction of the computational time cost. The model results showed that as the particle diameter increases from the fine mode the effect of the gravitational settling increases reducing the stratification in concentration created by the DV system and thus increasing the particle concentration at the breathing level of the exposed person. This effect remains until reaching a diameter where deposition on the floor opposing the DV principle acts as a removal factor. Nevertheless, the removal effect occurs for relatively high diameters and thus doesn't cover the critical inhalable range (particles of diameter below 10μm (3.28x10⁻⁵ft)). To overwhelm the gravitational effect, higher ventilation air flow rates are needed to satisfy good indoor air quality but unfortunately at a higher energy cost. To overwhelm the gravitational effect, flow rates higher than 100L/s (3.53ft³/s) with a recommended supply temperature of 22°C(71.6°F) are needed to satisfy good IAQ for the inhalable particle range and thermal comfort criteria.

INTRODUCTION

Occupants are one of the principal sources of contaminants. Particle behavior is significantly affected by their size and the air flow pattern imposed by the ventilation system. The displacement ventilation system (*DV*) is one of the widely used air conditioning systems in office spaces (Bjørn et al. 2002). DV results in stratification in temperature which increases from the floor to the ceiling level due to the rising plumes leading to lower energy consumption compared to the conventional mixing ventilation system. The particle concentration variation with respect to height is widely dependent on the relation between heat and infectious particles sources (Mundt, 1995). Usually the heat sources are simultaneously the contamination sources, therefore the upward buoyant plumes emanating from the heat sources result in an increase in the removal of contamination efficiency of the DV system (Brohus et al. 1996). To insure good IAQ not only tracer gases should be considered but also the inhalable range covering particles of diameter up to 15μm (4.92x10⁻⁵ft). With higher particle diameter, the gravitational settling effect increases acting against the vertical rising air flow decreasing the effectiveness of the *DV*. To date, the majority

Carine Habchi is a graduate student. Kamel Abou Ghali and Nesreen Ghaddar are professors in the Department of Mechanical Engineering, American University of Beirut, Beirut, Lebanon.

of the literature work on active particle is performed using computational fluid dynamics (*CFD*) in spite of the high computational time cost of this modeling technique. Hence, it is of primary concern to modify the simplified *DV* models developed for predicting passive particles distribution to account for active particles behavior by including the gravitational .effect and particle deposition on different walls orientation. The aim of this work is to develop a simplified model predicting accurately the transport of active particles generated from an infected person in an indoor environment served by a *DV* air conditioning system at a significantly less computational time compared to *CFD* simulations. The model will be used to properly design the DV system to minimize cross contamination between occupants.

PHYSICAL MODEL

The simplified model will resolve for thermal, airflow and concentration fields within an office conditioned by a *DV* system. The investigated system creates stratification in particle concentration and temperature in the vertical direction leading to the determination of variable vertical zones as illustrated in Figure 1. At the lower level, Zone 1 is characterized by upward circulation of the air surrounding the different thermal plumes. This zone is limited by the stratification height defined by the level at which the supplied flow rate equals the summation of the rising plumes' flows. Zone 2 is bounded by the stratification height and the critical height where the plumes expand. At the higher level, due to plume expansion a recirculation zone (Zone 3) is created. As horizontal gradients of the different variables are negligible compared to the vertical ones the space is discretized vertically into a number of layers. Horizontally, each layer is constituted by lumped regions: wall and heat sources thermal plumes regions, and circulating air region. The contaminant concentration is affected by different physical phenomena. Transport of particles within the plume and its surrounding is driven by convective buoyant flows and vertical diffusion. Laterally, interaction between the plume and its surrounding occurs by lateral diffusion due to concentration gradient at the boundary separating the plume from its surrounding and by entrainment of room air by the plume. As we are dealing with active particles, gravitational effect and particle deposition on walls affect largely the particle behavior and are included in the model.

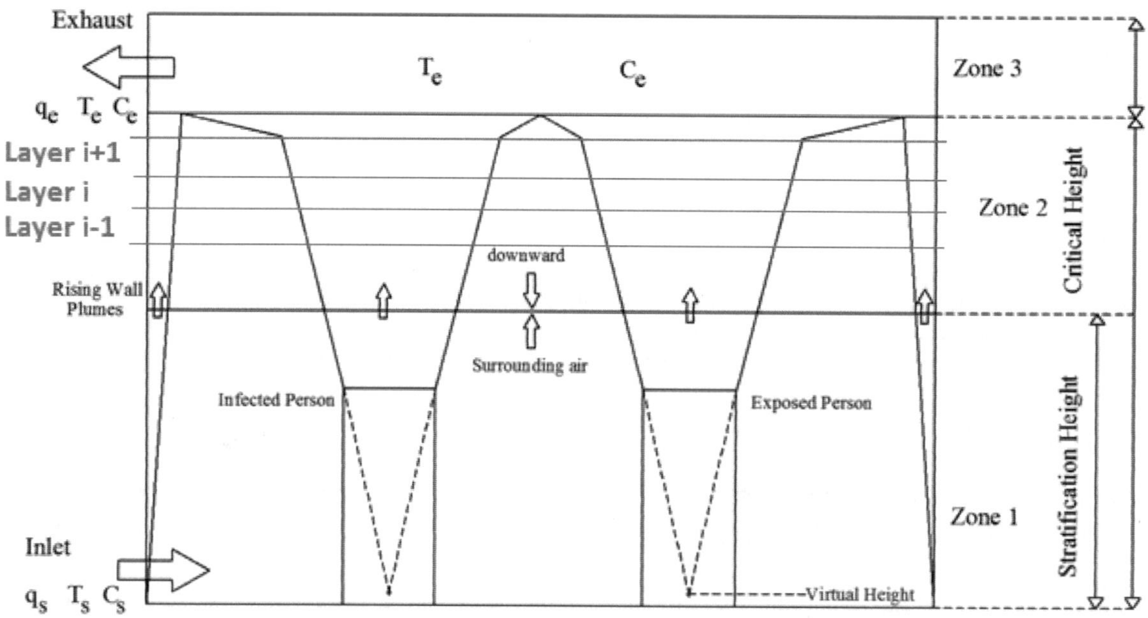

Figure 1: Representation of the studied domain

Energy and mass balances

The flow pattern plays a major role in particle distribution within the domain. Flow rates within the upward convective plumes of walls (M_w) and internal heat sources (M_{pl}) were obtained respectively from the empirical relations developed by Jaluria

(1980) and Mundt (1992). Applying a mass balance for each layer i (Figure 1) the expression of the mass flow rate of the circulating air $M_{a,i}$ at the interface between layers i and $i+1$. is obtained :

$$M_{pl} = 2.38*10^{-3} \rho Q_H^{3/4} \left(\frac{dT\infty}{dz} \right)^{-5/8} m_n \tag{1}$$

$$M_w = 2.87*10^{-3} \rho (\Delta T_{wa})^{1/4} z^{3/4} X \tag{2}$$

$$M_{a,i} = M_s - \sum M_{w,i} - \sum M_{pl,i} \tag{3}$$

The buoyant plumes are widely affected by the temperature variation in the studied domain. An energy balance for each layer i allows the determination of the thermal field within the domain:

$$\sum_i \dot{Q}_i + \sum_{in,i} M_{in,i} h_{in,i} - \sum_{out,i} M_{out,i} h_{out,i} = 0 \tag{4}$$

Where $\dot{Q}$ is the heat source rate, $h_{in,i}$ /$h_{out,i}$ and $M_{in,i}$ /$M_{out,i}$ are respectively the entering/exiting enthalpies and mass fluxes.

As mentioned before horizontal gradients in concentration and temperature are negligible compared to the vertical ones. Therefore, at the horizontal level the contaminant concentration is assumed to be uniform within the surrounding of the plume. In order to determine the interactive diffusion between the thermal plume and its surrounding a profile of the particle concentration should be assumed satisfying the physical boundary conditions of a maximal concentration at the centerline of the plume and continuity of the concentration at the plume boundary (Kanaan et al. 2010). The contaminant mass balance in the surrounding air is given by:

$$\begin{aligned}&\left[\max(M_{a,i-1};0)C_{a,i-1} - (\max(M_{a,i};0) + \max(-M_{a,i-1};0))C_{a,i} + \max(-M_{a,i};0)C_{a,i+1}\right] + \\ &\left[\rho v_s A_{a,i} C_{a,i+1} - \rho v_s A_{a,i-1} C_{a,i}\right] + \left[A_{a,i} D_p \rho \left(\frac{\partial C_a}{\partial z} \right)_i - A_{a,i-1} D_p \rho \left(\frac{\partial C_a}{\partial z} \right)_{i-1} \right] - \\ &\sum_{k=1}^{m} A_{\text{int},(i,k)} (D_p + D_t) \rho \left(\frac{\partial c_{pl,k}}{\partial r} \right)_{R_{(i,k)}, z_{mi}} - \sum_{k=1}^{m} M_{ent,(i,k)} C_{a,i} - \sum_{j=1}^{n} \rho v_{d(i,j)} A_{w,(i,j)} C_{a,i} = 0\end{aligned} \tag{5a}$$

The first term of Equation (5a) illustrates the particle vertical convection flux in the circulating air, the second term represents the gravitational effect, the third term is the vertical diffusion flux while the fourth term represents the lateral particle diffusion fluxes due to the interaction between the surrounding air and the buoyant plumes of the occupants. The fifth term is the entrainment fluxes of room air by the rising plumes and the last one presents the sum of deposition fluxes on walls within a layer i.

$$\begin{aligned}&\left[M_{pl,(i-1,k)} C_{pl,(i-1,k)} - M_{pl,(i,k)} C_{pl,(i,k)}\right] + \left[\rho v_s A_{pl,(i,k)} C_{pl,(i+1,k)} - \rho v_s A_{pl,(i-1,k)} C_{pl,(i,k)}\right] + \\ &\left[- A_{pl,(i-1,k)} D_p \rho \left(\frac{\partial C_{pl}}{\partial z} \right)_{i-1,k} + A_{pl,(i,k)} D_p \rho \left(\frac{\partial C_{pl}}{\partial z} \right)_{i,k} \right] + \\ &A_{\text{int},(i,k)} (D_p + D_t) \rho \left(\frac{\partial c_{pl}}{\partial r} \right)_{R_i, z_{mi}} + M_{ent,(i,k)} C_{a,i} + S_{p,(i,k)} = 0\end{aligned} \tag{5b}$$

Equation (5b) constitutes the particles mass balance within each rising plume. The first term of this equation is the plume upward convective flux of particles, the second one presents the gravitational effect, the third term is the particle vertical diffusion flux, the fourth term is the lateral diffusion due to interaction between the plume and its surrounding. The fifth term illustrates the entrainment flux and the last term presents the generation rate within the plume in layer i.

Since the behavior of active particles is studied deposition on surfaces is a determinant factor that should be properly modeled. Boundary deposition fluxes J_d of particles towards the walls can be expressed as:

$$J_d(y=0) = v_d C_\infty \quad (6)$$

Where C_∞ is the core concentration outside the concentration boundary layer (air concentration $C_{a,i}$ varying with height due to the stratification effect created by the DV+TP system). v_d is the deposition velocity which is a function of the wall orientation and particle characteristics and is modeled according to Nazaroff et al (2000) who accounted for turbulent and brownian diffusion in addition to the gravitational settling effect on particle deposition.

Boundary conditions and numerical methods

The space is discretized into n layers vertically (n=20 was enough to satisfy grid independence) and a number of lumped regions horizontally (determined by the number of non-interactive heat sources plumes). The model inputs include temperature and flow rate of the supplied air, generation rates of heat and contaminants sources, and heat fluxes from the walls. The finite volume technique developed by Patankar (1980) was used in the discretization of the coupled mass and energy equations. The convergence criterion for the residuals of the different equations was set to 10^{-6}. The predicted variables are the concentrations within the occupant plumes and surrounding air and temperatures within each layer.

Case study

The developed model was used to investigate the cross-contamination between infected and exposed persons in a typical office served by the DV system. The space of dimensions [4m (13.12ft)× 3.5m (11.48ft)× 3m (9.84ft)] is occupied by two sitting persons (75W (0.1hp) each); one of them is contaminated and generating particles. A lighting load of 100W (0.134hp) was considered and a load of 250W (0.335hp) was imposed from the external walls. The supply temperature was fixed at 22°C (71.6°F). The supply airflow will be varied until reaching good IAQ. In fact, a low intake fraction is required at the breathing level of the exposed person (particles concentration at the breathing level of the exposed person normalized to the concentration of generation by the infected person)for the inhalable range especially for particle of diameters smaller than 10μm ($3.28x10^{-5}$ft) which constitute the critical range because they present the largest potential of penetrating the airways and reaching the alveoli which increases the probability of cross-contamination between the occupants (Mangili et al 2005; Guha, 2008). Furthermore, for proper design of the DV system thermal comfort criteria should be satisfied. Therefore, the temperature in the occupied zone should be between 23°C (73.4°F) and 26°C (78.8 °F) (BS EN ISO 7730: 1995), and the temperature difference between the feet and the head should not be more than 2°C (3.6°F) to 3°C (5.4°F) (BS EN ISO 7730: 1995). The supply temperature was set to 22°C (71.6°F) as it is 3°C (5.4°F) lower than the set-point temperature of the space to avoid thermal draught (Engineering Guide Displacement Ventilation). For this temperature, the flow rate will be varied in order to satisfy the IAQ and thermal comfort criteria.

Results and discussions

The behavior of particles is largely dependent on the particle diameter and supply flow rate. Figure 2 represents the variation of the normalized particle concentration of the surrounding air zone with height for different diameters from the fine to the coarse mode [1μm ($3.28x10^{-6}$ft), 5 μm ($1.64x10^{-5}$ft), 10μm ($3.28x10^{-5}$ft) and 15μm ($4.92x10^{-5}$ft)] and variable flow rates [50L/s (1.77ft^3/s), 75L/s (2.65ft^3/s), and 100L/s (3.53ft^3/s)]. For a flow rate of 50L/s (1.77ft^3/s), the stratification height is very low at 1.1m (3.61ft) just at the breathing level which accumulate particles in the occupied zone increasing the risk of cross-infection as illustrated in figure3 which represents the variation of the intake fraction with particle diameter for different supply flow rates. Table 1 represents the effect of the supply flow rate and particle diameter on particle deposition. For the same flow rate, for small diameters turbulent diffusion and Brownian motion are the main phenomena affecting deposition while gravitation plays a minor role. Meanwhile, as the particle diameter increases the gravitational effect becomes more significant leading to the increase of particle deposition. Particle deposition is decreased as the flow rate increases because the settling effect is lowered. For example for a flow rate of 50L/s (1.77ft^3/s), total deposition is below 1% for particle of diameter 1μm but increases largely for heavier particles reaching a value of 79.55% for particles of 15μm ($4.92x10^{-5}$ft) with 99.99% depositing on the floor (Table 1). Floor deposition plays an important role in decreasing the probability of cross-contamination only for particle diameters larger than 10μm ($3.28x10^{-5}$ft). For particles above 5 μm ($1.64x10^{-5}$ft) the gravitational settling begins to act as a removal factor decreasing the concentration at the breathing level of the exposed person (Figure 3). Nevertheless, the removal effect doesn't cover the entire critical inhalable range for which relatively high intake fractions took place at the breathing level thus a higher flow rate is needed. For a flow rate of 75L/s (2.65ft³/s), the stratification height is shifted upward to a level of 1.3m (4.27ft) where the maximum normalized concentration of particles of diameter less than 5 μm (1.64×10^{-5}ft) takes place. In fact the stratification height is the level at which the velocity of circulating air is zero due to the downward motion of the surrounding air above this height and upward motion below it which make it a critical zone where particles accumulate. As the diameter of particle increases, the concentration at the breathing level of the non-infected person increases due to a decrease in the protection effectiveness provided by the ascending thermal plume of the exposed person (Figure 3). Thus as the diameter of the particle increases the performance of the DV system is weakened (Figure 2) due to the downward gravitational settling effect that acts against the upward motion of the surrounding air below the stratification height.

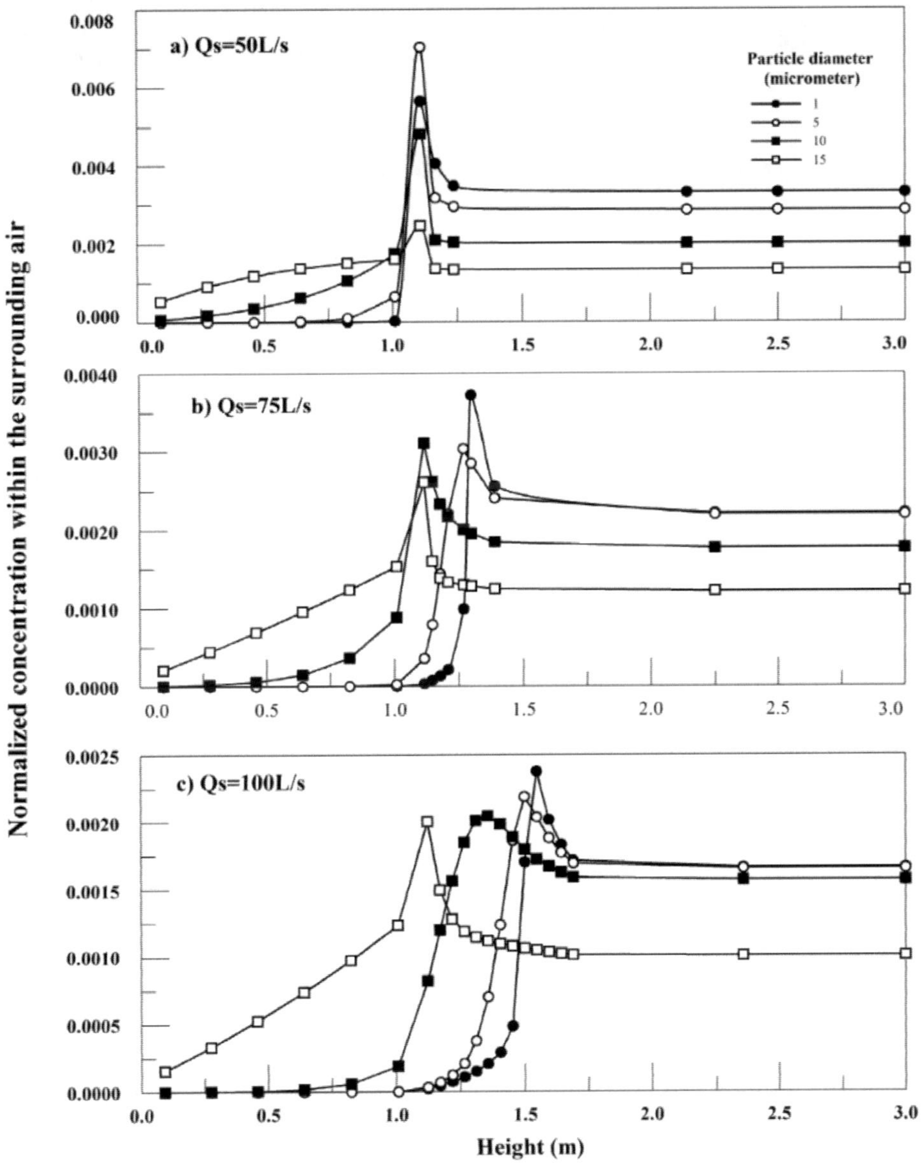

Figure 2: Variation of the normalized concentration within the surrounding air for different particle diameters for a supply flow rate of: a) 50L/s (1.77ft³/s); b) 75 L/s (2.65ft³/s); c) 100 L/s (3.53ft³/s)

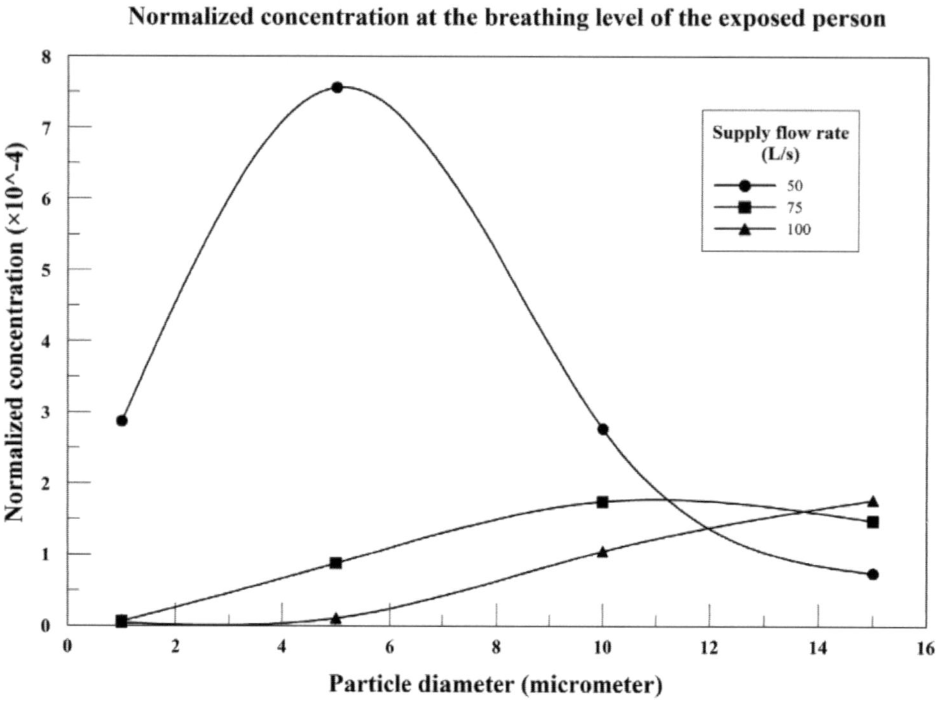

Figure 3: Effect of flow rate and particle diameter on particle distribution

Table 1. Effect of flow rate and particle diameter on particle deposition

flow rate (L/s); particle diameter (μm)	Total deposition fraction (%)	Deposition fraction on vertical walls (%)	Deposition fraction on the floor (%)
(50;1)	0.51	49.16	50.84
(50;5)	5.68	4.25	95.75
(50;10)	30.23	0.15	99.85
(50;15)	79.55	0.01	99.99
(75;1)	0.32	89.88	10.12
(75;5)	2.45	28.33	71.67
(75;10)	15.89	1.17	98.83
(75;15)	53.43	0.11	99.89
(100;1)	0.19	100	0
(100;5)	0.21	49.44	50.56
(100;10)	5.67	2.48	97.52
(100;15)	20.03	0.31	99.69

In order to insure a better indoor air quality (IAQ) covering the critical inhalable range, the air flow rate should be further increased to overcome the gravitational settling effect and higher the stratification level. By increasing the supply flow rate from 75L/s (2.65ft^3/s) to 100L/s (3.53ft^3/s) the stratification and critical heights are raised from 1.3m (4.27ft) to 1.55m (5.09ft) and from 1.5m (4.92ft) to 1.75m (5.74ft) increasing the removal of contamination efficiency leading to a significant decrease of particles concentration at the breathing level of the exposed person (Figure 3). Therefore, designing the DV system such that the stratification height is just above the breathing level (Brohus et al., 1996) is not enough to satisfy good *IAQ* for the critical inhalable range. For a flow rate of 100L/s (3.53ft^3/s) no particle within the critical inhalable range accumulates at the breathing level providing good IAQ with a maximum intake fraction of 1.05×10^{-4} (Figure 3). The required supply airflow conditions should not only satisfy the IAQ criteria but also the thermal comfort. For this reason, the thermal performance of the DV system for the different supply conditions investigated was studied. With the increase of the flow rate from 50L/s (1.77ft^3/s) to 100L/s (3.53ft^3/s) the thermal comfort is enhanced. In fact the difference in temperature between the feet and the head for a seated person decreases from 5 °C (9 °F) to 2.35 °C (4.23 °F) and the temperature range within the occupied zone is changed from [23.4°C (74.12°F); 28.4°C (83.12°F)] to [23.2°C (73.76°F); 25.55°C (77.99°F)]. Therefore, the two criteria of thermal comfort are satisfied for the supply conditions of 22°C (71.6°F) and 100L/s (3.53ft^3/s).

CONCLUSION

A simple analytical model was developed to predict active particle behavior in offices ventilated by *DV* systems. The model implemented deposition of particles on walls of different orientations, and the gravitational effect on particles spread. The simplified model was used to investigate cross-contamination between occupants in typical office spaces. A parametric study on the supplied flow rate was conducted to determine the design of the DV system satisfying both thermal comfort and IAQ criteria for the inhalable range specifically for particle of diameter lower than 10µm (3.28x10^{-5}ft) as they present higher probability of being inhaled by the exposed person. The obtained results revealed that as the particle diameter increases the gravitational effect increases reducing the stratification in concentration imposed by the DV system and hence increasing the particle concentration at the breathing level of the exposed person. For a flow rate of 50 L/s (1.77ft^3/s) this effect remains until reaching a particle diameter above 5µm (1.64x10^{-5}ft) where deposition on the floor opposing the DV principle acts as a removal factor. For the critical inhalable range of interest, gravitational settling accumulates particles in the occupied zone as the diameter increases leading to higher probability of cross-infection. In order to enhance the DV performance in terms of IAQ, higher ventilation air flow rates of the order of 100 L/s (3.53ft^3/s) are required to satisfy good IAQ with a recommended supply temperature of 22 °C (71.6 °F) to insure the thermal comfort. It was found that designing the DV system such that the stratification height is just above the breathing level is not enough for insuring good IAQ covering the inhalable range. The stratification height should be shifted upward to meet the IAQ criteria which requires the increase of the flow rate up to 100L/s (3.53ft^3/s) for a typical office loaded by 40 W/m^2 (4.98×10^{-3}hp/ft^2).

ACKNOWLEDGMENTS

The authors would like to thank the Lebanese National Council for Scientific Research (CNRS) for their financial support. The support of the University Research Board (URB) at the American University of Beirut is also acknowledged as well as the Shammas PhD Fellowship.

NOMENCLATURE

$A_{a,i}$ = cross sectional area outside the thermal plume at the interface between the layers i and $i+1$ (m^2)
$A_{int,i}$ = interaction area between the plume and its surrounding within layer i (m^2)
$A_{p,i}$ = cross sectional area at the interface between the layers i and $i+1$ inside the thermal plume (m^2)
A_w = wall surface area (m^2)
C = particle concentration (kg/m^3)
C_∞ = core concentration outside the concentration boundary layer (kg/m^3)
D_p = particle Brownian diffusion (m^2/s)
D_t = turbulent diffusion coefficient (m^2/s)
DV = displacement ventilation
h = enthalpy (J/kg)
IAQ = indoor air quality
J_d = deposition flux of particles towards a wall (kg/m^2.s)
M_a = mass flow rate within the surrounding air region (kg/s)
$M_{ent,I}$ = mass flow rate of room air entrained by the plume in layer i (kg/s)
M_{pl} = mass flow rate within the plume (kg/s)
M_s = supply mass flow rate (kg/s)
M_w = mass flow rate within the wall plume (kg/s)
m_n =non-dimensional parameter defined by Mundt
n :=number of walls
$\dot{Q}$ = heat source rate (J/s)

Q_H = heat source load (W)
$S_{p,i}$ = particle concentration source term in layer i inside the plume (Kg/m^3.s)
$T\infty$ = ambient temperature at the surrounding of the plume (°C)
ΔT_{wa} = difference in temperature between the wall and the surrounding air (°C)
v_d = deposition velocity (m/s)
X := wall width (m)
z_{mi} = mid-height of layer i (m)
ρ = air density (kg/m^3)

Subscripts

a = air
sur :=surrounding
i = layer number
in = inlet
j = wall number within a layer
k = occupant number
out = outlet
pl =plume
w =wall

REFERENCES

Bjørn, E. Nielsen, P.V. 2002. Dispersal of exhaled air and personal exposure in displacement ventilated rooms. Indoor Air, 12: 147–164.

Brohus, H., Nielsen, P.V. 1996. Personal exposure in displacement ventilated rooms. Indoor Air, 6: 157-167.

BS EN ISO 7730. 1995. Moderate thermal environments – Determination of the PMV and PPD indices and specification of the conditions for thermal comfort. British Standards Institute, London.

Guha, A. 2008. Transport and deposition of particles in turbulent and laminar flow. Annual Review of Fluid Mechanics, 40: 311–341.

Jaluria, Y. 1980. Natural Convection, Heat and Mass Transfer, Free boundary flows; Chapter 4, New York: Pergamon Press.

Kanaan, M., Ghaddar, N., Ghali, K. 2010. Simplified Model of Contaminant dispersion in rooms conditioned by chilled-ceiling displacement ventilation system. HVAC&R Research, 16: 765-783.

Lai, A.C.K., Nazaroff, W.W. 2000. Modeling indoor particle deposition from turbulent flow onto smooth surfaces. Journal of Aerosol Science, 31: 463-476.

Mangili, A., Gendreau, M.A. 2005. Transmission of infectious diseases during commercial air travel. Lancet, 365: 989–996.

Mundt, E. 1992. Convection flows in rooms with temperature gradients –Theory and measurements. Proceedings of Roomvent Third International Conference on Air Distribution in Rooms.

Mundt, E. 1995. Displacement ventilation systems-Convection flows and temperature gradients. Building and Environment, 30: 129-133.

Patankar, S.V. 1980. Numerical Heat Transfer and Fluid Flow, series in computational methods in mechanics and thermal sciences. Hemisphere Publishing Corporation

Price HVAC. Engineering Guide Displacement Ventilation. http://www.pricehvac.com/Catalog/Section J. Accessed February 1, 2014.

Facing the realities and Challenges of Sustaining Energy Performance and Improving Indoor Air Quality During Operations & Maintenance of Buildings

Om Taneja, Ph. D. PE,
Member ASHRAE

ABSTRACT

Buildings are high users of energy and other resources. There are opportunities to make the building equipment, systems and operations more efficient. It is believed that improvements in efficiency of buildings can be achieved at a relatively lower cost, by including efficiency measures during renovations and reconstruction. The process of improving efficiency entails that an effective integrated design, construction and operations approach be used, engaging architect, engineer, builder, tenants, information technology administrators and trained operating and maintenance staff. In addition, investment decisions need to focus on transforming existing buildings into "Smart and resilient Buildings". Behavioral changes, development of middle management and Buildings Managers to invoke operational excellence is valuable. This movement is driven not only from climate change concerns, but also due to legislative or business pressures of cutting waste, lowering costs, and adopting modern technology to bring greater transparency. . Concurrent with efficient operations, greater focus is also necessary on hardening the buildings and their infrastructure so as to be smarter and adaptive to inclement climatic conditions. However, without the staff developing competencies in energy management, technology, safety, and performance based leadership, such goals cannot be met. Structured training and development of Buildings Managers and Operators can help in immediate operational Improvements without significant capital needs. Building managers face many challenges in sustaining high performance of buildings. The purpose of this paper is to first address the causes of the problems and elements required for transformation to best in class operations. Benefits lie in integrating those elements to sustain high performance, over the life-cycle of a building. Missed opportunities of including efficiency measures can have adverse performance and cost consequences during the life time of the building. Embracing performance measurements and benchmarking, employing operational "best practices" and commitment to development of staff capacity can be a great asset. Energy use tracking and analytics can guide the operations and maintenance staff in performing their responsibilities and facilitate continuous commissioning. Facilities managers and IT system administrators need to collaborate to be able to dynamically manage the building operating parameters based on occupancy, time of day, and other human, environmental, and business and resilience considerations. Use of Green Procurement, recycling program, regular sensor and equipment calibration, and computerized maintenance management systems can further boost the performance.

INTRODUCTION

21st century is the digital era, with enormous growth of service sector, internet, social media, wireless communications and flexible workspace. This creates the need for more robust, high performance, resilient infrastructure. A day in the life of building "operations and maintenance staff" is very active not just for operating and maintaining buildings, but for other widely varying auxiliary duties. A building manager may be required to prepare and respond to incident reports, questions from an inspector general, using collaboration and social tools, parking, providing barriers for a planned protest, resolving a security breach, running a fire drill, special event set-ups and break-down, changing the power plug to a new copier , or moving people. With such a broad mix of duties, the day goes by fast, but it distracts from primary building operations and maintenance (O&M) functions. As building systems become complex and integrated, there are so many more meetings, conference calls and web searches that it is common for a building manager to voice his opinion, "I can't get my work done". With changing demographics, agglomerations, growth in services, urbanization trends, and buildings sector will expand and will continue to consume more resources. The world population is projected to reach 9 billion by 2050, an increase of 34% from where it is today. Buildings represent 32% of total final energy consumption. In terms of primary energy consumption,

Om Taneja is the Associate Director, Public Building Service, GSA, Region2, New York City buildings represent around 40% in most IEA countries. Consumer electronics and computer equipment now represent 15% of global residential electricity consumption. Energy used by these devices is expected to triple by 2030 unless actions are taken to increase energy efficiency. In the United States, buildings are responsible for 40% of annual energy consumption and up to 30% of all energy-related greenhouse gas emissions, 12% of fresh water and generate 40% of solid waste (IEA 2013, 2007). The buildings sector of today is energy and resource intensive and has an oversized ecological footprint (IEA 2013; USEPA 2008). 25% reduction in global energy use and greenhouse gases can be accomplished by cutting waste and employing basic efficiency measures (Figure 1). Awareness of social costs and benefits, Innovation, technology adoption, standardization, operational astuteness can make sustainable buildings an engine for socio-economic growth of the end users, the people (Kennedy et al, 2013).

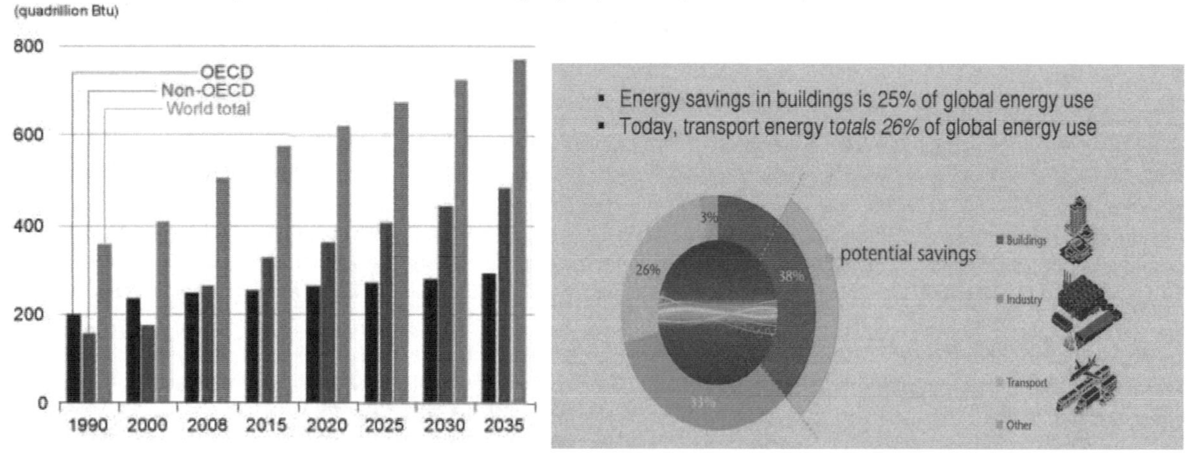

FIGURE 1 - World Energy Consumption 1990-2035 & Potential for Energy Savings in Buildings (Source IEA 2013 - Global Energy Trends)

BUILDING LIFE-CYCLE COSTS – A BUSINESS CASE FOR O&M

According to the International Facilities Management Association (IFMA), the operating life cycle costs of a facility are typically an 80/20 split in favor of O&M — O&M at 57% and capital renewal at 23% make up the 80%, while design and initial first-cost construction accounts for 20%. O&M of the elements included in buildings, structures and supporting facilities is complex and requires a knowledgeable, well-organized O&M team and a skilled, well-trained work force whether the functions are performed in-house or out-sourced. ASHRAE suggests that a building with good operations and maintenance practices that is poorly designed will often outperform a well-designed building with poor operations and maintenance (ASHRAE 2009). The objective of the O&M team should be to operate, maintain, and improve the facilities to provide reliable, safe, healthy, energy efficient and effective performance of the facilities to meet their designated purpose throughout their life cycle. To accomplish these objectives, O&M management must manage, direct, and evaluate day-to-day O&M activities and budget funds to support the organization's requirements. For the US Federal Buildings, full life cycle costing is a requirement for all new and existing buildings renovations (Executive Order-2004).

REALITIES & CHALLENGES OF OPERATIONS & MAINTENANCE OF BUILDINGSBACKGROUND

Facilities management has been an outgrowth of a system where design and construction staff completes a project and leaves it to operators and technicians to learn on the job to manage complex building systems with advanced sensors and controls. Focus of the O&M staff quickly shifts to reducing the number of complaints, and avoiding equipment breakdowns. The facilities staff pre-cools or pre-heats the building, lets system and lighting stay for longer duration and bypasses controls that limit their ability to make changes to satisfy the individual customer's needs. In such a scenario, energy is wasted and buildings create a larger carbon footprint, adversely affecting the environment. Seasonal weather changes compound the matter. Facility managers and O&M staff are reluctant to take any innovative actions that could disturb the prevailing balance. There is only a limited communication with tenants and senior staff.

Customer's requests for service are not responded in a timely fashion and preventive maintenance is done as and when possible, or when the equipment breaks down. The building systems slowly slide into disarray. The staff struggles to keep the things stable. There is not enough time to manage effectively the safety, security or emergency evacuation systems. This process is can waste up to 30% of energy and result in higher operating costs (PECI-1999).

Reality of Resources - There is more glamour in building new or renewing existing buildings. Once the building is handed over to the O&M section, not enough funds and resources are provided to maintain the building to the level and scope intended. The O&M budgets are usually the first budget line to be reduced and during times of budget squeeze, organizations require a zero-based budgeting that does not allow adjustments for inflation and emergencies. Furthermore, during value engineering sessions, the items that help O&M receive a lower priority and usually deferred. This process slowly leads to a back log of deferred maintenance, which if not funded, can significantly degrade the condition of the buildings. The Organization should set up a separate "building fund" by using income surplus that could be used to lower the deferred maintenance back log. Lack of funding is not the only cause of the poor state of buildings. Other causes are: lack of adequate training, adopting technologies that are not user-friendly and assigning staff lacking in education and competencies to operate and maintain an energy-efficient building (Federal Facilities Council 2010)

Challenge of Right People in Right Job - Focus on people and performance Expectations-An Organization can advance its values by defining leadership expectations, employee expectations and roles and responsibilities to be used for team building. Certain elements are essential, namely, organization and people must understand what is expected of them, the workforce must have the necessary skills and employee must be motivated for success. The mission cannot be accomplished without the right people, in right jobs and with the right attitude. Human capital strategy regarding performance management, employee development and employee recognition should be instituted.

Integrated Design, Construction and Operations - Synthesis of people, processes and Technology- It is very inefficient for different groups to operate in silos and not share information in a timely manner. To operate and maintain a high performance building efficiently, requires the synthesis of people, process and technology using a systems-thinking approach. This requires the extraction of knowledge from multiple areas of expertise, including but not limited to HVAC and control systems, energy and maintenance management, software and IT systems. In order to achieve the goals set by both government and private organizations, it is necessary to transition from commonly used silo-thinking to systems-thinking. Integrated design provides evaluation of the cross-component impact of multiple systems as they are changed or upgraded. A high-performance building cannot yield results unless its heating and cooling plants are designed in harmony with pumping, whole building distribution and control systems. For all new buildings' design and for energy efficiency projects for existing buildings, US General Services administration employs government's guiding principles that establish performance goals for energy, water, materials and indoor air quality along with other goals that need to be pursued throughout the life-cycle of the buildings (Federal leadership 2006) (figure 2)

***Systems-thinking* at the leadership level** - entails integration of functions and understanding of the people that make up the organization, as well as the values, structure, processes, policies, regulations and supervision within an organization. Systems-thinking requires goal setting, development, incentives, communication, reviews, rewards and accountability.

***Systems-thinking* when applied to infrastructure and building systems -** is the understanding of how all the components of a larger system interact with each other to meet the needs of the building occupant (El-Homsi, et al-2010). For example, using a systems thinking approach, when a technician is dispatched to respond to an occupant's complaint that a room is too cold, the technician will consider how the adjustment made to the diffuser or damper within the terminal box will impact the entire cooling system, often including the chillers and the cooling towers. As a second, broader example, systems-thinking is also necessary during major equipment and system replacements and upgrades. If more efficient lamps are installed in an office area, the cooling load will be reduced, which will reduce the load on the chiller. Depending on the operational parameters of the chiller, reduction of the amount of heat generated from the more efficient lamps and other factors, the chiller controls may need to be readjusted so that the chiller operates efficiently under new space conditions.

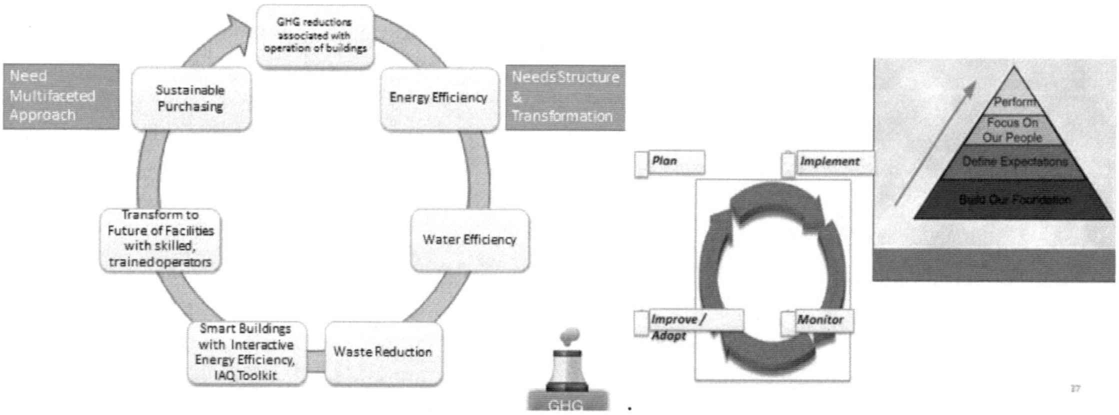

Figure 2 – Integrated approach and team Hierarchy

Expectation of Excellence in Operations –Closing the GAP between Design & Actual Performance

All too often a building's energy performance does not meet design expectations, particularly a new building's energy savings projection that overstates achievable performance (figure 3). Across the high-performing building industry, these unrealistic energy performance goals have come from, among other things, inadequate modeling and benchmarking practices, unreliable monitoring and equipment control systems, significant changes in space usage and processes during tenant improvements (Taneja, et al 2008). An important link toward delivering sustainability and energy savings targets that has been too often missing is the role and training of building operations staff. In order to get newer or renovated buildings operating at the efficiency level predicted in the design process, applying relevant monitoring, diagnostics and commissioning tools and active engagement of building operators are critical (Federal Facilities Council 2008).

Measures that O&M staff needs to take for bridging the gap from planned to actual performance are (Taneja, et al.- 2008): a.) better building operating plans, b.) simpler, user-friendly controls c.) Better understanding of customer needs d.) Measurements & verification tools, e.) Technology adoption and automation, f.) Benchmarking, g.) commissioning, h.) Design-build with third party guarantee, i.) Customer Feedback.

Figure 3 - Measures to Bridge the Gap between Planned and Actual Performance

LEVERAGE TECHNOLOGY TO MAKE BUILDING SYSTEMS BETTER INTEGRATED, AND "SMART"

Technology is all pervasive and subtle. All commercial class buildings today have physical building systems such as HVAC, lighting, power, elevators, AV equipment co-existing with IT infrastructure. It is conventional for buildings to a have a "Building Management System" and "IT infrastructure". The building mechanical, electrical systems and IT platform typically reside in separate silos and are not interconnected due to concerns with data security and reliability of operations. If these siloed systems can be integrated and optimized, the result can add a great value. To avail of the benefits of high speed internet of today, networking of the BMS and IT infrastructure with possibly cloud computing can add intelligence and interactivity of building systems and components. Integration of these systems provides a lofty jump in interactivity and sharing of data. Cloud computing treats computer resources as a service delivered over a network and financed through operating expenditures (Watson 2012). The amount of data available to the facility manager continues to increase, and use of emerging intelligent buildings' analytics tools can help organize and effectively convey the data in a useful form (Elkington, 2012 Ethan et al-2013, GSA-2013).

For mined data to be useful we must go beyond just information gathering. Adding greater intelligence to equipment and operations assigning attributes, understanding context, recognizing actions from analytics, navigating social networks, benchmarking, and ability to "shave energy" as conditions change, are important. If such an inclusive process is used, the incremental costs are small, but the owner reaps the benefits over the life-cycle of buildings. Sooner the building managers adopt the relevant technologies, better off they would be in their performance. Almost all building components come with embedded chips or sensors that can be networked to better integrate systems, equipment, environmental conditions, indoor air quality. In a pattern of growing urbanization, alternative work spaces, enhanced service expectations, and sustainable living, investing in more intelligent buildings and infrastructure has become a priority as a precursor to "Smart Living Buildings". A smart building has the sensors and monitoring equipment to provide the status of equipment, normalize data being generated from different systems, supplied by different vendors and perform building analytics , diagnostics

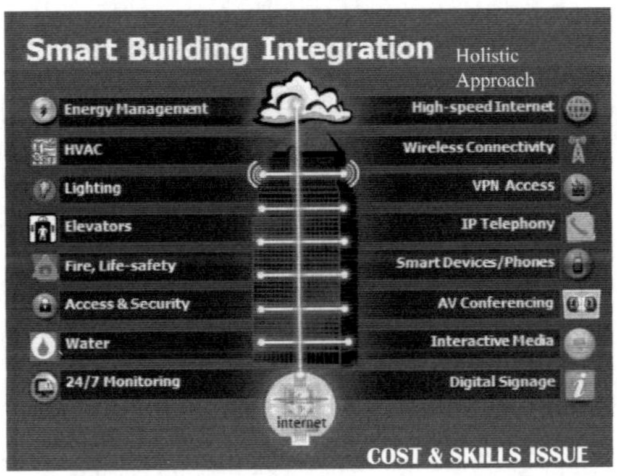

Figure 4 - Smart Buildings integrated function of Building systems.

benchmarking, and continual commissioning. It is generally expected that smart buildings provide simultaneously energy savings, comfort and safety of the users. Users want intelligent workplace but they would like to control their workplace environment.

Deploy Specific Technology Features for Strengthening Operations

Data is the gateway for managing building performance. Technology can help towards mining, analyzing and developing Key Performance Indicators (KPI). Employing Computer Maintenance Management System (CMMS) to track maintenance and customer complaints improves service. . An O&M manual should be developed that in addition to general items, should list industry best practices so that O&M staff can emulate to improve. Level of technology introduction can vary from a simple digital controls and spreadsheets to enterprise networked building monitoring, and diagnostics systems. Integrated design provides evaluation of the cross-component impact of multiple systems as they are changed or upgraded (Figure 4).

Smart Building should have simplified actionable data system to monitor, in real-time, the building performance and stream data to the building and a central facility, with easily usable dashboards displays. Each system component should have an attribute or contextual IP address. Features that help O&M, reporting and analytics are:

a.) Advanced meters and Sub-meters Technologies to monitor energy use -Application of smart meters helps with "measurements and Verification (M&V) data, helps optimize performance, enhance occupant awareness and behavior (Pike Research-2013 GSA-2013); b.) Employ or upgrade sensors, and Building and Energy Management Systems configured to provide a system, building or enterprise level performance; c.) Provide Analytics and Diagnostics applications; d.) It is important to provide reliable measurements and verification (M&V) tools; e.) Continual or periodic benchmarking and commissioning applications, including sensor calibration. This way energy performance of all assets can be compared. Best performing buildings should be recognized and opportunities to enhance performance of low performing buildings or building with sliding performance can be reviewed and addressed; f.)Applications to manage equipment sequencing and hours of operation consistent with occupancy; g) *Generate recommended action items* on discovery of variance from optimal conditions; h) Use computerized applications to *track and improve service delivery*; i) Include web and social media applications and tools for e-learning, collaboration and sharing of best practices; j) Employ CMMS to strengthen services and maintenance.

Innovative measures for central heating and cooling plant and auxiliary systems should be employed to achieve highest possible efficiency, not only during full load, but also for part-loads. Most buildings operate at peak load less than 300 hours/year which is less than 5% of time (Frank Stern 2013). Therefore, part-load and after-hours efficiency is the driving force. O&M staff can use "Low-cost, no-cost Shave Energy Program" to trim energy use with changing occupancy and schedule a "Demand Response Program" for load reduction during high utility grid demands, by shutting down non-essential equipment (Koomey et al 2002). The goal of these systems is to keep the load as flat as possible. (Stern 2013;. Mathieu, et al April 2011). Energy efficiency measures can lower the demand and help avoid overload on the grid. For this, buildings should deploy innovative HVAC and lighting technologies, such as, LED lighting, decoupling heating and cooling from ventilation, dedicated outdoor air ventilation, demand control ventilation, displacement air conditioning, heat recovery and humidification systems, evaporative and desiccant cooling, geothermal heat pumps, cogeneration systems, thermal storage, ductless cooling, chilled beams, radiant heating and cooling, variable refrigerant flow systems. Resetting chilled water temperature higher and condenser water temperatures lower during seasonal transitions or lower demand periods can improve performance.

Strengthening O&M BY Transformation to Future of Facilities (FOF)

Facility managers are now dealing with advanced technology, new building systems and components, increased level of building complexity, and pressure to lower energy use and improving IAQ. Future of Facilities requires structured administrative framework policies, programs, processes and people to maintain and operate the high performance buildings. Future of Facilities is both a philosophy and strategy towards the way we design, build and operate buildings. It is no longer purely a technical matter of equipment and Operational efficiency improvements. Many interrelated technical, environmental, social, cultural and economic subjects need to confluence. Green thinking takes the long-term view and demands concrete actions to integrate Internet, Web tools, social media, and the human dimension to be inclusive of all stakeholders. The traditional way of constructing and operating buildings has become outdated, where facilities managers had a limited input to strategic planning, design and construction and their input was not appreciated. The days when the facilities management or engineering group was housed in a basement office, next to the mechanical rooms are long gone. Facilities Managers have to become responsible and accountable for maintaining a healthy and safe place with flexible workspace solutions, deep environmental, energy and cost savings, greater tenant satisfaction and fostering a best in class workforce.

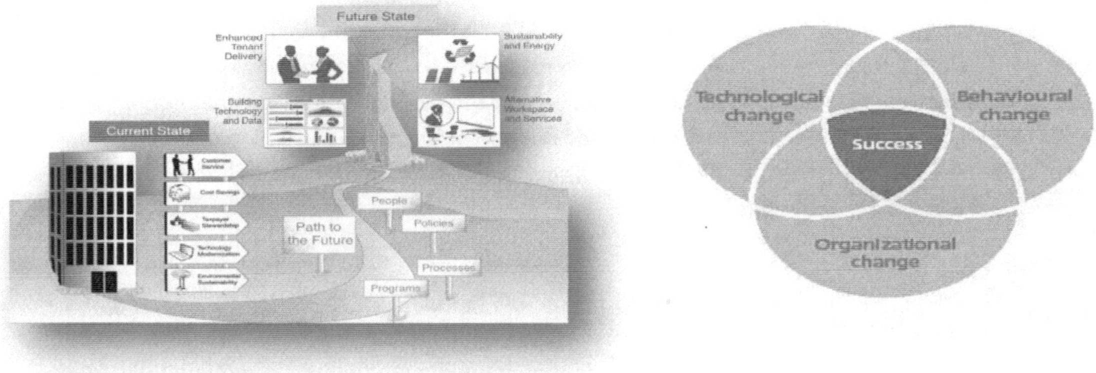

Figure 5 Pathways to Future of Facilities, balancing technological, behavior and organizational changes

Strengthened Operation to Manage, Track & Communicate Indoor Air Quality

The indoor environment is where people spend 90 % of their time and so as the building is upgraded for energy efficiency, it must simultaneously improve indoor environmental quality. It is postulated that improvements in IAQ lead to greater comfort and productivity of staff (Executive Office 2010). It has been estimated that in 75 percent of the cases maintenance deficiencies are a cause of IAQ problems. Maintenance problems caused by building structure, heating, ventilating, and air conditioning (HVAC) systems, general housekeeping, and the lack of a comprehensive maintenance plan have all been identified as causes of poor IAQ (US EPA 2008). Operational problems include improper or inadequate control of HVAC systems, changes in use and occupancy levels beyond those allowed for in the building design, and the introduction of new building material or furnishings that produce indoor air pollutants. O&M staff can contribute to IAQ by timely maintenance, replacing filters, cleaning coils calibrating sensors and monitoring complaints (USDOE 2010; USEPA 2008).

Incorporate Technology to Facilitate Climatic Resilience

Electrical, water, IT infrastructure, and building structures and systems need to adapt to and quickly recover from inclement weather conditions. Climate change represents one of the greatest challenges of our time. Climate change is no longer a distant threat; we are already feeling its impacts across the country and the world. As we act to curb the greenhouse gas pollution that is driving climate change, we must also prepare for the impacts that are too late to avoid. US Federal agencies are required to create a virtual climate, resilience toolkit, and best practices for "shelter in place", evacuation, remote access and ability to reconstitute after a challenging weather condition. US GSA seeks to provide robust, climate-ready workplaces for the Federal customer agencies and to secure the Federal real property investment. It involves not only coping with immediate problems more efficiently, but also establishing location of critical equipment above the flood plain, maintaining a reserve, and quick hookup capabilities for generators, chillers, boilers etc. In essence, climate change adaptation requires building a robust capacity to manage risk (UNDP 2012, Kennedy-2013, and Executive Office-2013). The process of adaptation is not new; but the idea of incorporating future climate risk into policy making is.

CONCLUSION

Public and private sectors are shifting their strategic direction towards collaboration, integrating smart building technologies and training of operations and maintenance staff into widely varying aspects of buildings operations and services. Buildings need to be built and operated so as to harmonize with the natural environment. They can evolve from a drain on our resources, to a vehicle for economic growth and improved productivity. To reap the full benefit of technologies, organizations must transform their commitment, structures, processes, and cultures. They will need to become more open and nonhierarchical and to invigorate a culture of trust. Ultimately, the power of system and social technologies hinges on the full and enthusiastic participation of employees who are not afraid to share their thoughts and trust that their contributions will be respected. Creating these conditions will be far more challenging than implementing the technologies themselves. Improved communication and collaboration through social technologies could strengthen interaction amongst workers and enhance trust. By leveraging use of technology, buildings can become more intelligent and optimize performance under changing occupancy, weather or other situations. Skilled O&M staff shaves 5% to 20% of energy use by pursuing dynamic adjustments (FEMP 2010, PECI 1999). In addition, by building capacity of facilities managers with requisite skills for operating high performance buildings, buildings can further sustain high performance over their life cycle. Innovative ideas and belief in delivering excellent services to the tenants, coupled with "tried and tested" industry best practices and technologies offer hope for climate change mitigation and resilience. Greater Resilience for buildings to adapt to increasing frequency of debilitating climate needs to be included in buildings design, equipment locations and local and remote and operations

REFERENCES

Adam Elkington ,"New Smart Building Technology to increase Federal Buildings Energy Efficiency", US GSA Press Release, May 14, 2012 (http://www.gsa.gov/portal/category/100731).

ASHRAE, "The Decision-Maker's Guide to Energy Efficiency in existing buildings", 2009", Atlanta, Georgia

Executive Office of the US President, "Climate Action Plan"", June 26, 2013

Executive Order 13327 Federal Real Property Asset Management, February 4, 2004

Federal Facilities Council of the National Research Council, "Core Competencies for Federal Facilities Asset Management Through 2020: Transformational Strategies", 2008, http://www.nap.edu/catalog/12049.html

Federal Energy Management Program, "Operations & Maintenance Practices", Release 3.0; a Guide to Achieving *Operational Efficiency,* August 2010.

El-Homsi, A.; J.slutsky, "Corporate Sigma Optimizing the Health of your Company with Systems Thinking", 2010, CRC Press, Boca Raton

Ethan A. Rogers, R. Neal Elliott, Sameer Kwatra, Dan Trombley, and Vasanth Nadadaur, "intelligent Efficiency Opportunities, Barriers and solutions", American Council for Energy Efficient Economy, October 2013

Federal Leadership in high Performance Buildings – inter-Agency Memorandum of Understanding, February 2006

Frank Stern, "Peak Demand and Time-Differentiated Energy Savings Cross-Cutting Protocols", Navigant Consulting Subcontract Report, Report NREL/SR-7A30-53827, April 2013.

International Energy Agency (www.IEA.Org), World Energy Outlook, 2012 and 2013

Johanna L. Mathieu, Phillip N. Price, Sila Kiliccote, Mary Ann Piette, "Quantifying Changes in Building Electricity Use, with Application to Demand Response", Lawrence Berkeley Laboratories, April 2011

Jonathan Koomey* and Richard E. Brown, LBNL, "the role of building technologies in reducing and Controlling peak electricity demand", Lawrence Berkeley Nation Laboratory, Report 49947, September 2002.

Kevin Kennedy and Kristin Meek, "By the Numbers, The Economic Benefits of a National Climate Action Plan" Insights, World Resources Institute, Washington, DC, (insights.wri.org), 2013

Om Taneja and Adam Hinge, "Life Cycle Assessment Measures of Energy Efficient Projects" International Conference on Improving Energy Efficiency in Commercial Buildings, IEECB08, Frankfurt, Germany.

Pike Research, Navigant Consulting, "Smart meters for commercial and industrial markets, May 2013.

Portland Energy Conservation Inc. (PECI), "Operation & Maintenance Assessments", September 1999, www.peci.org

Todd Watson, "IBM's 2012 Tech Trends Report: Skills, Skills, and More Skills!" Internet Evolution, Dec 2012

UNDP Perspective, Resilience: the importance of prioritizing disaster risk reduction – Hopkins Lecture, University of Canterbury, Aurora Centre, Burnside High School, Christchurch, August 15, 2012

US Department of Energy, Energy Information Administration, Energy & Environment Data, see EIA's Environmental Web Site: http://www.eia.gov/environment , 2010, and 2013

US EPA, "EPA's Report on the Environment 2008; highlights of the National Trend", 2008 and 2011 update, www.usepa.gov/roe

US General services Administration, "sustainability Strategic Planning, FY14/FY15", August 2013

US General Services Administration, "GSA's Smart Buildings & advanced Metering Technologies", 2013

Slashing Energy Use in Labs and Commercial Facilities While Still Providing Superior IEQ

Gordon P. Sharp
Member ASHRAE

ABSTRACT

Laboratory facilities are energy intense building types due to the vast amounts of 100% outside air required. With today's concerns over high energy expenses, reducing carbon footprints, plus efforts to make facilities' green and provide a better indoor environment, reducing both new and existing lab facility energy expenses has become a critical challenge particularly in hot and humid climates where the cost of conditioning this vast amount of outside air can be prohibitive. The primary reason behind many labs' high energy expenses is the minimum ventilation or air change requirements that often dominate the amount of outside air required by these facilities. This paper will present a proven Demand Based Control solution to this problem. In this approach that is referred to in the ASHRAE handbook, rather than use a fixed air change rate of 6 to 12 ACH (Air Changes Per Hour), real time measurements of the actual lab indoor environmental quality are used to vary the air change rates from as low as 2 to 4 ACH to upwards of 8 to 16 ACH based on the cleanliness of the lab room air. By safely cutting the lab air change rates often times by as much as 50% or more for about 98% of the time, this approach is often cited as the single largest energy conservation approach for many lab facility designs. Furthermore in new designs significant net reductions in first cost may also be achieved through reductions in the sizing of the HVAC systems. Finally by significantly increasing air change rates when contaminants are sensed, improved Indoor Environmental Quality (IEQ) can be achieved. This paper will describe this successful approach for labs and vivariums plus discuss the technology used to implement it cost effectively. Additionally, the results will be shown for the largest study ever done of lab IEQ conditions covering over 1.5 million hours of lab operation. Furthermore a variation of this concept is often applied to other types of facilities such as office buildings, educational facilities, public assembly buildings such libraries and student centers, as well as healthcare facilities. This approach has been used in over 400 buildings including well over 200 lab facilities and two case studies will be highlighted such as a full lab building retrofit of the 2006 R&D Lab of the year at Arizona State University which is saving over $1 million a year in annual energy costs from this concept. Finally, a highly sustainable, "near zero carbon" lab project in the often hot and humid climate of Abu Dhabi, UAE will also be discussed.

INTRODUCTION

Heretofore very little objective data has been available on the environmental and energy savings impact of both reducing and varying air change rates in labs and vivariums. This article attempts to address this data gap with the results of a major research study that generated a significant amount of objective data on the indoor environmental quality (IEQ) conditions of labs and vivariums that are using dynamic control of air change rates.

Specifically, this paper will focus on the results of research that represents what is believed to be the largest study ever

Gordon P. Sharp is the Founder & Chairman of Aircuity, Newton, MA, USA.

done of laboratory IEQ conditions covering over 1,500,000 hours of lab operation from labs and vivariums that employ Demand Based Control of air change rates or real time sensing and control of Lab IEQ. In total over 20 million sensor values were collected and analyzed including data on TVOC's, Particles of a size range of 0.3 to 2.5 microns, carbon dioxide, and dewpoint (absolute humidity). This study of over 300 lab areas at 18 different facilities indicates that although lab events occur on a regular periodic basis requiring an increase in air change rates to purge levels, on average, air flow can be maintained at minimum levels of 2 to 4 ACH for over 97 to 98% of the time while still providing excellent IEQ.

METHODOLOGY

In a large majority of labs (particularly life sciences labs) and vivariums the air flow is often dictated by the minimum air change rate for the space which might be 12 to 20 ACH in a vivarium or 6 to 12 ACH in a lab room. Although, high thermal loads, or the heavy use of fume hoods or animal racks with heavy makeup requirements can sometimes drive the room airflow rates, often times it is the minimum ventilation rate that determines the air flow. However, if the air in these rooms is "clean" or free of any harmful or irritating contaminants that the minimum air change rate is intended to dilute, then a high air change rate is not needed, at least for when the air is clean.

As such one approach that has been shown to effectively and safely vary air change rates in labs and vivariums is to sense the quality of the air for such contaminants as volatile organic compounds (VOC's), ammonia, plus some other chemical vapors and odors, as well as particulates. When the room air is free of these contaminants then the air change rate in the vivarium can be reduced to for example 4 to 6 ACH, or if it is a lab room to for example 4 or even 2 ACH.

In order to economically and reliably accomplish this sensing of environmental conditions in many labs and vivarium rooms within a facility a novel sensing architecture known as multiplexed sensing was used in this study. With this approach, one central set of sensors is used in a multiplexed fashion to sense not one but many different rooms or areas. Thus instead of placing multiple sensors in each of the rooms to be sensed, this networked system routes packets or samples of air sequentially in a multiplexed fashion to a shared set of sensors. Every 40 to 50 seconds a sample of air from a different area is routed through a common air sampling backbone consisting of a hollow structured cable to the centralized set of sensors, known as a sensor suite, for measurement. These sequential measurements are then "de-multiplexed" for each sampled area to create distinct sensor signals used for traditional monitoring and control. Typically 15 to 20 areas can be sampled with one set of sensors approximately every 15 minutes depending on the requirements for those spaces. A variety of different types of sensors can be used to analyze the air samples for multiple air parameters. Figure 1 shows an example of the architecture of the multiplexed sensing system used to implement the study.

In addition to dramatically reducing the number of sensors needed to implement this concept such as by a factor of 15 to 20 this multiplexed sensing concept has another advantage over using individual sensors. Typically for controlling the lab room space airflow and IEQ it is best to look at the contaminant levels in the room differentially. In other words we typically want to subtract the contaminant levels seen in the supply airflow from the exhaust or room levels. Thus if the supply airflow particle levels were high for example then we would not erroneously call for more supply flow which would just pump more particles into the room from the supply system. Dilution ventilation only works to dilute the contaminants generated in the room itself. A significant benefit of using the multiplexed sensing approach is that it can measure differential contaminant or parameter levels much more accurately and reliably than individual sensors. This is because using two different sensors, one for the room or exhaust and another for the supply levels can actually double the sensor drift errors since one sensor could drift negative while the other could drift positive, thereby doubling the error. With a multiplexed sensing system the same centralized sensor is used to measure both the supply contaminant levels as well as the room levels. Any offset drift error of the sensor will be the same for both measurements since the sensor is the same for both measurements. Therefore when the differential is taken and the supply level is subtracted from the room level, the offset drift error of each measurement is cancelled out. Thus the multiplexed sensing architecture can generate much more accurate differential measurements compared to using individual sensors.

This research study was conducted using environmental data obtained from the above multiplexed sensing system from 18 different lab and vivarium sites representing over 300 lab and vivarium rooms across the US and Canada where dynamic control of air change rates was employed. These sites consisted primarily of life sciences and biology related areas as well as a smaller amount of chemistry and physical sciences lab areas. Three of the above sites involved animal facilities which are not reported on in this paper.

Regarding the scope of the study, approximately 1,500,000 operating hours of lab data and about 100,000 hours of vivarium data were analyzed. Put in other words, if only one lab room was studied vs. over 300, this amount of operating hours would have spanned over 18 decades. The data from the various sites was for different lengths of time depending on when the site came online. In total over 20 million sensor values were collected and analyzed including data on TVOC's, and particles of a size range of 0.3 to 2.5 microns which are both reported on in this paper as well as carbon dioxide, and dewpoint (absolute humidity).

All particle and TVOC measurements were taken as differential measurements of the room minus the environmental conditions of the supply air feeding the lab or vivarium. This was done to significantly reduce potential effects of any sensor drift as well as to subtract out any impact of the outdoor conditions on the measured room conditions. Since all measurements were taken using a multiplexed sensing system, the measurements of both the room conditions and the supply air feeding these rooms were taken with the same sensor, thereby creating an accurate differential measurement.

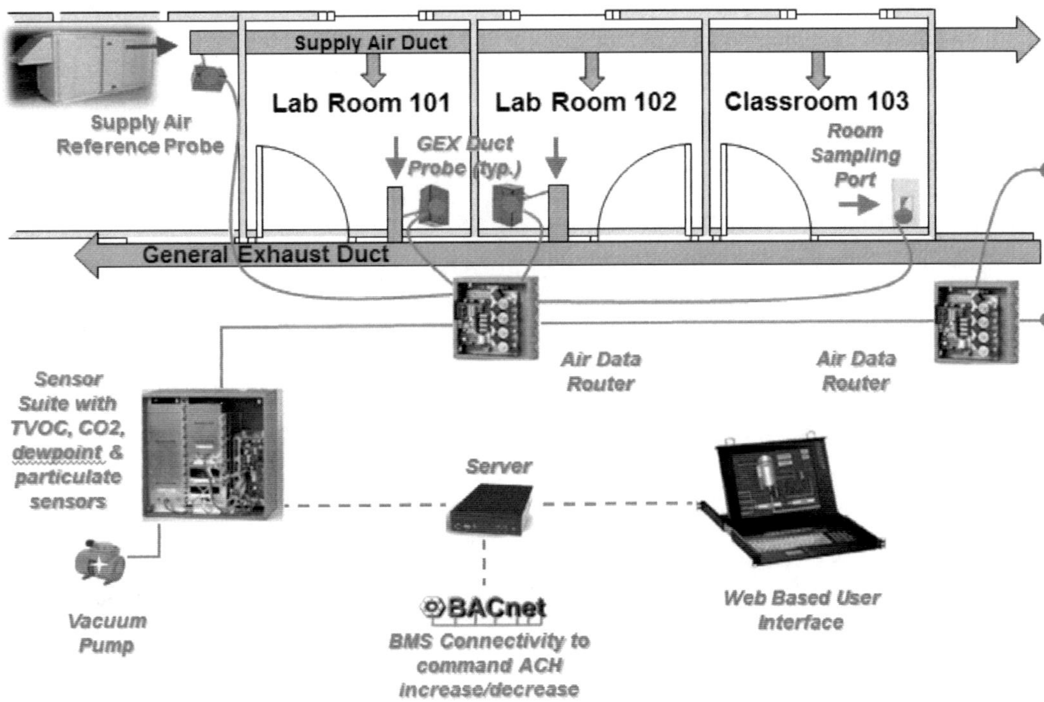

Figure 1 *Multiplexed sensing architecture for Demand Based Control.*

To simply the analysis, all sensor data was put into bins representing the number of counts or times that a parameter exceeded a specific threshold level corresponding to that bin. The data was then normalized based on the total number of data points or counts to generate the percent of time the data exceeded the bin value thresholds. When these values are graphed they form a cumulative graph showing the percent of time that each bin value was exceeded.

RESULTS

Figure 2 shows a graph of the average TVOC levels over all of the lab locations representing about 1,500,000 hours of operating data. As mentioned previously, this is a cumulative graph so that the value of 0.84 % at 0.10 PPM means that on average this is the amount of time that a lab location has a TVOC value greater than or equal to 0.1 PPM. Since this represents the average, some locations can be much higher than this and others potentially near zero. However, the average gives a good idea of the potential energy savings across all these different locations.

As can be seen in Figure 2, labs are typically "clean" of most chemical contaminants about 99.2% of the time. As such this means that energy can be saved by operating at reduced minimum air change rates up to about 99.2% of the time in labs at least with respect to the TVOC sensor. Looking at this same data in another light, on average TVOC events occur for about 1.5 hours a week or over 3% of a typical 40 hour work week.

To show the variations in this data between different sites, Figure 3 shows the same TVOC graph but with each of the lab sites shown as a separate line with the average curve shown by the black dotted line. Note that even at the site with the greatest amount of TVOC activity, the dynamic control concept can still save energy about 97% of the time with almost 6 hours of time during which the minimum room ACH rate is safely increased to respond to TVOC based events.

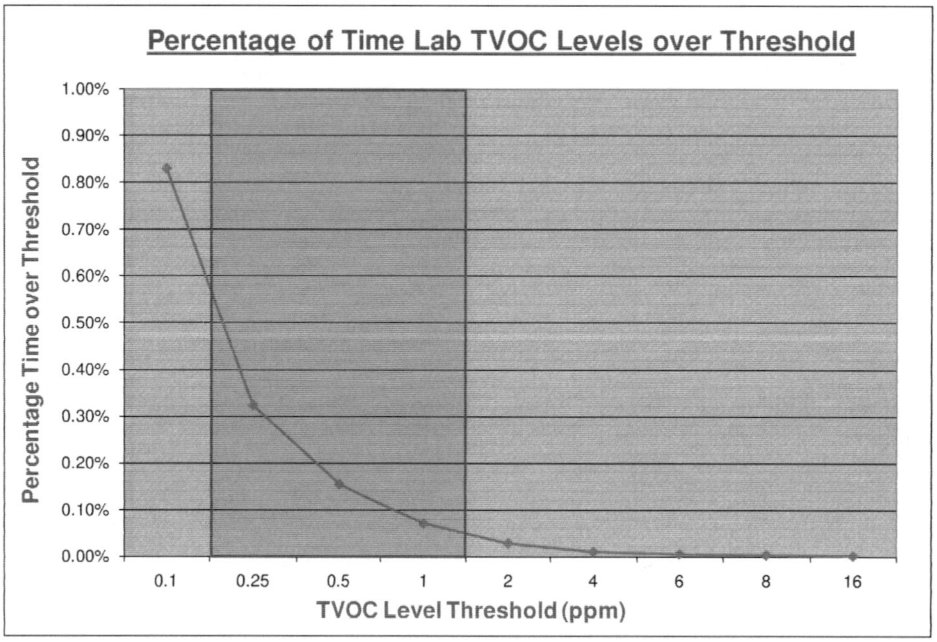

Figure 2 *Average TVOC level percentages over threshold (1.5M hours of lab operation.*

Another parameter that can cause an increase in the minimum air change rate is an increase in particles in the lab due for example to a reaction that may go out of control or an acid spill that causes an evolution of smoke or an aerosol into the lab room. Figure 3 shows a graph of the average level of 0.3 to 2.5 micron particle counts (PM2.5) that exceeded a background level of the lab room's supply air for all the different sites of the study. Typically about 1.0 million particles a cubic foot (pcf) is used as the threshold for increasing the minimum air change rate.

As can be seen in Figure 4, the average lab room, indicated by the dotted black line, is above the 1M pcf threshold almost 0.5% of the time or about 30 minutes a week on average. The individual sites show a range of values from near zero up to about 1.5 % of the time that flow should be increased based on a particle event. If this amount of time is added to the time that TVOC's are above the control threshold, this comes to total of about 1.2% of the time on average or for some sites 2 to 3% of the time. In other words minimum air change rates of between 2 to 4 ACH can be achieved from 97% to in excess 99.0% of the time due to the presence of either TVOC or particle events occurring on average up to about 5 hours a week.

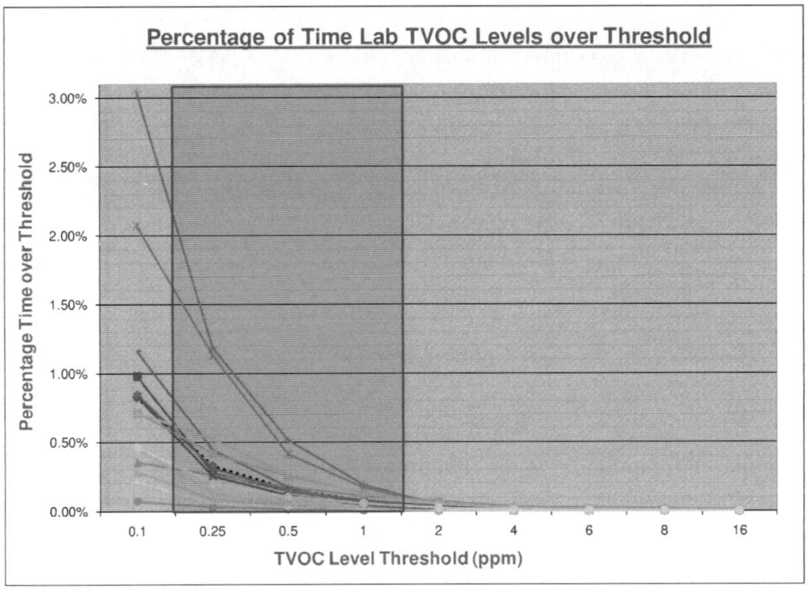

Figure 3 *Average TVOC level percentages for multiple lab sites.*

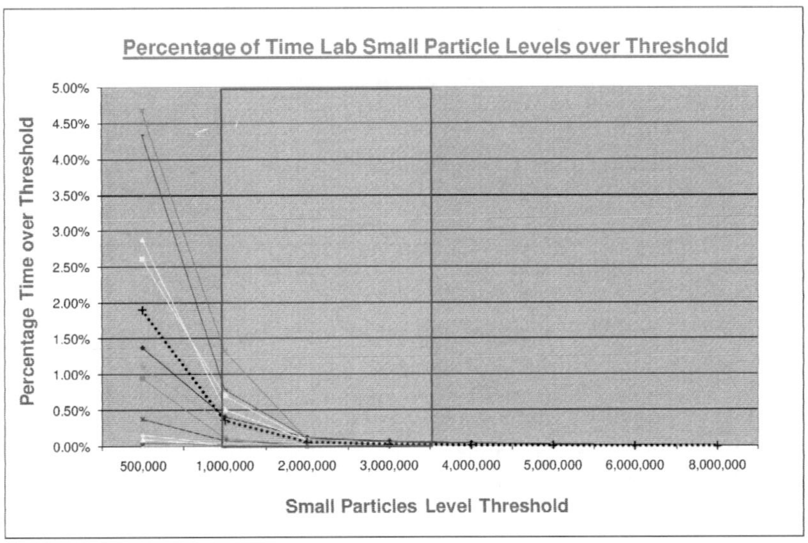

Figure 4 *Percentage of time that small particle levels exceed threshold.*

CASE STUDIES

One case study of Demand Based Control and multiplexed sensing is the Biodesign A & B buildings at Arizona State University in Phoenix, Arizona. These highly acclaimed facilities were R&D Magazine's Lab of the Year in 2006 and also achieved USGBC's (US Green Building Council) LEED Platinum certification. The building was initially designed with a minimum ventilation rate of 12 ACH. It was decided to reduce the air change rate by a factor of 3 down to 4 ACH using demand based control when the lab air is clean, and to increase the airflow to about 16 ACH when contaminants are sensed in the lab. This was successfully tested in a pilot project in 2007 and then implemented in 2009 in over 200 lab spaces and another 90 vivarium spaces. The net result of this 2009 retrofit was an energy savings of approximately $1 million annually for this approximately 35,000 gross square meter facility. Figure 5 shows the reduction in airflow that was achieved upon implementing the Demand Based Control in about 10 lab areas at ASU Biodesign Institute. Some variation in total airflow can be seen from day to day but the total average savings is quite significant with an excellent payback.

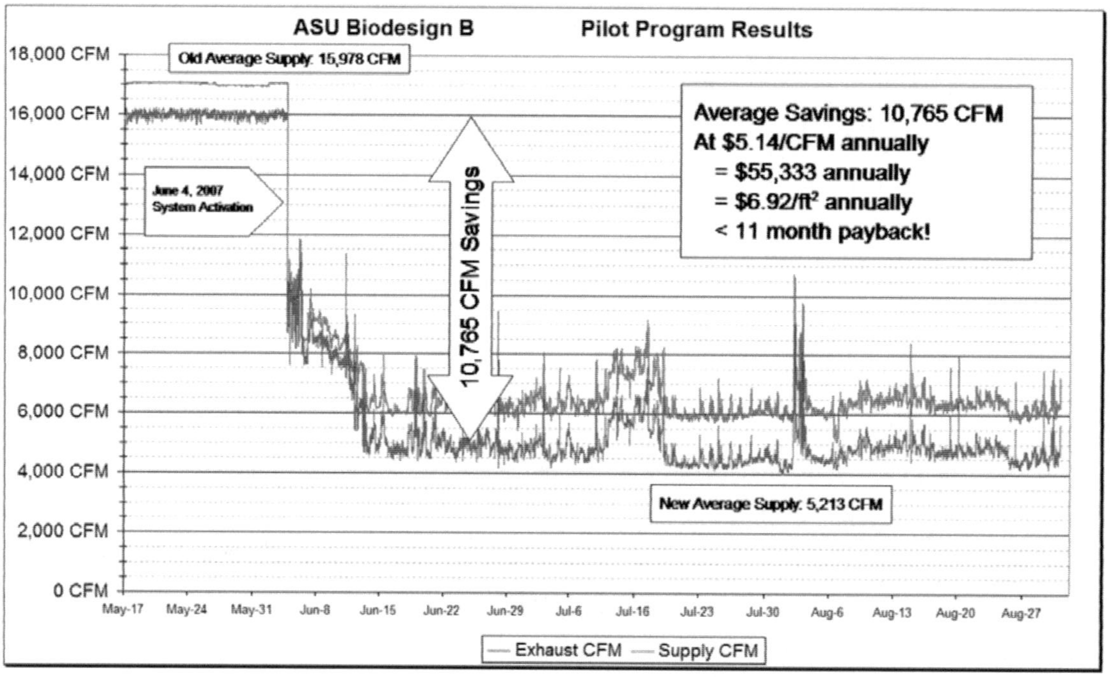

Figure 5 *Representative flow reduction from Demand Based Control at the ASU Biodesign Institute*

Furthermore not only was energy saved but as shown in Figure 6 the Indoor Environmental Quality (IEQ) of the Biodesign Institute was improved as well due to the increase in the purge airflow rates when contaminants were present in the lab areas. This can be seen in the lower amounts of time that various TVOC thresholds were exceeded when Demand Based Control (Dynamic ACH Control) was used versus a constant minimum airflow rate of 12 ACH. Thus Demand

Based Control proved both significant energy savings and improved IEQ.

In addition to safely saving energy in laboratory and vivarium facilities, multiplexed sensing technology is also being used in non-lab facilities at Arizona State University such as in their offices, classrooms, library, sports arena, student center and many other buildings where the demand control of outside air is important to save energy. In total Arizona State University has implemented this technology approach in over 25 lab and non-lab buildings.

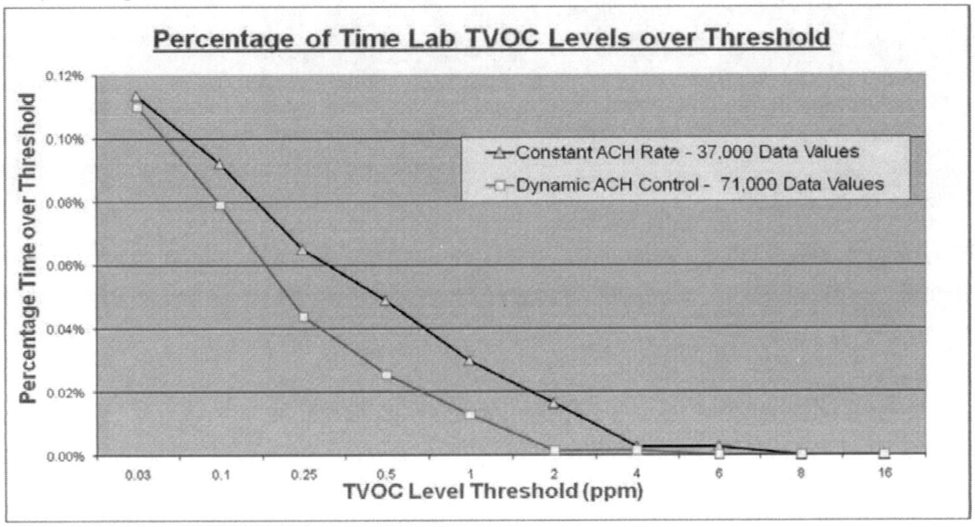

Figure 6 *Improved IEQ in the form of lower TVOC levels from Demand Based Control (dynamic ACH) versus a constant airflow*

A second case study of Demand Based Control is the Masdar Institute of Science and technology or MIST which is located in Masdar City in Abu Dhabi, UAE. Designed to be one of the world's most sustainable facilities with a near net zero carbon footprint, the MIST 1A and 1B buildings comprise mixed use lab, office, classroom, and residential space over about 150,000 sq. m. The MIST 1A part of the facility is shown in Figures 7 & 8. The region experiences severe climate with both very high temperatures and very high humidity levels.

Figure 7 *MIST1X views - Courtesy of Fosters and Partners.*

MIST used Demand Control in its classrooms and office areas to reduce ventilation based on determining occupancy by sensing carbon dioxide. In the MIST labs, Demand Based Control is being used in conjunction with either chilled beams or fan coil units to reduce lab air change rates to 2 ACH day and night when lab air is sensed to be clean. When contaminants are sensed purge airflow rates of as high as 14 ACH are commanded to provide greater dilution than a fixed minimum ACH rate of 6 to 8 ACH for safer operation. The lab's VAV (Variable Air Volume) fume hoods were initially designed to go as low as 150 l/s when the sash was closed, but due to the new ANSI Z9.5-2012 standard a VAV fume hood min of 45 l/s was to further increase energy savings.

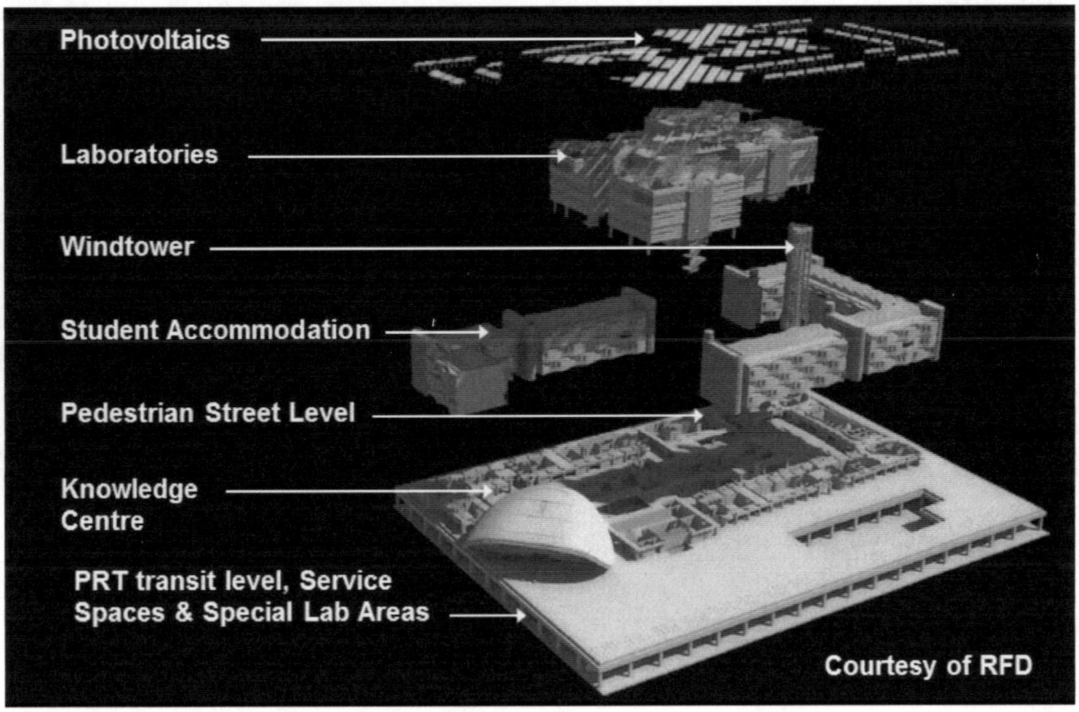

Figure 8 *Further detail of MIST 1A lab spaces.*

Demand Based Control is providing high impact energy savings for the MIST 1A and 1B facilities of approximately 9,000 MWh per year while also enhancing lab and non-lab area IEQ. Additionally, the reduction in lab and office airflow volumes reduced the project's HVAC capacity requirements creating a significant reduction in first or capital costs. This was achieved though downsizing the main HVAC capital equipment such as air handlers, exhaust fans, chillers, heat recovery systems, etc. Finally, the reduction of about 9,000 MWH/year of energy use translated into 4 MW less of solar photovoltaic panels required to approach net zero energy capacity. This solar photovoltaic panel capital savings was alone in excess of $20 million!

CONCLUSIONS

The largest and most comprehensive study to date of the impact of varying air change rates on lab and vivarium IEQ conditions and energy savings was completed using a multiplexed sensing approach in buildings that implemented demand based control. This study showed that that although particle and TVOC events occur upwards of a few hours a week and require high ventilation rates, much lower flow rates can be utilized with an automatic ACH rate control system for well over 97% of the time. With the current challenges many organizations are facing concerning reducing their carbon footprint and their usage of energy, this research and the related case studies provide ample evidence of the significant contribution that the demand based control of lab air change rates can make towards safely meeting these goals.

REFERENCES

American Industrial Hygiene Association (AIHA). 2012. Laboratory ventilation. ANSI/AIHA Standard Z9.5-2012. American Industrial Hygiene Association, Fairfax, VA.

American Society of Heating, Refrigeration, and Air-Condition Engineers (ASHRAE). (2011). ASHRAE handbook - HVAC applications Chapter 16, Laboratories (pp. 16.1 to 16.22)., Atlanta, GA: ASHRAE, Inc.

A New Full Scale Multi-Physics Model to Predict Hydrodynamics and Heat and Mass Transfer Performance of a Water Cooled Evaporator

Iyad Fayssal
PhD, ME Department, American University of Beirut

Fadl Moukalled
Professor, ME Department, American University of Beirut

ABSTRACT

In a recent article Moukalled et al. (2011) reported on the performance of a roof-top air conditioning unit by numerically predicting the thermal and hydrodynamic fields on the air flow side of the unit. This paper extends the work reported by Moukalled et al. (2011) and predicts the performance of the unit by analyzing the refrigerant flow side simultaneously with the air flow. The problem involves a turbulent multi-phase flow situation with condensation on the outer surface of the evaporator coil. Further, the heat transfer between the air flowing on the outer surface of the coil and the refrigerant flowing inside the pipes necessitates solving a conjugate heat transfer problem. Solutions are generated via a collocated unstructured finite volume method. . A conformal mesh was generated at the common interface regions separating fluid and solid zones to account for the continuity of heat fluxes at these interfaces. Two sets of results are presented in this manuscript simulating a wet and a dry coil. Detailed numerical predictions are compared with a developed lumped analytical model and with available experimental data for a similar coil model (Moukalled et al., 2011) and are found to be in good agreement. Complex turbulent flow fields were obtained in regions located upstream and downstream of the evaporator coil.

INTRODUCTION

In the last few decades, the world has witnessed a great revolution in many sectors including industrial, commercial, and residential fields. Unfortunately, this development was at the expense of available energy resources required to cover certain demands causing a depletion of these resources and increasing environmental pollution. Shifting towards renewable energy resources is among the suggested alternatives; in addition, recycling, reusing, and reducing energy consumption are still considered widely applicable strategies for energy management and pollution mitigation in various applications. For instance, in building applications, many trends and technologies are being under research and development to save energy during cold and hot seasons. These include the exploration of new wall materials, low-e windows, shading devices, etc. However, these structures still rely on integrated HVAC systems to cover their cooling and heating needs. Therefore, to achieve a global optimization of reducing power and energy consumption in residential and non-residential buildings, a sight at the HVAC units should be taken effectively into consideration. To improve the performance of HVAC units, it is first essential to explore deeply their function and fully understand the hydrodynamic and thermal behavior of the flow in such units. Accordingly, experimental procedures are necessary for improving performance of building end-use machines; however, high costs are sometimes associated with building experimental setups beside some difficulties in retrieving inaccessible data. Fortunately, in the last few years, computational power has witnessed an important improvement through the development of high performance computer machines capable of handling extensive large scale simulations for many engineering applications, thus giving a great opportunity to build numerical experiments which models the real physical situation. Such "numerical experiments" which are based on governing physical laws help in extracting much information for a specific physical phenomenon under investigation while experiments could sometimes lack this ability. Computational fluid dynamics, implemented in this study, is a finite volume based method (Patankar, 1980) derived to solve simple to complex type problems involving single and multiphase flow phenomena to multi-physics situations involving conjugate heat and mass transfer. Few studies have been reported for investigating air flow hydrodynamics and thermal characteristics in AC units following experimental and numerical procedures. Z.G.Xu et al. (1996) investigated the turbulent incompressible flow in an industrial air conditioning unit following a two-dimensional numerical analysis for simplified coil geometry. In their work, fine modeling of the turbulent flow through the evaporator coil tubes was considered by selecting a central symmetrical plane at the coil to predict the pressure losses. Gotham et al. (1993) modeled artificially the turbulent flow in an air conditioning unit by adopting the porous medium approach to model the evaporator coil. However, this situation is limited to a uniformly distributed flow resistance where the porous factor and Darcy coefficient are defined. Another experimental study was conducted by Z.G.XU et al. (1998) in an attempt to validate the solution of the turbulence models for an air conditioning unit for an incompressible isothermal flow. Experimental investigations were also done by Aganda et al. (2000) and Timoney and Foley (1993) on heat exchangers to capture the effects of maldistribution in the air flow and report its impacts on the heat exchanger performance. While previous studies have lacked many features including simplifying geometry, neglecting condensation, and introducing artificial drag models, recently Moukalled et al. (2011) developed a more sophisticated model for predicting and enhancing the performance of a roof-top air conditioning unit operating under actual indoor and outdoor environmental conditions. In their study, Moukalled et al. integrated a new feature for computing the water vapor condensation at the surfaces of the coil following Fick's Law (M.P.P. Valencia, 2004). Moreover, the physical model was considered in its real geometry, and thus no drag terms were introduced while building the CFD model for the unit. Results were generated for the base case and parametric models and validated with available experimental data for the unit. New designs were then suggested in an attempt to enhance the performance of the machine by incorporating different geometric modifications on evaporator coil. Based on the obtained numerical simulations, results have shown an enhancement in several suggested designs, highlighting to the fact that no added manufacturing costs are required to achieve this purpose. The current research extends the work of Moukalled et al. by combining the refrigerant circuits to the air side and simulating a full multi-physics conjugate heat transfer model. This research work aims at developing a numerical tool to help manufacturers optimize the performance of HVAC units. Details of the geometry, physical models, and predicted results are discussed though the rest of this manuscript.

MATHEMATICAL MODEL

Figures 1(a) and 1(b) show the details of the evaporator coil geometry. The coil specifications are the same as in (Moukalled et al., 2011).

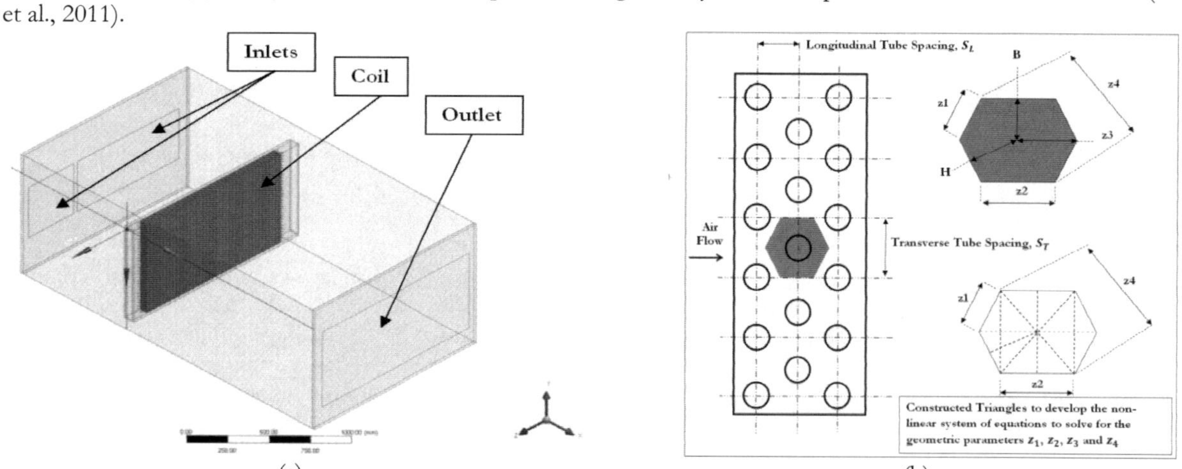

(a) (b)

Figure 1: (a) Model Geometry; (b) Flat plate finned staggered tube heat exchanger geometric parameters

Governing Equations

The problem possesses a muli-physics nature where fluid-solid thermal interactions are resolved by solving the set of governing mass, momentum, energy, turbulence, species and volume fraction equations. In addition, the multiphase mixture model is adopted to resolve the distribution of humid air species where condensation is accounted for at the evaporator coil surface. Details of the integrated governing equations for the mixture model are described briefly in what follows:

Continuity

$$\frac{\partial \rho_m}{\partial t} + \nabla.(\rho_m \boldsymbol{u_m}) = 0; \quad \boldsymbol{u_m} = \frac{\sum_k \alpha_k \rho_k u_k}{\rho_m}; \quad \rho_m = \sum_k \alpha_k \rho_k \tag{1}$$

Momentum

$$\frac{\partial(\rho_m \boldsymbol{u_m})}{\partial t} + \nabla.(\rho_m \boldsymbol{u_m} \boldsymbol{u_m}) = -\nabla p + \nabla.(\bar{\bar{\tau}}_{ij}) + \rho \boldsymbol{g} + \boldsymbol{F} + u_{dr,k} \sum_k \alpha_k \rho_k u_{dr,k} \tag{2}$$

Energy

$$\frac{\partial}{\partial t}(\sum_k \alpha_k \rho_k E_k) + \nabla.[\sum_k u_k(\rho_k E_k + p)] = \nabla.(K_{eff} \nabla \mathrm{T}) + S_h; \quad S_h = \dot{Q}_{lat}/volume = \dot{m}_{cond} L\,/volume \tag{3}$$

Species

$$\frac{\partial \rho Y_v}{\partial t} + \nabla.(\rho u Y_v) = \nabla.\left(\rho D_{v,m} + \frac{\mu_t}{SC_t}\right) \nabla Y_v + S_v \tag{4}$$

Volume fraction

$$\frac{\partial(\alpha_k \rho_k)}{\partial t} + \nabla.(\alpha_k \rho_k \boldsymbol{u_m}) = -\nabla.(\alpha_k \rho_k u_{dr,k}) + S_\alpha \tag{5}$$

Turbulence model

The realizable k-ε two equation model was implemented. The governing equations for $k$and ε are:

$$\frac{\partial(\rho k)}{\partial t} + \nabla.(\rho u k) = \nabla.\left[\left(\mu + \frac{\mu_t}{\sigma_k}\right) \nabla k\right] + G_k + G_b - \rho \varepsilon + S_k \tag{6}$$

$$\frac{\partial \rho \varepsilon}{\partial t} + \nabla.(\rho u \varepsilon) = \nabla.\left[\left(\mu + \frac{\mu_t}{\sigma_\varepsilon}\right) \nabla k\right] + \rho C_1 S_\varepsilon - \rho C_2 \frac{\varepsilon^2}{k + \sqrt{v\varepsilon}} + C_{1\varepsilon} \frac{\varepsilon}{k} + C_{3\varepsilon} G_b + S_\varepsilon \tag{7}$$

Condensation Model

The condensation rate at the coil surface is modeled as a source term (S) appearing in the species and volume fraction

$$\dot{m}_{cond} = -\rho C_v (D_{v,m} + D_t) \frac{Y_{v,cell} - Y_{sat,well}}{Distance_{centroid\ cell,centroid\ wall}} \tag{8}$$

NTU Method

A modified lumped heat transfer model based on the NTU Method was developed for the aim of comparing some numerical predictions obtained from the CFD model for a dry coil assumption, while the wet coil predictions were compared with published data of similar coil construction. The mathematical model of the NTU method is presented next. The refrigerant side heat transfer coefficient, h_r, is computed by

$$h_r = \frac{Nu_r K_r}{D_i}, \qquad Nu_r = \begin{cases} 4.36 & if\ Re_r < 2300 \\ 0.023 Re_r^{0.8} Pr^{0.4} & if\ Re_r > 2300' \end{cases} \qquad Re_r = \left(\frac{\rho U D_i}{\mu}\right)_r \tag{9}$$

The overall air side surface efficiency (η_o) for a multiple row staggered tubes plate-fin-and-tube heat exchanger is given by

$$\eta_o = 1 - \left(\frac{A_f}{A}\right)(1 - \eta_f), \qquad \eta_f = \frac{\tanh(m r_t \varphi)}{m r_t \varphi}; \qquad r_t = r_0 + \delta_t \qquad m = \sqrt{\frac{2h_a}{k_{fin}\delta_t}}, \quad L_{fin} \gg \delta_t \tag{10}$$

The fin efficiency parameter for a circular fin (φ) is calculated using

$$\varphi = \left(\frac{R_e}{r_t} - 1\right)\left(1 + 0.35 \ln\left(\frac{R_e}{r_t}\right)\right) \tag{11}$$

The equivalent circular fin radius (R_e) is defined as

$$R_e = \frac{P_{hex}}{2\pi}, \qquad P_{hex} = 4z_1 + 2z_2 \tag{12}$$

A four set of non-linear equations are developed to solve for the unknown geometric parameters z_1, z_2, z_3 and z_4 (Figure 1-b)

$$\begin{cases} z_1^2 = (z_3 - z_2/2)^2 + B^2 \\ z_4^2 = z_2^2 + (2B)^2 \\ z_3 = \dfrac{z_4}{2} \\ \left(\dfrac{z_4}{2}\right)^2 = \left(\dfrac{z_1}{2}\right)^2 + H^2 \end{cases} \qquad \begin{cases} B = S_L, & S_L < S_T/2 \\ B = S_T/2, & S_L \geq S_T/2 \end{cases} \qquad H = \frac{1}{2}\sqrt{\left(\frac{S_T}{2}\right)^2 + S_L^2} \tag{13}$$

The outer heat transfer coefficient for a staggered bundle of pipes (h_0) is computed as

$$h_0 = \frac{Nu_{D,NL}k_a}{D_0}; \quad Nu_{D,NL} = FNu_D; \quad Nu_D = CRe_D^m Pr^n (Pr/Pr_s)^{0.25} \tag{14}$$

The overall heat transfer coefficient for a dry coil surface is then computed

$$\frac{1}{UA} = \frac{1}{\eta_0\left[\left(\frac{N_{fins}-1}{1/h_oA_{bf}}\right) + \left(\frac{N_{fins}}{1/h_fA_f}\right) + \left(\frac{1}{1/h_oA_{uf}}\right)\right]} + \frac{\log(D_0/D_i)}{2\pi(L_{pipe} - N_{fins}d_{fin})k_{pipe}} + \frac{1}{h_rA_i} \tag{15}$$

The number of transfer units (NTU) is given by the following dimensionless parameter

$$NTU = \frac{UA}{C_{min}}; \qquad C_{min} = min\left[C_a = \left(\dot{m}C_p\right)_a, C_r = \left(\dot{m}C_p\right)_r\right] \tag{16}$$

For the cross-flow heat exchanger, the effectiveness (ε) is given by

$$\varepsilon = 1 - exp\left[\left(\frac{1}{C_r}\right)(NTU)^{0.22}\left\{e^{[-C_r(NTU)^{0.78}]} - 1\right\}\right]; \; C_r = \frac{C_{min}}{C_{max}}; \; C_{max} = max\left[C_a = \left(\dot{m}C_p\right)_a, C_r = \left(\dot{m}C_p\right)_r\right] \tag{17}$$

BOUNDARY CONDITIONS

The boundary conditions (tables 1 and 2) include the hydrodynamic, humidity ratio and thermal conditions for each of the air and refrigerant zones based on industrial data (*Moukalled et al., 2011*).

Table 1. Inlet boundary conditions for air and refrigerant zones

	Velocity (m/s)	Temperature (K)	w (kg/kg d.air)	Volume fraction (secondary phase 1)	Volume fraction (secondary phase 2)
Air Zone	3.99	300	0.0111	0	0
Refrigerant Zone	0.3233	279	-	0	1

Table 2. Outlet boundary conditions for air and refrigerant zones

	Pressure (Pa)	Backflow Temperature (K)	w (kg/kg d.air)	Volume fraction (secondary phase 1)	Volume fraction (secondary phase 2)
Air Zone	10	290	0.0055	0	0
Refrigerant Zone	557480	284	-	0	1

SOLUTION ALGORITHM AND BASE CASE RESULTS

A hybrid grid of total number of 30 million control volumes was generated to model the evaporator unit. The multiphase mixture model with species transport was integrated with one primary phase and two secondary phases. In the primary phase, dry air and water vapor were defined while in the secondary phases water condensate and refrigerant (water) are defined. User defined functions (for serial and parallel computations) were developed using C codes to compute the mass transfer rates (due to condensation of moist air) at the coil surface and computing the sensible, latent, and total heat transfer rates for the evaporator coil. The pressure and velocity coupling was resolved by the segregated SIMPLE-Corrected algorithm. While only the air side zone was simulated by Moukalled et al. (2011), in the current work, a multi-physics conjugate heat transfer problem is solved for ≈ 30x10 million variables including, pressure, x, y, and z velocities, energy, volume fraction, mass fraction, etc... by coupling the air and refrigerant zones with the solid regions (fins and pipes). Steady state computations were then simulated on a 60 core cluster machine. Results describing the flow properties were then analyzed at different planes defined at specific locations in the unit in terms of temperature, humidity ratio and velocity distribution. Figure 2 shows the temperature contours predicted at different (y-z) planes located upstream, crossing, and downstream the evaporator coil. In figure 3, velocity vectors colored by temperature were plotted at different (x-y) planes locations showing variations of hydrodynamic and thermal flow field between the inlet and outlet sections of the evaporator unit. Complex flow nature is induced at regions located in the upstream vicinity of the evaporator coil. This complex nature of the air flow field is also depicted in figures 4 and 5 showing some recirculation regions induced at the top and sides of the evaporator unit. This physical realization gives a future opportunity for optimizing the design of the unit to reduce such turbulent features in the flow field. Due to the condensation phenomenon occurring at the evaporator coil surface, lower water vapor fractions were predicted at different (y-z) planes (figure 6); however, non-uniform water vapor distribution in the air mixture was revealed for the same plane location due to the lower coil surface temperature at regions located near the refrigerant inlet sections, while water vapor contents increases with regions approaching the exit of refrigerant sections.

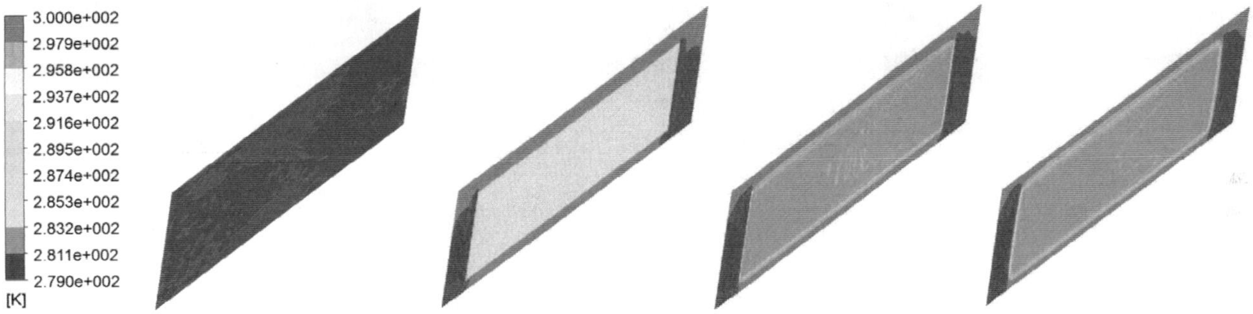

(a) (b) (c) (d)

Figure 2: Air zone temperature (K) contours at (y-z) planes. (a) x = -10cm; (b) x = 2cm; (c) x = 6cm; (d) x = 8cm

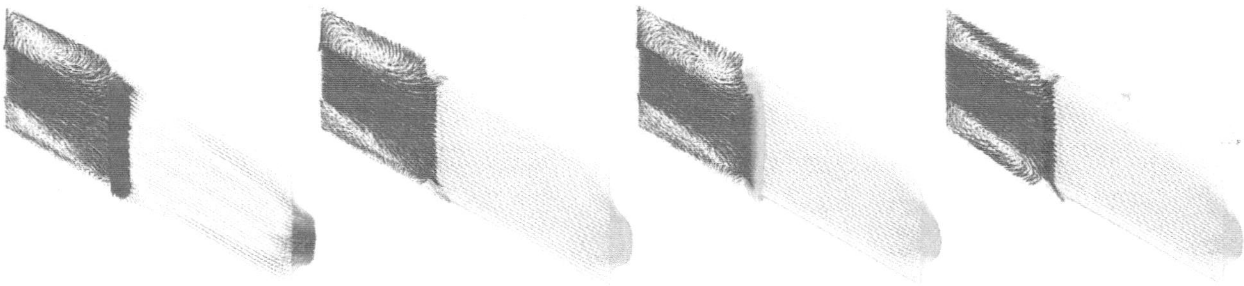

(a) (b) (c) (d)

Figure 3: Velocity vectors colored by temperature at (x-y) planes. (a) z = 0cm; (b) z = -10cm; (c) z = -40cm; (d) z = -70cm

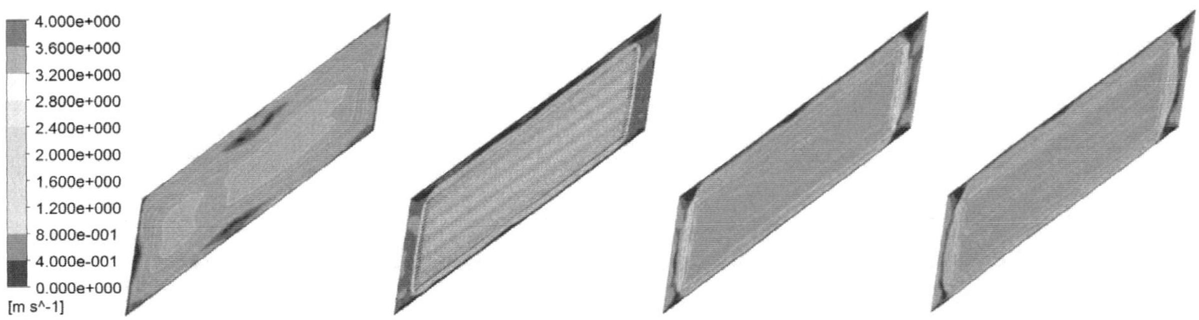

(a) (b) (c) (d)

Figure 4: Air zone velocity magnitude (m/s) contours at (y-z) planes. (a) x = -10cm; (b) x = 2cm; (c) x = 6cm; (d) x = 8cm

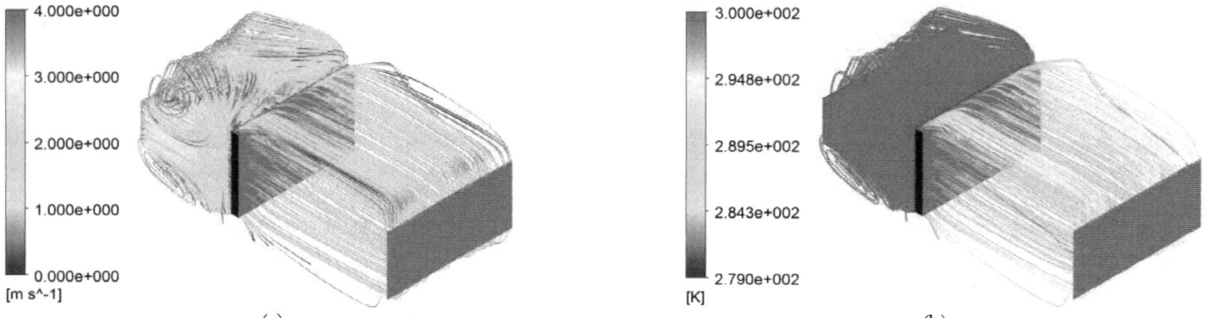

(a) (b)

Figure 5: Air zone streamlines colored by: (a) velocity magnitude; (b) temperature

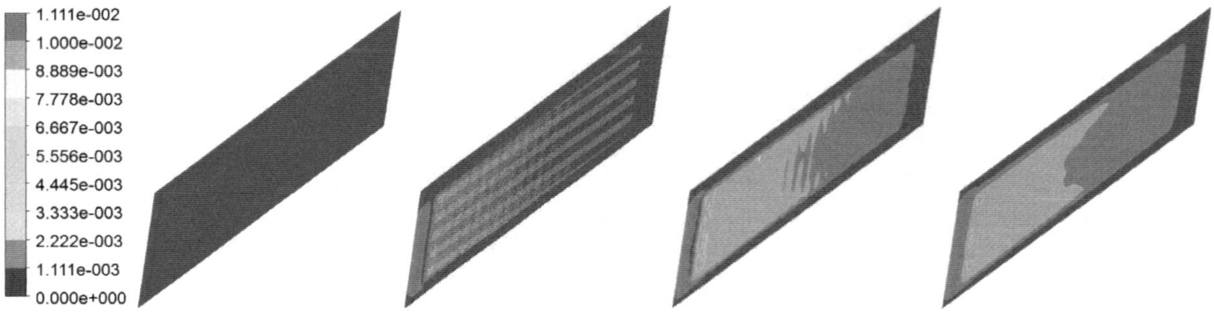

(a) (b) (c) (d)

Figure 6: Air zone humidity ratio (kg/kg) contours at (y-z) planes. (a) x = -10cm; (b) x = 2cm; (c) x = 6cm; (d) x = 8cm

Figures 7(a) and 7(b) show the streamlines of refrigerant colored by temperature. Variations in refrigerant's temperature are significant among the evaporator pipe rows where the first row of pipes reveals a higher temperature distribution and it decreases for the second and third rows. This is due to the hotter air entering the evaporator coil, and as the air flow crosses the coil staggered pipes, its temperature decreases gradually; thus reducing the temperature variation between the external coil surface and inner refrigerant. Accordingly, lower heat transfer rates occur at the last row of pipes, allowing for a refrigerant exit temperatures range between 282 and 285 K (Figure 7-c).

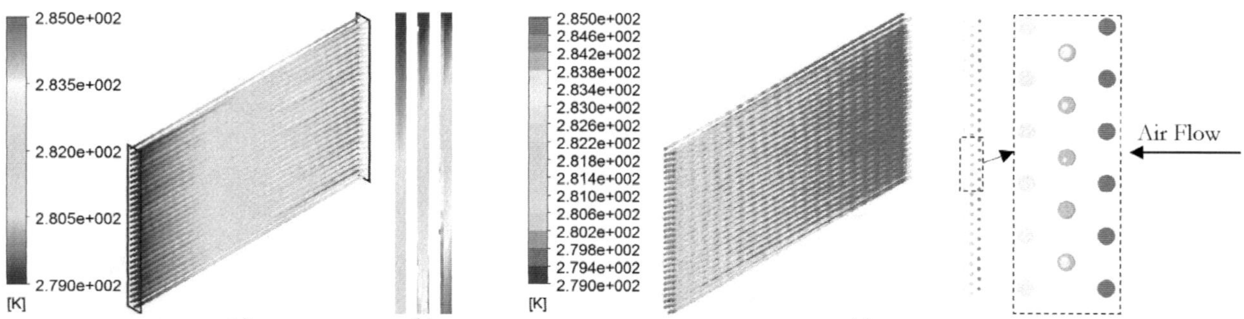

(a) (b) (c)

Figure 7: Refrigerant zone streamlines colored by temperature: (a) 3D view; (b) top view; (c) Refrigerant wall and outlets temperature

Discrepancies in water vapor species distribution were predicted at the evaporator coil surface (Figure 8-a). Higher values of water vapor mass fractions were revelaed at the tip of the fins and at entrance regions of the coil. These values decreases gradually as flow crosses axially the evaporator coil. However, discrepancies are also revealed in the z-direction showing lower values of water vapor mass fraction at regions near the refrigerant inlet sections and it increases at farther locations. Comparisons of coil surface temperaure for wet and dry coils are shown in figures 8-b and 8-c respectively. The analysis of temperature distribution at the coil surface is similar to the one reported for the water vapor mass fraction distribution.

However, it should be noted the increase of coil surface temperature for the case of a wet coil as compared to a dry coil. The computed area weighted average of the evaporator surface temperature was 284.6 K and 285.59 K for the dry and wet coils respectively. Table 3 shows some reported predictions obtained for a wet and dry coil cases.

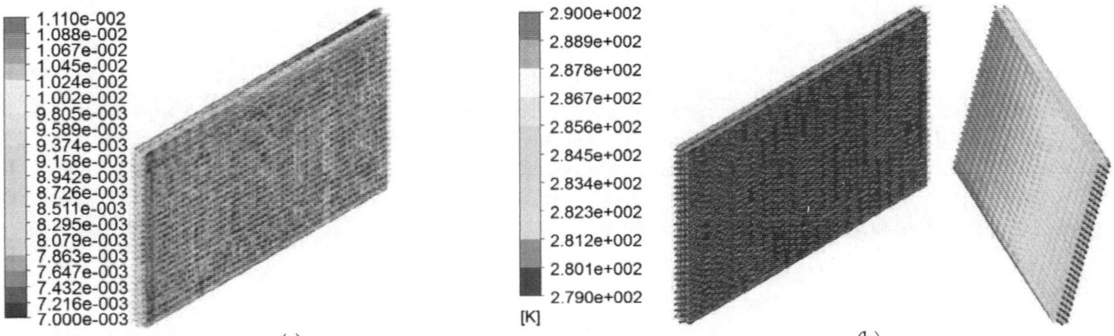

(a) (b)

Figure 8: (a) Humidity ratio contours on wet coil; (b) Temperature distribution on: (b) wet coil (3D view)

Table 3. Variation of air mass weighted average outlet temperature (K) and humidity ratio (kg/kg d.air), sensible, latent and heat transfer rate (W) and condensation rate (kg/s) for a wet and dry coil

	Base Case (dry coil)	Base Case (wet coil)
T_Outlet_air, (K)	288.5	291.2
w_Outlet_air, (kg/kg d.air)	-	0.0092227
$Q_{Sensible}$, (W)	27221	17678
Q_{Latent} (W)	-	11046.94
Q_{Total}, (W)	27221	28724.94
Condensation rate, (kg/s)	-	0.004445

MODEL VALIDATION

The numerical CFD model was compared with the lumped NTU analytical method (for the dry coil case) and with some existing experimental data (for the wet coil case) for a similar coil construction (Moukalled et al., 2011). For the wet coil simulation, the computed latent heat was 11046 W; whereas the latent capacity predicted by Moukalled et al. (2011) was 10000 W for a similar coil. This 9% difference is interpreted by the fact of the constant surface coil temperature assumption taken by Moukalled et al. (2011) where a model constant, C_v (Eq.8), was defined in computing the condensation rate at the coil surface; while in this study, no model constant was accounted for and condensation rate varies at different coil surface locations. Table 4 shows a comparison between numerical and the developed analytical model for a dry coil assumption. A maximum temperature difference of 0.5 K was obtained and a 4% error was reported for the sensible heat capacity between the two models. This discrepancy is due to the limited capabilities of the analytical model to capture the variations in temperature among the refrigerant rows; thus causing a slight over-prediction in the computations.

Table 4. Comparison of outlet air temperature (K) and sensible heat transfer rate (W) between CFD and NTU method for a dry coil

	Dry Coil (Numerical, CFD)	Dry Coil (Analytical)	% diff
T_Outlet_air (K)	288.5	288	0.17 %
$Q_{sensible}$ (W)	27221	28315	4 %

CONCLUSIONS

A numerical investigation was conducted to predict the performance of a water cooled evaporator unit using a finite volume based scheme. The current study extended the work of Moukalled et al. (2011) by integrating the refrigerant flow as an additional feature. Thus the analysis considered the hydrodynamic and thermal characteristics of the air and refrigerant zones simultaneously, causing a higher complexity in the modeling process. A full scale multi-physics CFD model was constructed to model the conjugate heat transfer phenomenon at the evaporator coil. The mass transfer mechanism occurring due to the condensation of water vapor at the coil surface was accounted for by adopting the mixture model with multi-species transport. Moreover, the model accounted for turbulence effects with the $\mathrm{k}-\varepsilon$ realizable model integrated to predict the chaotic behavior induced in the flow field. Two sets of results were presented in this manuscript simulating a wet and a dry coil. The numerical model was capable of accurately capturing the physical variations between the dry and wet coil assumption where lower temperature values were predicted at the dry coil surface, while a higher temperature was obtained at the wet coil surface. This phenomenon is due to the latent heat released from the condensation at the wet coil surface; thus leading to increase slightly the coil surface temperature. The simulated CFD model for the wet and dry coil assumptions was then compared with a developed lumped analytical NTU model and with available experimental data for similar coil construction (Moukalled et al., 2011); a good agreement was achieved. This research work aims to aid manufacturers in improving the design of HVAC units through building 'numerical experiments' capable of exploring detailed features of thermal and flow fields.

NOMENCLATURE

$\rho =$	Density (kg/m^3)	$D =$	Mass diffusion coefficient (m^2/s)	$\delta_t =$	Fin thickness (m)
$\boldsymbol{u} =$	Velocity (m/s)	$K =$	Thermal conductivity $(m/W\ K)$	$L =$	Fin Length (m)
$\alpha =$	Volume fraction	$Y =$	Species mass fraction	Subscripts	
$t =$	Time (s)	$\mathrm{T} =$	Temperature (K)	$dr, k =$	Drift velocity
$p =$	Pressure (Pa)	$k =$	Turbulent kinetic energy (m^2/s^2)	$k =$	Phase k
$\bar{\bar{\tau}}_{ij} =$	Stress tensor (N/m^2)	$\varepsilon =$	Turbulent dissipation rate (m^2/s^3)	$a =$	Air
$\boldsymbol{F} =$	Volumetric body force (N/m^3)	$\mu =$	Viscosity $(Pa\ s)$	$r =$	Refrigerant
$E =$	Total energy (J/m^3)	$Nu =$	Nusselt number	$f =$	Fin
$\sigma =$	Turbulent Prandtl number	$Re =$	Reynolds' number	$i =$	Inner
$G_k =$	Generation of turbulence kinetic energy due to mean velocity gradients $(N/m^2 s)$	$h =$	Heat transfer coefficient (W/m^2K)	$o =$	Outer
$G_b =$	Generation of turbulence kinetic energy due to buoyancy $(N/m^2 s)$	$A =$	Area (m^2)	$v =$	Water vapor

REFERENCES

F. Moukalled, S. Verma, M. Darwish, 2011. *The use of CFD for predicting and optimizing the performance of air conditioning equipment*, International Journal of Heat and Mass Transfer 54, 549–563.

Patankar, S.V. 1980. *Numerical Heat Transfer and Fluid Flow, series in computational methods in mechanics and thermal sciences.* Hemisphere Publishing Corporation.

Z. G. Xu, D. H. T. Gotham and M. W. Collins, J. E. R. Coney, C. G. W. Sheppard and S. Merdjani, 1996. *A numerical and experimental study of turbulent flow through the evaporator coil in an air-conditioning unit*, Int J. Refrig. Vol. 19, No. 6, pp. 369-381.

Gotham, D. H. T., Xu, Z. G., Collins, M. W. 1993. *Modelling the air flow through the EHA 5 air-conditioning system,* Research Memorandum RM 186, Thermo-Fluids Engineering Research Centre, City University, London.

Z.G. XU, D.H.T. GOTHA,, M.W. COLLINS, J.E.R. CONEY, C.G.W. SHEPPAR AND S. MERDJANI, 1998. *VALIDATION OF TURBULENCE MODELS IN A SIMULATED AIR-CONDITIONING UNIT*, Int. J. Numer. Meth. Fluids 26: 199–215.

A.A. Aganda, J.E.R. Coney, C.G.W. Sheppard, 2000. *Airflow maldistribution and the performance of a packaged air conditioning unit evaporator*, Applied Thermal Engineering 20, 515-528.

D.J. Timoney, P.J. Foley, 1993. *Some effects of air flow maldistribution on the air performance of a compact evaporator with R134a.* Eurotherm Seminar No. 26.

M.P.P. Valencia, 2004. *Condensation of water vapor and acid mixtures from exhaust gases*, Doctoral Dissertation, Institut Fur Energietechnik, Fachgebiet Energie-, Impuls- und Stofftransport, Berlin.

Schmidt, Th.E., 1949. *Heat transfer calculations for extended surfaces*, Refrigeration Engineering, April, 351-357.

Body Cooling Vest for Improving Endurance under Hot Environment

Haneen Hamdan

Nesreen Ghaddar, PhD
ASHRAE Member

Kamel Abou Ghali, PhD
ASHRAE Member

ABSTRACT

This work models a passive method of cooling the body to increase the work periods in hot environment. Cooling is performed by wearing a cooling vest containing packages of phase change materials (PCM) made of salt mixtures that melt at a fixed temperature to prevent increase in skin temperature thus maintaining comfort. A dynamic mathematical model of the heat and moisture transport in clothing layers with PCM material is developed. The fabric-PCM model is validated by experiment on clothed heated cylinder in wind tunnel under controlled environment. The internal air layers confined between the PCM and the heated cylinder (microclimate air) and between the PCM and the outer clothing (macroclimate air) are measured. Good agreement is found between experimental temperatures and predicted values by the fabric-PCM model with maximum errors of 4.5% and 2.5 % respectively.

INTRODUCTION

Workers in hot environment are exposed to heat strain that will reduce their working endurance. Under these conditions, the body will try to maintain its normal core temperature of 37 °C (310.15 K) by diverting the blood to the skin, rather than the muscles, in order to dissipate excess heat. If increased blood flow to the skin and the evaporative skin sweating are not enough to dissipate heat, the human core temperature will rise increasing the likelihood of thermal stress. Therefore, passive cooling methods such as clothing ventilation and cooling vests have been used to combat heat stress (Ghaddar et al., 2011).

Different types of personal cooling vests exist depending on blower fans, circulating liquid or phase change material (PCM) to cool the body. Cooling vests with incorporated phase change materials (PCM), ice, paraffin or salt, are considered as the most promising techniques. The very low melting temperature of ice may cause erythema leading to discomfort (House et al., 2013). The salt PCM is superior to ice since it can be incorporated in the fabric structure in microcapsules or used as packs with the possibility of controlling the melting temperature. There are different types of PCMs that are non-toxic, have high melting temperatures and large latent heat of fusion (Mondal, 2007). Cooling vests with incorporated salt packets have caught the attention of many researchers that tried to find the optimum design of this type of garments that suits a specified work and does not affect the wearer's mobility (Tate et al., 2008). Reinertsen et al. (2008) studied experimentally the performance of PCM material and the effect of its position on thermal comfort and sensation under simulated work situations. Published experiments on human subjects showed that the lower the melting temperature the stronger is the cooling effect on the torso and the mean skin temperature (Gao et al., 2011). Zhao et al. (2012) showed that the cooling vest is effective only in the hot humid case. Other than experimentation, numerical models were also performed to study the effect of wearing a PCM cooling vests (Yifen et al., 2011).

In this work, a dynamic thermal model is developed for a PCM cooling vest to simulate heat and mass transfer through the clothing layers with PCM packets integrated within the air layers. An experiment on a clothed heated cylinder is performed

Haneen Hamdan is a graduate student. Nesreen Ghaddar and Kamel Abou Ghali are professors in the Department of Mechanical Engineering, American University of Beirut, Beirut, Lebanon in a controlled chamber to validate the temperatures of the air gaps trapped between the PCM packets and the fabric layers.

FABRIC-PCM MATHEMATICAL MODEL

To mathematically model the effect of varying the location, weight and melting point temperature of the PCM materials on the air layers temperatures of the clothing system, a fabric-PCM model is needed. This model calculates the temperature and the moisture content of the different fabric layers and the associated air gaps between them at given ambient conditions and at constant skin temperature and for known number of PCM packets and their thermophysical properties along with the type of clothing and their dry and evaporative resistances.

The vest is made of a low evaporative resistance fabric to facilitate moisture transfer to the environment and it contains pockets to hold the PCM packs. Figure 1 depicts schematic of (a) side view of the skin and the different clothing, PCM, and air layers and (b) front view of the cooling vest in which PCM packets are placed on the upper and lower segments of the vests. The thin inner layer of cotton is necessary for preventing any direct contact of the PCM with the skin. Above this layer is the inner air layer followed by the distributed PCM material of known quantity and known melting point temperature. An outer protective fabric layer then covers the PCM packs. There is also an air layer separating the PCM and the outer fabric. For this arrangement, the PCM packet material is hanging between the inner and outer layers without being bonded to either layer but it is closer to the underlying layer because of the Velcro fasteners.

In work conditions, the heat and moisture from the body surface is transferred to the microclimate air whose temperature depends on the amount of the transported heat and the enclosed PCM material. The macroclimate air temperature is modeled as lumped air layer with a uniform temperature between the PCM packets and behind the packets as shown in the Figure. 1. In the regions between the PCM packets and the underlying clothing, the microclimate temperature is different than the surrounding macroclimate air since Velcro fasteners are used to bring the PCM material closer to the body.

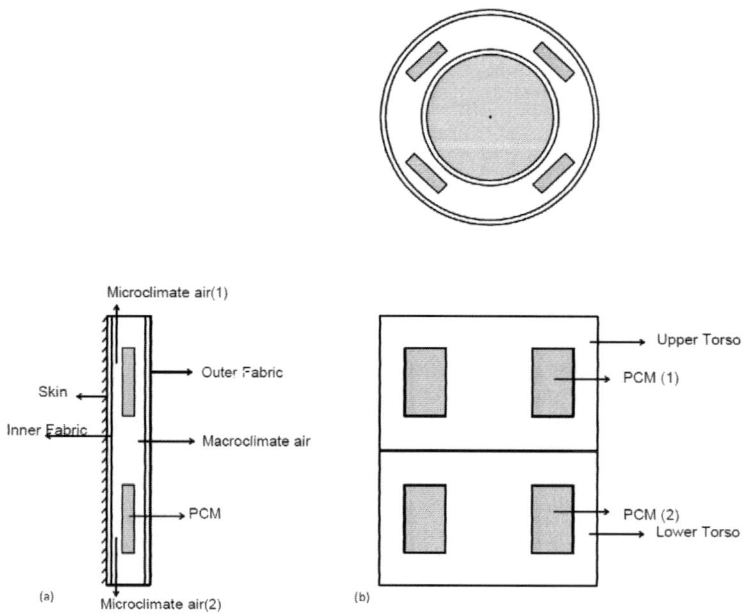

Figure 1: Schematic of (a) side view of the skin and the different clothing, PCM, and air layers and (b) front and top view of the cooling vest in which PCM patches are placed on the upper and lower segments of the vests

Governing Equations

Mass and energy balance equations for heat and mass transport are written for each fabric and PCM layer. The underlying clothing layer exchanges heat and water vapor transfer with the skin and the microclimate air layer. The region of microclimate air layer between each PCM piece and the underclothing layer is considered as lumped. However, each region has its own conditions depending on the PCM melting temperature and its position. On the other hand, the macroclimate air layer along the vest is basically at a lumped air temperature.

The Mass and energy balances for the underlying fabric layer are respectively given by:

$$\rho_{if} * e_{if} * \frac{dR_{if}}{dt} = \frac{P_{skin} - P_{if}}{\frac{R_{e,if} * h_{fg}}{2} + \frac{1}{2 * h_m}} + \frac{A_f - \sum A_{pcmj}}{A_f} * \frac{P_{air} - P_{if}}{\frac{R_{e,if} * h_{fg}}{2} + \frac{1}{2h_m}} + \frac{\sum A_{pcmj}}{A_f} * \frac{P_{aj} - P_{if}}{\frac{R_{e,if} * h_{fg}}{2} + \frac{1}{2h_m}} \tag{1}$$

$$\rho_{if} * e_{if} * (C\frac{dT_{if}}{dt} - h_{ad}\frac{dR_{if}}{dt}) = \frac{T_{skin} - T_{if}}{\frac{R_{d,if}}{2} + \frac{1}{2h_f}} + \frac{A_f - \sum A_{pcmj}}{A_f} * \frac{T_{air} - T_{if}}{\frac{R_{d,if}}{2} + \frac{1}{2h_f}} + \frac{\sum A_{pcmj}}{A_f} * \frac{T_{aj} - T_{if}}{\frac{R_{d,if}}{2} + \frac{1}{2h_f}} \tag{2}$$

Tair is the temperature of the air layer (macroclimate air) sandwiched between the outer fabric, the PCM, and the inner fabric as shown in Figure 1, Taj is the air temperature trapped between the PCM and inner fabric, Apcm,j is the surface area of the PCM packet and n is the number of PCM packets distributed inside the cooling vest.

When there is no phase change (Tpcm ≠ PCM melting temperature), the PCM is treated as a dry intermediate layer and its energy equation is then similar to fabric layer equations as follows:

$$m_{pcmj} * C_{pcmj} * \frac{dT_{pcmj}}{dt} = h_c * A_{pcmj} * (T_{aj} - T_{pcm}) * \frac{r_{if}}{r_{pcm}} + h_c * A_{pcmj} * (T_{air} - T_{pcm}) \tag{3}$$

The ratio r_{if}/r_{pcm} takes into account the increase in the resistance due to difference in the areas of the different layers due to the cylindrical geometry. Figure 2 shows the radii of the different layers

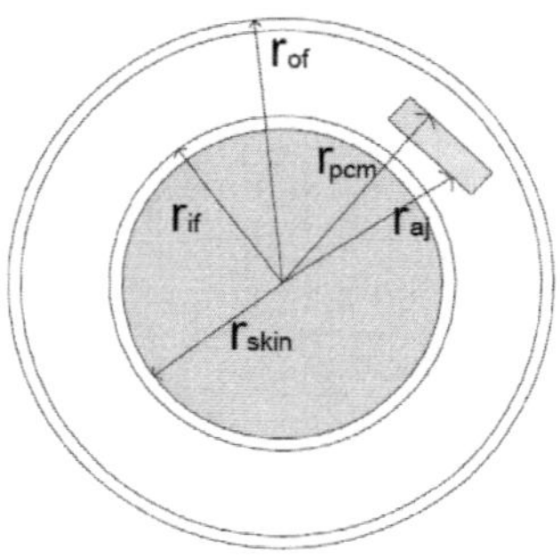

Figure 2: Radii of different Layers

During the phase change, Tpcm = constant and it is equal to the PCM melting temperature and the heat transport equation is given by:

$$h_c * m_{pcmj} * \frac{d\alpha_j}{dt} = h_c * A_{pcmj} * (T_{aj} - T_{pcm}) * \frac{r_{if}}{r_{pcm}} + h_c * A_{pcmj} * (T_{air} - T_{pcm}) \tag{4}$$

The mass and energy balances for the outer fabric layer are respectively given by:

$$\rho_{of} * e_{of} * \frac{dR_{of}}{dt} = \frac{P_{env} - P_{of}}{\frac{R_{e,of} * h_{fg}}{2} + \frac{1}{2h_m}} + \frac{P_{air} - P_{of}}{\frac{R_{af} * h_{fg}}{2} + \frac{1}{2h_m}} * \frac{r_{if}}{r_{of}} \tag{5}$$

$$\rho_{of} * e_{of} * (C\frac{dT_{of}}{dt} - h_{ad}\frac{dR_{of}}{dt}) = \frac{T_{env} - T_{of}}{\frac{R_{d,of}}{2} + \frac{1}{2h_f}} + \frac{T_{air} - T_{of}}{\frac{R_{d,of}}{2} + \frac{1}{2h_f}} * \frac{r_{if}}{r_{of}} \tag{6}$$

Microclimate and Macroclimate Air Modeling. Assuming that the gap width between the PCM pack and the underlying clothing layer to be small, the air flow in this region that is triggered by buoyancy forces caused by temperature differences and can be assumed to follow the Poiseuille flow regime due to the small gap width. The mass flow rate of air is represented by Bejan (1994) as:

$$\dot{m}_a = \rho_{air} * g * \beta * \frac{T_{air} - T_{aj}}{12\upsilon} * d^3 * w \tag{7}$$

The mass and energy balances for microclimate air are respectively given by

$$\rho_{air} * d * \frac{dw_{aj}}{dt} = \sum \frac{P_{if} - P_{aj}}{\frac{R_{e,} * h_{fg}}{2} + \frac{1}{2h_m}} * \frac{r_{if}}{r_{aj}} + \dot{m}_a * (w_{air} - w_{aj}) * \frac{1}{d * w} \tag{8}$$

$$\rho_{air} * d * A_{pcn,j} * C\frac{dT_{aj}}{dt} = A_{pcmj} * h_c * (T_{pcm} - T_{aj}) + A_{pcmj} * h_c * (T_{if} - T_{aj}) * \frac{r_{if}}{r_{aj}} + \dot{m}_a * C_{air} * (T_{air} - T_{aj}) \tag{9}$$

The mass and energy balances for macroclimate air are respectively given by

$$\rho_{air} * e_{air} * \frac{dw_{air}}{dt} = \frac{P_{of} - P_{air}}{\frac{R_{e,} * h_{fg}}{2} + \frac{1}{2h_m}} \sum \dot{m}_a * (w_{aj} - w_{air}) * \frac{1}{d * w} + \frac{A_f - \sum A_{pcmj}}{A_f} * \frac{P_{if} - P_{air}}{\frac{R_{e,} * h_{fg}}{2} + \frac{1}{2h_m}} * \frac{r_{if}}{r_{air}} \tag{10}$$

$$\begin{aligned} \rho_{air} * e_{air} * (C * \frac{dT_{air}}{dt} - h_{fg} * \frac{dw_{air}}{dt}) &= h_c * (T_{of} - T_{air}) + \sum \frac{A_{pcmj}}{A_f} * h_c * (T_{pcmj} - T_{air}) + \\ \frac{A_f - \sum A_{pcmj}}{A_f} * h_c * (T_{if} - T_{air}) + h_{fg} * \frac{P_{of} - P_{air}}{\frac{R_{e,} * h_{fg}}{2} + \frac{1}{2h_m}} &+ h_{fg} * \frac{A_f - \sum A_{pcmj}}{A_f} * \frac{P_{if} - P_{air}}{\frac{R_{e,} * h_{fg}}{2} + \frac{1}{2h_m}} * \frac{r_{if}}{r_{air}} \\ + \sum \dot{m}_a * C_{air} * (T_{air} - T_{aj}) \end{aligned} \tag{11}$$

A first order numerical model will be developed using standard numerical integration techniques. The equations will be solved algebraically using the explicit Euler forward method providing the initial conditions such as the temperature and the pressure of all the different layers of clothing and air, as well as the initial PCM temperature.

EXPERIMENTAL METHADOLOGY

The aim of the experiment is to validate the results of lumped temperatures of the microclimate air layer (adjacent to human body) and the macroclimate air layer (adjacent to environment) as predicted by the fabric-PCM model in the melting phase. A commercially available PCM packet is inserted between a heated cylinder and a fabric placed in a controlled chamber. The PCM used is made of a non-toxic, non-flammable salt mixture sealed inside an aluminum wrapper. The melting temperature of the PCM used in the experiments is 28°C (301.15 K).

The experimental setup is placed in a climatic chamber to provide controlled environmental conditions at 28 ± 0.5 °C and relative humidity of 50 ± 2% (Ghadar et al., 2005). In order to hold the PCM packet and the fabric, three concentric metallic mesh cages placed around the heated cylinder were used allowing the formation of two air layers, the microclimate air confined between the PCM packet and the heated cylinder and the macroclimate air confined between the PCM packet and the fabric. Figure 3(a) shows a schematic of the clothed heated cylinder experimental setup and other components. The heated cylinder is wrapped with metallic resistance heaters Omega-KH-1012 to produce constant heat flux condition per heater at the cylinder surface. The heat fluxes from the heaters are regulated independently such that an almost uniform mean steady temperature of 35 °C ± 0.5 °C is attained for each strip heater due the high conductivity of the surface heaters.

The temperature of each air layer is measured by two thermocouples mounted on the mesh cages at two positions in the region adjacent to the PCM packet as shown in Figure 3(b). The PCM was kept at 20 °C (293.15 K) to solidify all night and at the onset of the experiment, its temperature was 20 °C (293.15 K). The average value of the two readings was compared against the predicted one. The thermocouple readings, the climatic chamber conditions are monitored in time by a data acquisition system of national instruments SCXI 1121, and Labview software. The temperatures were sampled at 10 reading/s and are averaged and recorded every 60 seconds with an accuracy of ±0.1°C.

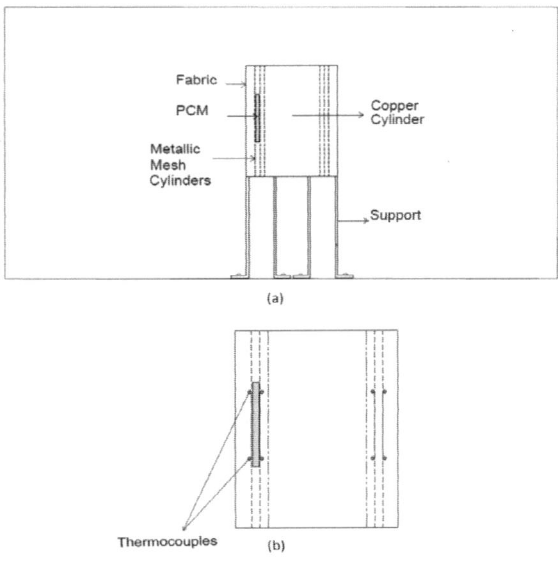

Figure 3: Schematic of (a) the clothed heated cylinder setup in the chamber and (b) the location of the thermocouples

Experimental Results

The clothed heated cylinder with the PCM geometry was simulated at the same experimental environment temperature of 28°C (301.15 K) to validate model predictions and experimental measured values of microclimate and macroclimate air temperatures. Figure 4 presents the averaged experimental readings and the model predictions of (a) the microclimate air layer temperature and (b) the macroclimate air layer temperature during the melting phase. A good agreement is found in general between the measured and predicted values of microclimate and macroclimate air temperatures. Examining the microclimate air, which resembles the air gap between the PCM and the heated cylinder, the maximum error was found to be about ±4.5%, corresponding to 1.1°C. However, the maximum error for the macroclimate air, which is the air trapped between the PCM and the outer fabric exposed to the environment, was around 2.5% during the melting phase which corresponded to ±0.5°C. The errors are easily explained by the lumping assumption of the PCM and the consideration of a uniform temperature at all positions throughout the melting phase when the PCM reaches melting point. In real applications, the inner surface of the PCM melts first. Therefore, in the microclimate air, the experimental value of temperature was lower than the predicted one because the melting at the inner surface of the PCM in the experiment started before the predicted one by the model, which assumes uniform temperature across the whole mass. As the melting process continued, the experimental temperature of the microclimate air did not show the slight decrease similar to the predicted values because of the slight increase of the PCM inner surface facing the heated skin. On the other hand, the experimental values of the macroclimate air temperature was greater than that of the predicted one by the model at the beginning of the experiment because it is affected by the outer surface of the PCM packet that experienced a delayed onset time of melting compared to the inner surface.

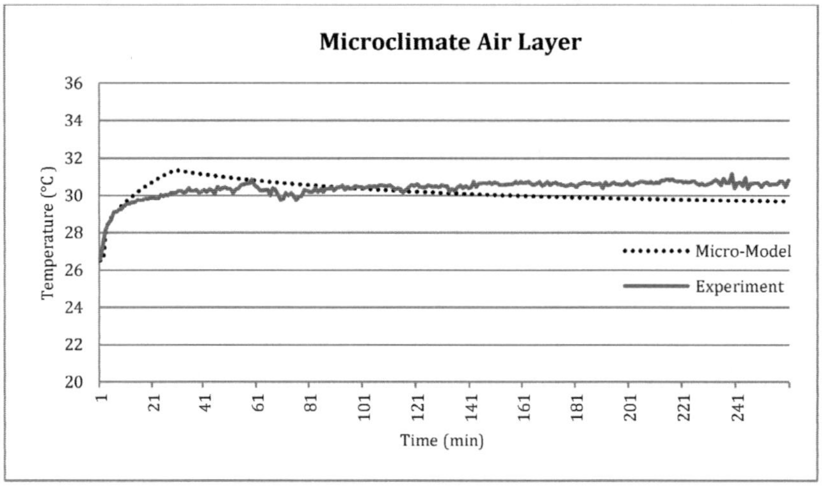

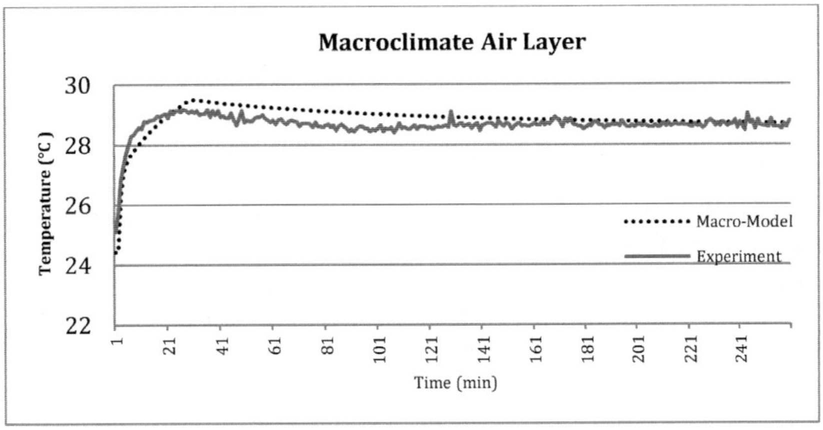

Figure 4 (a-b): Plots of the averaged experimental readings and the model predictions of (a) the microclimate air layer temperature and (b) the macroclimate air layer air layer temperature during the melting phase.

CONCLUSION

A mathematical model is developed and is validated by experiment to simulate heat and mass transfer in a cooling vest containing salt PCM packets. The model predictions of the air layers temperatures show a good agreement with the experimental findings with slight differences due to the lumping assumption of the air layers and the PCM taken by the model. In future work, the validated fabric-PCM model may be integrated with a multi-node segmental and transient bio-heat model capable of predicting accurately the segmental skin and core temperature of all body segments (Karaki et al., 2012). Therefore, the skin temperature obtained from the bioheat model can be used in the fabric-PCM model so that both continuity of flux and skin temperature are observed at each time step. Thus the skin and the core temperatures of the segments where the PCMs will depend on the microclimate air temperature calculated by the fabric-PCM model. This integration between the two models may be used latter as an optimization tool to reduce the weight of the vest while maintaining its performance.

NOMENCLATURE

A = area (m^2, ft^2)
C_p = thermal capacitance (J/kg.k, Btu/lb.K)
e = thickness (m, ft)
g = gravitational acceleration (m/s^2, ft/ s^2)
h_{fg} = heat of evaporation (J/kg, Btu/lb)
h_{ad} = heat of adsorption (J/kg, Btu/lb)
h_m = mass transfer coefficient (kg/m2·kPa·s, lb/ft^2.kbars.s)
h_f = heat transfer coefficients (kg/m2·kPa ·s, lb/ft^2.kbars.s)
hc = convective heat transfer (W/m2·K, $Btu/hr.ft^2.K$)
m = mass (kg, lb)
p = vapor pressure (kPa, kbars)
r = radius
Rd = dry Resistance of the fabric (m2·K/W, ft^2.K.hr/Btu)
Re = evaporative Resistance of the fabric(m2·kPa/W, ft^2.K.hr/Btu)

R = fabric regain

w = humidity ratio (kgw/kgair, lbw/lbair)

Greek Symbols

α = fraction of melted PCM

β = volumetric thermal expansion (°C-1)

ρ = density (kg/m3, lb/ft^3)

ν = kinematic viscosity of air (m2/s, ft^2/s)

Subscripts

air = macroclimate air between the PCM and the environment

aj = microclimate air between PCM and skin

env = environment

of = outer fabric

if = inner fabric

pcm = phase change material

REFERENCES

Bejan, A. 1994. Convection Heat Transfer, 2nd Edition, Wiley-Interscience. **(ISBN:** 0471579726)

Gao, C., K. Kuklane , K. ,Holmér, I. 2010. Cooling vests with phase change material packs: the, Ergonomics. Vol. 53, no. 5, pp. 716-723.

Gao, C., K. Kuklane , K. ,Holmér, I. 2011. Cooling vests with phase change materials: the effect of melting temperature on heat strain alleviation in an extremely hot environment. Eur J Appl Physiol 111:1207-1216

Ghaddar, N., Ghali, K., Harathani, J., and Jaroudi, E., 2005. Ventilation Rates of Micro-climate Air Annulus of the Clothing-skin System under Periodic Motion. Int. J Heat Mass Transfer, 48(15), pp. 3151-3166.

Ghaddar N., K. Ghali, S Chehaitly, Assessing thermal comfort of active people in transitional spaces in presence of air movement. 2011. Energy and Buildings, accepted for publication, 10.1016/j.enbuild.2011.06.040.

House, J.R., Lunt, H.C. and Taylor, R. 2013. The impact of a phase-change cooling vest on heat strain and the effect of different cooling pack melting temperature. Eur J Appl Physiol . 113:1223-1231

Karaki W., Ghaddar, N., Ghali, K., Kaleve, K., Holmer, I. and Vanguard, L.L. 2012. Human Thermal Response with Improved AVA Modeling of the Digits. *International Journal of Thermal Sciences*, accepted for publication

Mondal, S. 2007. Phase change materials for smart textiles-An overview. Applied Thermal Engineering

Reinertsen, R.E., Fæ̈revik, H. and Holbǿ, K. 2008. Optimizing the Performance of Phase-Change Materials in Personal Protective Clothing System. International Journal of Occupational Safety and Ergonomics (JOSE), vol. 14, no. 1, p. 43–53.

Tate, M., Forster, D. and Mainwaring, D.E. 2008. Influence of garmet design on elite athlete cooling. Sports technol. 1, No. 2-3, 117-124

Yifen, Q., Nan, J., Weia, W., Guangweia, Z. and Baoliang, X. 2011. Heat Transfer of Heat Sinking Vest with Phase-change Material. Chinese Journal of Aeronautics, 24(6): 720–725.

Zhao, M., Gao, C., Wang, F., Kulkane, K. and Holmer, I. 2012. The Torso cooling of Vests Incorporated with Phase Change Materials (PCMs): A Sweat Evaporation Perspective. Textile Research Journal 0: 1–8

Predicting Ventilation Rates of Clothed Human Segment

Nagham Ismail

Nesreen Ghaddar, PhD
Member ASHRAE

Kamel Abou Ghali, PhD
Member ASHRAE

ABSTRACT

Most comfort indices for indoor environment are based on static ensemble clothing insulation and assumed sedentary activity level at low air velocities. However, to consume less energy, research has recently focused on increasing velocity to bring thermal comfort at elevated air temperatures by enhancing ventilation through clothing. This work aims to implement an integrated heat transport model of clothed human body subject to external wind. Each clothed segment of the human body is modeled as a vertical annulus of a heated cylinder surrounded by a permeable cylinder, subject to cross uniform wind with open or close end to the environment, depending on the aperture configuration, in the presence of natural convection. The flow and heat transport characteristics are obtained by solving the steady state mass and energy balance equations of body segments 'microclimate air annulus numerically. Experiments were performed in a low speed wind tunnel in which an isothermally heated vertical clothed cylinder with open bottom aperture was placed in uniform cross wind. Good agreement was found between the model predictions and experimental measurements of temperature at different angular and vertical location in the microclimate air layer. It is found that clothed segments opened from bottom increases ventilation rate by 40 % when compared to clothed segments opened from top. Furthermore, an increase of wind speed by 1 m/s (3.28 feet/s) leads to an increase of about 36 % of ventilation rate. Permeability also plays another role in enhancing ventilation and total heat loss from the clothed segment especially when natural convection is important. At wind velocity less than 2 m/s (6.56 feet/s), the increase in ventilation and heat loss by moving from semi permeable clothing to permeable clothing were around 20% and 19 % respectively.

INTRODUCTION

In order to consume less energy in buildings, researchers have recently focused on increasing indoor air velocities to enhance clothing ventilation. Thus, thermal comfort is reached at elevated air temperature (as well as localizing higher air flow to upper body segments enhancing associated ventilation (Melikov, 2004; Cermal et al., 2006; Khalifa et al., 2009; Ghaddar et al., 2011). Clothing ventilation is predicted by two ways: empirical using correlations or by modeling. Indeed, predicting local ventilation by modeling has advantages in studying the effect of clothing apertures, permeability, wind conditions, air layer size, in increasing ventilation and reducing energy. Clothing microclimate ventilation is essentially related to the flow characteristics in the air layer between skin and clothes (Ghali et al., 2002), the clothing permeability (Lotens, 1993) and on the open clothing apertures (Ghaddar et al., 2005, 2010$_b$).The flow characteristics in the air layer are the results of two phenomena: natural convection associated with warm body skin and forced flow induced by wind effect that causes penetration of air through clothing and induced flow through the opening driven by pressure difference between ambient and air layer air (Sobera et al., 2003; Kind et al., 1991; Leong et al., 2006). This modeling approach was implemented by

Nagham Ismail is a PhD student. Nesreen Ghaddar and Kamel Ghali are professors in the Department of Mechanical Engineering, American University of Beirut, Beirut, Lebanon

Ghaddar et al. (2010a, 2010b) to predict the local ventilation for the trunk. However, they used empirical correlations to estimate the ventilation through clothing covering the limbs. Thus, it is of interest to expand the ventilation model to clothed limbs which normally have the clothing aperture open at the bottom especially that the literature does not have the effect of mixed convection on open bottom clothed cylinder which represents the major clothed segments of human body. This work aims to develop a local clothing ventilation model to predict clothed human ventilation and heat loss subject to external wind. Local ventilation rates for clothed limbs will be estimated based on modeling of mixed convection in microclimate air layer of clothed heated cylinder with bottom open aperture. The clothed heated cylinder model with bottom open aperture will be validated by experiments in a wind tunnel.

PHYSICAL MODEL

Fig. 1 (a) represents the physical configuration of this study. It consists of two concentric cylinders of radius R_s and R_f and height L. An isotropic fabric layer of permeability $\alpha$and thickness e_f covered the outer porous cylinder. The inner solid cylinder is maintained at temperature T_{skin}. A microclimate air annulus of thickness $Y = R_f - R_s$ is trapped between these two cylinders and considered small compared to L allowing the transformation from 3-D to 2-D problem and the use of plane Poiseuille flow in axial and angular directions. Thus, the radial direction is lumped. The top end of the annulus is closed adiabatically, while the bottom end is open to the environment at temperature T_{amb}. The configuration is placed in perpendicular to an air flow at V_∞. The outer fabric cylinder will be modeled using the three-node-fabric model of Ghali et al. (2002), which uses a lumped layer of two fabric nodes (outer and inner) and an air void node to represent thin fabric medium as shown in Fig.1 (b). The fabric outer node, which represents the exposed surface of the yarns, is in direct contact with the penetrating air in the air void node between yarns. However, the inner fabric node represents the inner portion of the solid yarn surrounded by the fabric outer node.

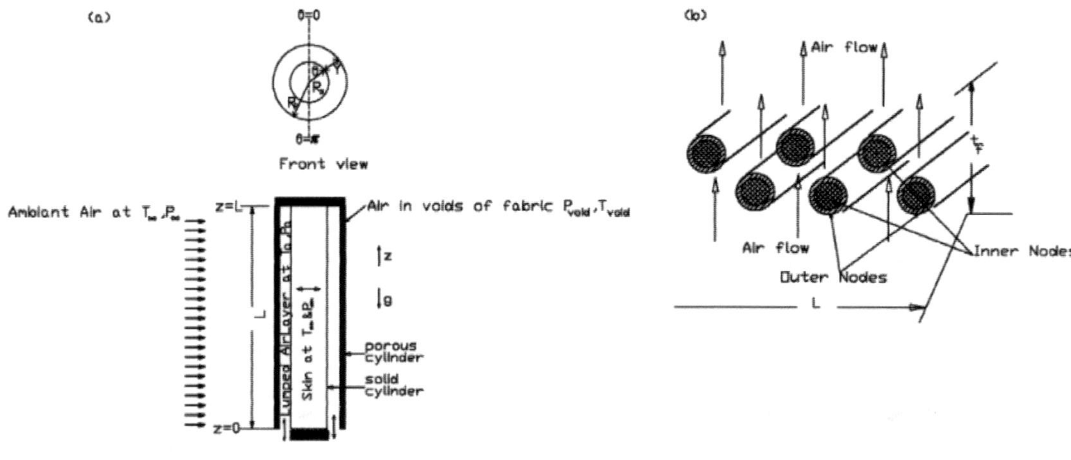

Figure 1 *(a) physical configuration of the clothed vertical cylinders with open end, (b) cylindrical representation of three-node fabric layer*

GOVERNING EQUATIONS

The mathematical formulation starts with the air annulus mass conservation equation given by:

$$\frac{\partial(Y\dot{m}_{az})}{\partial z}+\frac{\partial(Y\dot{m}_{a\theta})}{R_f\partial\theta}-\dot{m}_{aY}=0 \quad (1)$$

$$\dot{m}_{aY}=\frac{\alpha\rho_a(P_s-P_a)}{\Delta P_m} \quad (2)$$

$$\dot{m}_{a\theta}=-\frac{Y^2}{12R_f\nu}\frac{dP_a}{d\theta} \quad (3)$$

$$\dot{m}_{az}=-\frac{Y^2}{12\nu}\frac{dP_a^*}{dz} \quad (4)$$

$$P_a^*=P_a+\rho_a gz \quad (5)$$

$$\dot{m}_{an}=\rho\frac{g\beta Y^2}{12\nu}(T_a-T_{void}) \quad (6)$$

The radial, angular, and upward mass flow rates of equations. (2), (3), (4) and (6) are substituted in equation (1) to get the mass conservation equation in pressure and temperature form as

$$\frac{\alpha\rho_a(P_a-P_s)}{\Delta P_m}+\frac{\partial}{R_f^2\partial\theta}(\frac{Y^3}{12\nu}\frac{\partial P_a}{\partial\theta})+\frac{\partial}{\partial z}(\frac{Y^3}{12\nu}\frac{\partial P_a^*}{\partial z})+\frac{\rho g\beta Y^3}{12\nu}\frac{d(T_a-T_{void})}{dz}=0 \quad (7)$$

The dry energy balance of the air layer is then given by

$$h_{c\,skin-air}(T_{skin}-T_a)+h_{c-o}(T_o-T_a)+\max(\dot{m}_{ay},0)c_pT_{void}-\max(-\dot{m}_{ay},0)c_pT_a-\frac{\partial(Y.q_{\dot{m}_{az}})}{\partial z}$$

$$-\frac{\partial(y.q_{\dot{m}_{a\theta}})}{R_f\partial\theta}+\frac{k_a}{R_f^2}\frac{\partial}{\partial\theta}(Y\frac{\partial T_a}{\partial\theta})+k_a\frac{\partial}{\partial z}(Y\frac{\partial T_a}{\partial z})=0$$

(8)

$$q_{\dot{m}_{az}}=\dot{m}_{az}.c_p.T_a$$

$$q_{\dot{m}_{a\theta}}=\dot{m}_{a\theta}.c_p.T_a$$

(9)

The dry energy balance of the fabric void node is given by

$$q_{\dot{m}_{ay}}+H_{co}\times(T_o-T_{void})=0 \quad (10)$$

Where $q_{\dot{m}_{ay}}=\dot{m}_{ay}\times cp_{fabric}\times(T_{void}-T_a) \qquad if \quad P_s<P_a$

$$q_{\dot{m}_{ay}} = \dot{m}_{ay} \times cp_{fabric} \times (T_{amb} - T_{void}) \quad if \quad P_s > P_a \tag{11}$$

The dry energy balance of the fabric outer node is then given by

$$h_{coae} \times (T_{amb} - T_o) + H_{co}(T_{void} - T_o) + \frac{h_r}{2}(T_{skin} - T_o) + \frac{h_r}{2}(T_{amb} - T_o) = 0 \tag{12}$$

The total ventilation rate is calculated as the positive flow of air into the air annulus integrated per unit area of the clothed surface as follows:

$$\dot{m}_a = \frac{1}{\Pi L} \int_0^L \int_0^{\Pi} max(0, \dot{m}_{ay}) d\theta \, dZ \tag{13a}$$

The segmental sensible heat loss Q_s from the skin to the microclimate air layer used in the clothed cylinder model and bioheat model is given by

$$h_{coae} \times (T_{amb} - T_o) + H_{co}(T_{void} - T_o) + \frac{h_r}{2}(T_{skin} - T_o) + \frac{h_r}{2}(T_{amb} - T_o) = 0 \tag{13b}$$

BOUNDARY CONDITIONS AND NUMERICAL METHODS

The static pressure at the exit and the stagnation pressure at the inlet are equal to zero gauge pressure while the inlet fluid temperature is considered uniform and ambient and the gradient of the of the fluid leaving temperature is set equal to zero (Anil et al. 2009). The coupled pressure and energy equations of the clothed cylinder with associated boundary conditions (equations 1 to 13) are discretized using a finite volume methodology, where the air layer zone is divided into $N\Theta \times Nz$ grids of size Δz and $Rf.\Delta\theta$ with thickness Y. Since the numerical methodology is similar to the methodology used by Ghaddar et al.11, the discrete formulation of the coupled pressure and heat balances will not be repeated here. Central differencing is used for second order terms in the air layer pressure and energy equations. For an assumed skin temperature, an initial guess for the air temperature and pressure values for all nodes is assumed as ambient conditions and iterated until convergence is achieved when both coupled pressure and temperature equations converge such that overall mass and energy are conserved in each grid and values of pressure and temperature to a minimum relative error of 10-5. Once the pressure and temperature distribution in the air layer and the fabric void and the fabric outer node are obtained, the mass flow rates of all directions are calculated to estimate ventilation rates and to predict the sensible heat loss from the inner cylinder surface. The numerical solution is repeated for grid sizes to ensure that a grid-independent solution is obtained. In the presented simulations, the number of grid points were 400 with a maximum $\Delta\theta$ of 1° and minimum Δz of 0.1 cm (0.04inches) (close to the opening.

EXPERIMENTAL METHODOLOGY

An open loop wind tunnel was used to provide the flow. This tunnel used a blow type variable speed three-phase axial fan with 0.7 m (2.29 feet) diameter and peak power of 1.6 KW (1.611 Btu/s). The dimensions of the square test cross were 0.8 x 0.8 m^2 (2.29 x 2.29 $feet^2$). In order to mimic the clothed arm, a 40 cm (1.31 feet) cylinder of 25 mm (0.98 inches) inner diameter is inserted in the test section part. A cotton fabric, with 0.05 m/sec (3.28 feet/s) permeability covered a metallic mesh that is 0.5 cm (0.2 inches) away from the cylinder. The uniform crosswind experiments over the clothed cylinder were conducted at low-speeds. The air temperature was controlled to be 25 °C while the inner cylinder was kept at a constant temperature throughout the experiment to approach the skin temperature. This was assured by controlling the temperature of the heating rod that was inserted inside in the cylinder. Seven thermocouples at different positions in the axial direction were used to measure the temperature inside the cylinder. The cylinder was rotated at each time to measure the temperature variation with respect to several angles. Data logger was used in order to measure all the temperature values.

RESULTS AND DISCUSSIONS

Experiment validation

Fig.2 shows the predicted and experimentally measured values of (a) angular-averaged air annulus temperature as a function of height, and (b) axial-averaged air annulus temperature as a function of angle at wind speed of 1 m/s (3.28 feet/s). A good agreement is found between the experimental measurements of the air temperature and model predictions of the air layer temperature with small maximum error of less than ±0.2°C.

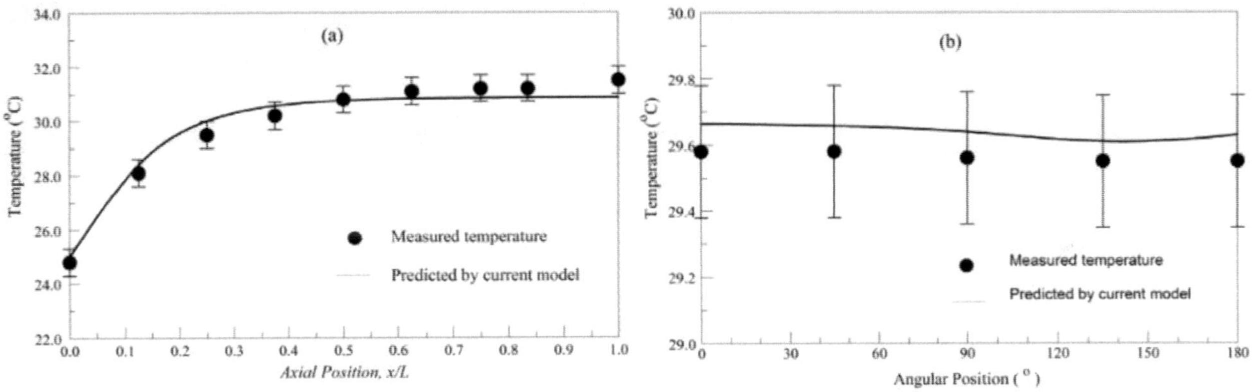

Figure 2 *(a) validation of the axial temperature distribution, (b) validation of the angular temperature distribution*

Simulation results

To understand the effect of mixed convection due to opening for clothed limbs, we analyze the cylinder model simulation results of ventilation rates and heat loss from a clothed vertical cylinder in three cases of open bottom, open top and closed end clothed and heated vertical cylinders. The simulation below is done at an ambient temperature of 25 °C at dry conditions. The skin is taken at constant temperature of 34°C. The dimensions are taken to be those of the clothed arm (L=40 cm (15.74 inches) and Rf =2.5 cm (0.98 inches)). Fig. 3 shows the angular-averaged microclimate air temperature along the axial distance. In the first case where the opening is at the top, the flow is only entered through fabric and is driven upward and leaves the cylinder at its end. The temperature decreases because of the high pressure at the bottom and the low pressure at the top that also draws in higher air flow in the radial direction contributing to the decreasing top temperature. In the open bottom cylinder, it is obvious that the temperature increases with the axial distance. The air enters from the bottom end and through fabric and leaves from the top. In the closed top and bottom cylinder, the air enters through fabric and leaves also through fabric; the temperature is slightly increasing axially. Clearly, the heat losses are highest in the case of open bottom cylinder at total heat loss followed by the closed top and bottom cylinder at total heat loss and then open top cylinder at total heat loss. In fact, the entering air at ambient temperature causes the high heat loss since ambient temperature is low compared to the skin temperature. However, in the case of open top annulus, air reaching the microclimate layer is penetrated through fabric only which heat the air and allow it to enter at high temperature relatively to ambient conditions. Thus, heat loss related to the difference of temperature between the skin and the microclimate air layer decreases. The top opening in this case does not enhance heat loss because ambient air cannot enter the microclimate layer from this opening. Because penetrating mass flow rate through the fabric is caused by the difference in pressure between the microclimate and the outer fabric surface external pressure around the cylinder (eq.2), thus, this mass flow rate is maximum in the case of the open bottom annulus followed by the closed ends followed by the open top annulus (fig.4).Table 1 summarizes the comparison between these three different configurations in order to emphasize the importance of the opening location in increasing the total heat loss and the ventilation rate.

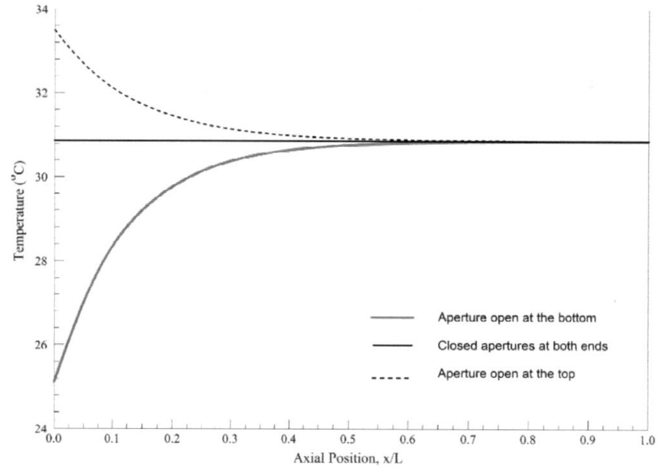

Figure 3 *comparison between angular-averaged temperature profiles variation in three cases*

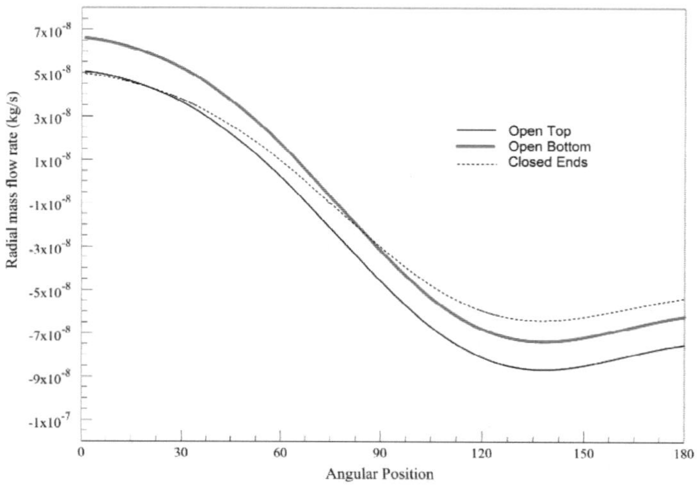

Figure 4 *comparison between angular-averaged temperature profiles variation in three cases*

Natural convection effect is removed from the mixed convection model and the decrease of ventilation rate is recognized. It is obvious that at zero wind speed, when only natural convection affects the ventilation, the decrease in ventilation is maximal (80 %). This percentage decreases when wind speed increases. For wind speed above 2.5 to 3m/s (8.2

to 9.4 feet/s) natural convection can be neglected. By comparing this result with Ghaddar et al. (2010$_b$) result, we deduce that at low velocities, the natural convection is more significant in the case of open bottom annulus than in the case of top annulus.

Table 1. Comparison of total heat losses and ventilation rate of a cylinder subjected to uniform cross wind in three geometrical cases

	Open top	Open bottom	Closed top and bottom
Q (W/m^2, $Btu/s.inches^2$)	44.03, 26.45×10^{-6}	56.41, 33.88×10^{-6}	48.01, 28.84×10^{-6}
$\dot{m}_a \times 10^{-2}$(l/min, $inches^3/min$)	13.83, 843.85	20.82, 1270.51	14.73, 898.87

CONCLUSION

A mathematical model derived from the first principles is developed and is validated by experiments to simulate the coupled ventilation and heat convection in a vertical open bottom end air annulus between concentric fabric cylinder and heated solid cylinder placed in perpendicular to airflow representing the clothed limb. When subjected to uniform cross wind, the open bottom cylinder presents significant ventilation rate when compared to the open top and closed ends cylinder. The reason is that the opening bottom end allows the ambient air to be sucked and then to ventilate the heated cylinder representing the skin.

In a future work, the clothed cylinder model will be integrated with a detailed segmental bioheat model to accurately determine the skin temperature associated with the clothed cylinder ventilation. The construction of total ventilation from segmental ventilation values of the trunk and limbs will lead to accurate prediction of total ventilation for wind speeds less than 2 m/s.

NOMENCLATURE

C_p : specific heat at constant pressure (J/kg.K , Btu/lb.K)

g : gravitational acceleration (m/s^2, inches/s^2)

$h_{cskin-air}$:heat transport coefficient from the skin to the trapped air layer (W/m^2.K, Btu/s. inches^2.K)

h_{coae} : heat transport coefficient from the fabric to the trapped air layer(W/m^2.K, Btu/s. inches^2.K)

h_r : linearized radiative heat transfer (W/m^2.K, Btu/s. inches^2.K)

H_{co} : convection heat transfer coefficient between the outer node and air flowing through fabric (W/m^{2}K, Btu/s. inches^2.K)

K_a : thermal conductivity of air (W/m·K, Btu/s. inches.K)

Z : axial direction

Y : air layer thickness (m, inches)

R_f : fabric cylinder radius (m, inches)

R_s : segment cylinder radius (m, inches)

L : fabric length in Z direction (m, inches)

e_f : fabric thickness (m, inches)

$\dot{m}_{aY}$: mass flow rate of air in radial direction (kg/m^2.s, lb/inches^2.s)

$\dot{m}_{a\theta}$: mass flow rate of air in angular direction(kg/m^2.s, lb/inches^2.s)

$\dot{m}_{aZ}$	: mass flow rate of air in axial direction (kg/m^2.s, lb/inches^2.s)
$\dot{m}_a$	: total ventilation rate (kg/m^2.s, lb/inches^2.s)
P_a	: pressure of the microclimate air (kPa, kbars)
P_s	: pressure at the external surface of the fabric (kPa, Kbars)
T_a	: temperature of the microclimate air (°C)
T_{amb}	: ambient temperature (°C)
T_{skin}	: skin temperature (°C)
T_o	: temperature of the fabric outer layer (°C)
T_v	: temperature of the fabric void layer (°C)
V□	: velocities of the environment cross wind (m/s, feet/s)
Q	: total heat losses (W, Btu/s)
Max	: maximum

Greek symbols:

α	: fabric air permeability (m^3/m^2.s, inches/s)
β	: the volumetric thermal expansion(°C-1)
ν	: kinematic viscosity of air (m^2/s, inches^2/s)
θ	: angular coordinate
ϱa	: density of air (kg/m^3, lb/inches^3)

Subscripts

a	: conditions of air in the annulus
o	: fabric outer node
void	: fabric void node

REFERENCES

Anil, L. X., and C. Reji. 2009. Numerical prediction of natural convection in vented cavities using Restricted Domain Approach. International Journal of Heat and Mass Transfer, 52:724-734

Cermak, R., Melikov, A.K., and L. Forejt. 2006. Performance of personalized ventilation in conjunction with mixing and displacement ventilation. HVAC&R Research, 12:295–311.

Ghaddar, N., Ghali, K., and J. Harathani. 2005. Modulated Air Layer Heat and Moisture Transport by Ventilation and Diffusion from Clothing with Open Aperture. ASME Heat Transfer Journal, 127 (3): 287-297.

Ghaddar, N., Ghali, K., and M. Othmani. 2010a. Effect of Moisture Transport on Mixed Convection in Vertical Annulus of a Heated Clothed Vertical Wet Cylinder in Uniform Cross Wind. Proceedings of 14th International Heat Transfer Conference (IHTC14): IHTC14-22342.

Ghaddar, N., Ghali, K., and S. Chehaitly. 2011. Assessing thermal comfort of active people in transitional spaces in presence of air movement. Energy and Buildings, 43 (10): 2832-2842.

Ghaddar, N.,Ghali, K., Othmani, M., Holmer, I., and K. Kuklane. 2010b. Experimental and Theoretical Study of Ventilation and Heat Loss from Clothed Vertical Isothermally-Heated Cylinder in Uniform Flow Field. ASME Journal of Applied Mechanics, 77 (3) :1-8.

Ghali, K., Ghaddar, N., and B. Jones. 2002. Multi-Layer Three-Node Model of Convective Transport within Cotton Fibrous Medium. Journal of Porous Media, 5(1):17-31, 2002.

Kind, R.J., Jenkins, J.M., and F. Seddigh. 1991. Experimental Investigation of Heat Transfer through Wind-permeable Clothing. Cold region Science and technology, 20:39-49.

Khalifa, H.E., Janos, and J.F. Dannenhoffer III. 2009. Experimental investigation of reduced-mixing personal ventilation jets. Building and Environment, 44:1551–1558.

Lotens, W.A. 1993. Heat transfer from humans wearing clothing. s.l : Doctoral thesis.
Melikov, A.K. 2004. Personalized ventilation. Indoor Air, 14 (7): 157–167.
Sobera, M. P., Kleijn, C.R., Brasser, P., and H. E. A. Van den Akker. 2003. Convective Heat and Mass Transfer to a Cylinder Sheathed by a Porous Layer. AIChE Journal, 49: 3018-3028.

Efficacy of Using Phase Change Materials (PCMs) As Thermal Energy Storage for Building Applications in the Mediterranean Region

Alissar Yehya
Student Member ASHRAE

Hassan Naji, PhD
Fellow ASHRAE

Laurent Zalewski, PhD
Fellow ASHRAE

ABSTRACT HEADING

In the past decade, the economic situation started to change due to the raise in energy prices, hence researchers began to give more attention to phase change materials (PCM) applications for heating and cooling of buildings. PCM can be used for temperature regulation and their high storage capacity can reduce energy consumption in buildings. The development and testing are usually conducted for prototypes of PCM wallboard and PCM concrete systems to enhance their thermal energy storage capacity, with main concern in peak load shifting and solar energy utilization. However, for a more adequate integration, the industry needs to optimize the performance of PCM. In this work we present a review on the thermal energy storage with phase change materials (PCM) in buildings, and focus on the evaluation of its applications in the Mediterranean region. The efficacy here is related to the availability of PCM with suitable fusion temperatures for the studied climate and the period by which the savings in energy will compensate the relatively high prices of PCM. The studies undergone in Cyprus and other Mediterranean countries are promising and show the importance of using PCM in improving building energy performance and maintaining low heating and cooling loads. Furthermore, the increasing number of studies on PCM in many Mediterranean countries should therefore set the subject seriously into discussion in other neighboring countries like Lebanon, where energy saving is vital due to the increasing price of fuel. On the other hand, we highlight on the importance of the characterization of the thermo-physical properties of PCM through efficient experimental and numerical methods for their adequate integration in building applications. This study thus presents a short overview of basic material characteristics and necessary energy performance data based on available experiences to evaluate the efficacy of future applications of PCM in Mediterranean countries.

INTRODUCTION

The increasing world population provoked by an increase in energy demand implies a significant environmental crisis. This promotes a major question whether the resources, possessed by our planet, are enough to satisfy such a demand for raw materials and primary energies. Figure 1a shows that the world primary energy consumption grew by a below-average 1.8% in 2012 according to the *"BP Statistical Review of World Energy 2013"*. Growth was below average in all regions except Africa. While oil remains the world's leading fuel, accounting for 33.1% of global energy consumption. Hydroelectric and other renewables in power generation both reached record shares of global primary energy consumption 16.7% and 1.9%, respectively. Moreover, as inferred from figure 1b, the Middle East, including Lebanon, has the highest consumption for oil

Alissar Yehya is a PhD student in Université Lille Nord de France and Université d'Artois, Hassan Naji and Laurent Zalewski are Professors in Université Lille Nord de France and Université d'Artois and natural gas, with a negligible contribution of hydroelectric energy and no renewables. This proves the lack of the presence of efficient strategies to mitigate the high consumption of the present or imported raw energy materials.

The use of energy in buildings accounts for a large share of the total end use of energy (IEA 2008). The average consumption of buildings is about 50% of all raw materials possessed in the world (Hegger et al. 2008). This percentage would differ depending on the region, and is equal to 40% in Europe, 30% in USA, and reaches 70% in non-developed countries like Lebanon. Energy consumption for thermal comfort in buildings has grown a lot in few years because of increasing users demand for comfort conditions and the associated market penetration of more cooling systems. This increase of energy consumption and the increase of the fuel price and CO_2 emissions are promoting a new policy of more sustainable buildings (Castell et al. 2008).

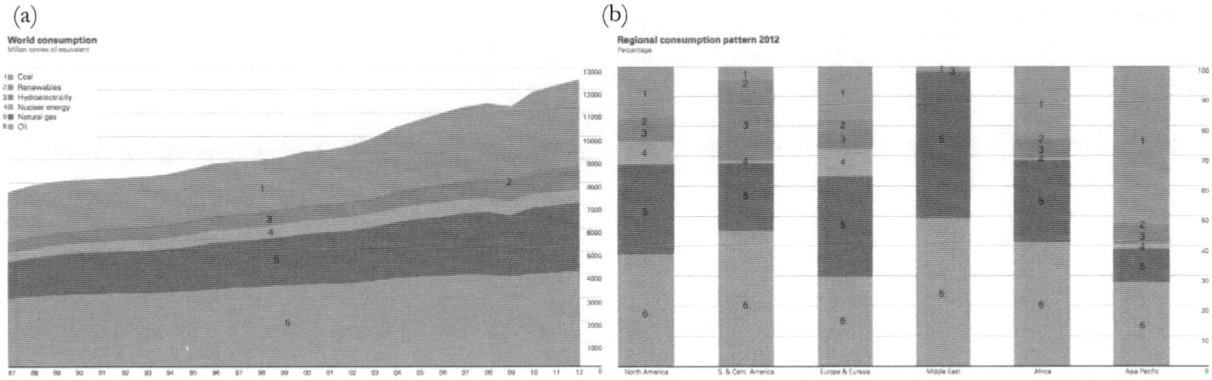

Figure 1 (a) World Consumption for different resources 2012 (b) Regional consumption pattern 2012 in %

In an attempt to conserve energy and reduce dependency on fossil fuels, it has become crucial to find effective means of reducing peaks in power consumption and to shift portions of the load from periods of maximum demand. Hence, it is vital today to drive the research towards energy-efficient buildings to reduce its heating and cooling demands. Using proper insulation in buildings is one of the effective ways of energy conservation in residential and commercial applications. Nevertheless, nowadays buildings are of lighter construction and have less thermal inertia. Hence, this implies less thermal comfort. So incorporating PCM into buildings designs, as part of the thermal mass, can help improving the thermal capacity and thus decrease the overall energy demand of buildings. PCMs increase the thermal capacity and regulate temperature to insure comfort.

As PCMs being part of Latent Heat Storage (LHS), they are based on the heat absorption or release when a storage material undergoes a phase change from solid to liquid or vice–versa. They can be used to store energy during day/summer to use it at night/ in winter, and can moreover be implemented in energy storage devices for heating/cooling applications. Briefly, Latent heat storage systems can be divided into applications for temperature control and applications for storage and supply of heat or cold with small temperature change (Mehling 2008).

PHASE CHANGE MATERIALS: APPLICATION IN BUILDINGS

Phase change materials are incorporated into buildings designs as part of the thermal mass. During the day, they absorb heat while melting, and consequently reduce the indoor temperature. At night, when the room temperatures are low, the PCM rejects the stored heat, thus preventing the temperature in the room from getting too cold. This material gives structures improved thermal storage capabilities with minimal change to the existing building design.

The use of PCM in buildings began in the mid-1940s (Lane 1983) and its application as a type of thermal storage was one of the first studied together with thermal storage tanks (Zalba 2003). Advancements in technology and research have since pushed studies into some new areas such as integration with walls (Cabeza 2007), gypsum wall boards (Banu 1998), and micro encapsulation in concrete mix (Bentz 2007), thermal solar storage (Lane 1983) and thermal control of photo voltaic (Telkes 1975).

The importance of PCM lies is referred to first their capability to narrow the gap between the peak and off-peak loads of electricity demand so as to save operative fees since the cost of electricity at night is lower than that during the day. Second, they have the ability to utilize solar energy continuously, storing it during the day and releasing it at night. In addition to the ability to use natural ventilation at night in summer, store the cold, and release it to decrease the room temperature, thus reducing the cooling load of air conditioning (Hasnain 1998). In Figure 2b, we can see a comparison between different sensible heat materials with respect to PCM. 1cm of PCM is equivalent 24cm of concrete and 51cm of gypsum. This proves the enhancement in thermal inertia by the usage of PCM.

(a)

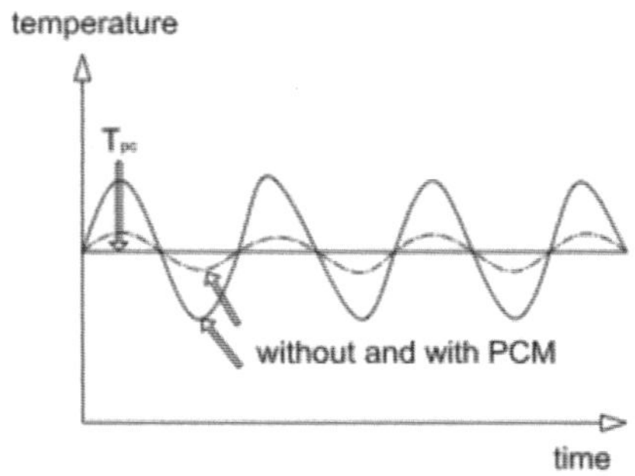

(b)

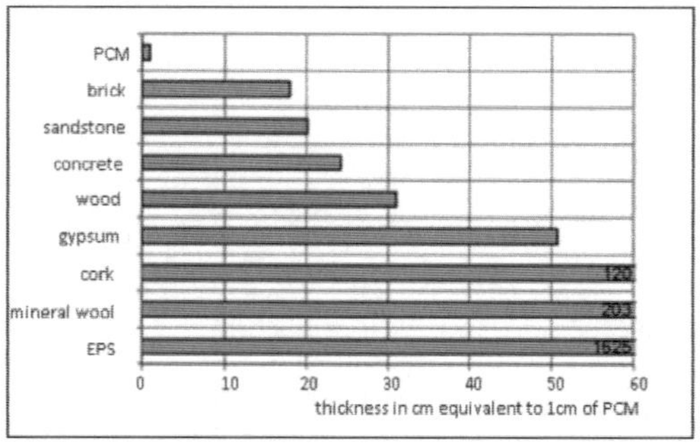

Figure 2 (a) PCMs with melting temperature at the average buffer temperature fluctuations (b) Heat capacities and heat stored 4K of interval for different building materials compared to PCM

Phase change materials (PCMs) can be incorporated in building materials in different forms. Gypsum plasterboards, Plaster, or concrete with microencapsulated paraffin are examples on these forms. On the other hand, building components can be equipped with PCM like in suspended ceilings; in this case, the PCMs are salt hydrates as seen in Figure 3a. Or can be incorporated in the blinds as in Figure 3b. Applications of phase change materials are divided in passive and active systems; with passive systems the regenerations of the thermal storage medium after melting occurs solely by means of free ventilation and natural cooling, there is no actively moved heat transfer fluid (Kaltenbach 2005).

(a)

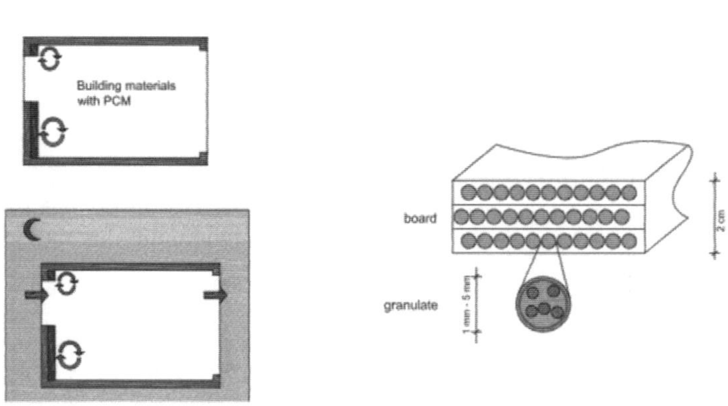

(b)

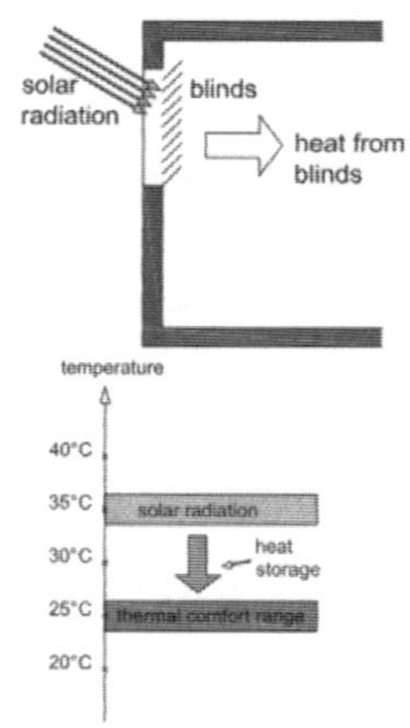

Figure 3 (a) The general concept for cooling with PCM (microencapsulated) integrated in building materials (b) Internal blinds absorb solar radiation and release the heat into the room, in order to cool the space by reducing the solar heat input of PCM

An example of passive systems is the PCM solar walls. However, in active systems, the airflow must be directed along the surface of the PCM in a way that heat transfer is by forced convection instead of free convection. The active solidification of the melted PCM can be achieved with a minimum of energy: by running water-bearing pipes through the substance, for example, or by blowing cool air over the surface of the elements with the aid of small fans (Kaltenbach 2005). Active thermal systems PCM can be integrated in ceiling, walls, or floors (Figure 4a and 4b).

A large number of PCMs are available in any required temperature range from −5 up to 190 °C. There exist some PCMs effective within the human comfort range which is 20–30°C. They store 5 to 14 times more heat per unit volume than conventional storage materials such as water, masonry or rock (Figure 2b). Among these we can mention the paraffin and non-paraffin compounds as organic materials, and salt hydrates as inorganic materials, and there are also the eutectic compounds.

(a) (b)

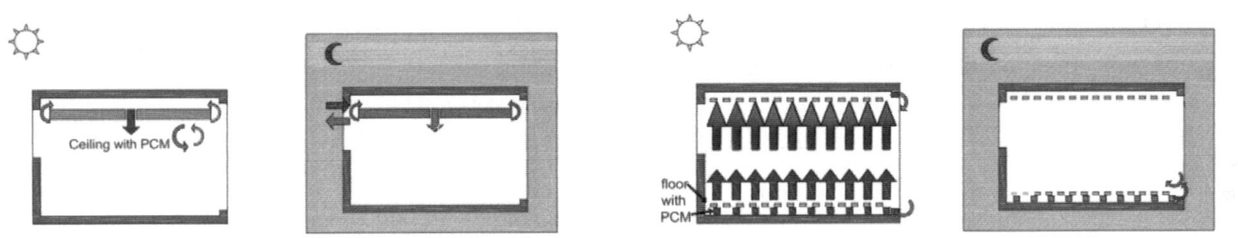

Figure 4 (a) The concept for cooling with PCM integrated in ceiling, excess heat from the room is stored in the PCM during day and released to the cold night air (b) The concept for cooling with PCM integrated in floor

The phase change material should possess a melting temperature in the desired operating temperature range with high latent heat of fusion per unit volume, high specific heat, and high density and high thermal conductivity. In addition, the material should possess high nucleation rate to avoid supercooling of the liquid phase and high rate of crystal growth, so that the system can meet demands of heat recovery from the storage system. However, the PCM should have relatively low cost to have an efficient integration in buildings applications.

Experiences of PCM application in different countries

To evaluate the performance of PCM construction materials in terms of thermal comfort, Athientis et al. fielded an experimental and numerical simulation study in 1997. It was shown that the utilization of PCM gypsum board in a passive solar building may reduce the maximum room resultant temperature by 4°C during the daytime and can reduce the heating load at night significantly. The experimental results showed a significant reduction in the room radiant temperature due to the absorption of solar gains in the PCM board. It can therefore be concluded that PCM applied over a large surface area in a passive solar building is effective for storage of solar gains and improvement of thermal comfort.

Furthermore, a comparative study was done by Zalewski et al. 2013 of a common mortar and a micro- encapsulated phase change material embedded in a mortar. Results demonstrated that the investigated PCM-M (BASF's microencapsulated PCM Micronal PCM) could provide important advantages for thermal storage systems as well as heat transfer applications. The energy stored and released in the phase change temperature range for the PCM-M is 41% higher than the energy stored (released) in the same temperature range for the conventional mortar. Moreover, this percentage can be enhanced when considering a temperature range around the melting temperature.

Experiences of PCM application in Mediterranean countries

In some Mediterranean countries different applications of PCM were implemented. Among those an experiment was conducted for a Mediterranean climate in Cyprus (Ozdenef 2012). The study showed a decrease of 1.6°C in temperature for

15mm PCM thickness with a reduction of 44.12% in cooling energy for a low thermal capacity simple building. While those if higher thermal capacities showed less significant reduction. This is due to the fact that low thermal capacity buildings have faster response to the changes in shortwave radiation, which allow PCM to be more effective.

Mandilaras et al. in 2012 conducted experiments on an unoccupied typical house in mid-western part of Greece, which contains no energy system. The walls contained multi-layers of insulation materials with gypsum plasterboard panels containing PCMs (15mm of Knauf Smartboard). The thermal characterization of the house's walling system was implemented. The thermal mass was enhanced during late spring, early summer, and autumn. This implied a decrease in decrement factor by 30 to 40% and increase in the time lag of approximately 100 minutes.

In the experimental study of Castell et al. 2008 on cubicles, of conventional and alveolar brick walls placed in Lleida Spain, macro-encapsulated PCM RT-27 paraffin and SP-25 A8 salt hydrate were added to the walls, with the presence of domestic heat pump heating and cooling. In summer the energy consumption decreased by 0.86% for set points higher than 20°C but a significant decrease was attained when the set points were close to the fusion temperature of PCM and reached 15%. In winter, an insulation role of PCM was recognized, with preserving the warmth of the inside with an increase of 0.4°C in the indoor temperature.

In the work of Panayiotou et al. 2013, the effect of the application of macro-encapsulated PCM on the thermal behavior of a test cubicle in Cyprus was theoretically investigated using a suitable model of TRNSYS software. The results indicate that the application of such material in Cyprus is very attractive since they operate very well in the specific weather conditions of Cyprus. 28.59% of energy savings was obtained by an optimum application of a PCM with melting temperature of 29°C on outer surface of a wall and inner side of the concrete slab just behind the plaster. Nevertheless, an increase in energy savings reached 66.17% when combining the use of PCM with an appropriate thermal insulation. Moreover, PCM showed a better behavior during the summer conditions than during winter where the role of PCM was insulating which confounded previous discussed experimental studies.

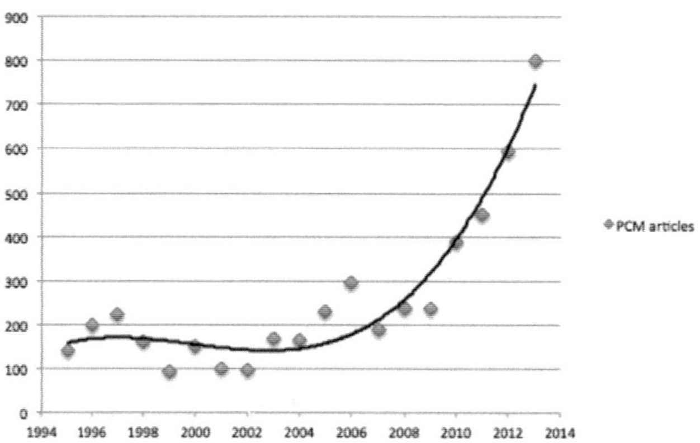

Figure 5 The increasing number of articles on PCM shows the rise in the researchers' interest towards LHS

Life-Cycle Assessment (LCA) for PCM

The energy consumption and CO_2 emissions are produced not only during the operational phase, but during the manufacturing and disposal of the materials, as well. The process used to quantify and evaluate the environmental impacts is known as life-cycle assessment (LCA) (Gracia et al. 2010). The LCA performed by Gracia et al. 2010 focused on the three experimental cubicles studied in Castell et al. in Spain. The energy consumption is registered to determine the energy savings achieved. The aim was to analyze if these energy savings are large enough to balance the environmental impact originated during the manufacturing of PCM. It was observed that the PCM must operate as long as possible during the year to reduce the operational impact. The benefits of the PCM were found to increase in locations where the weather conditions are similar all the year long, since the PCM would perform more cycles. Moreover, the use salt hydrates was found to be better than that of paraffin in order to reduce the manufacturing/disposal impact. From the lifetime analysis of the real scenario, the pay-back time of global impact was determined as 25 and 61 years for the hydrated salts and paraffins, respectively.

On the other hand, Panayiotou et al. 2013 economically evaluated, using LCA, the optimum PCM layer case of their previously discussed study. They defined the payback period and IRR (Internal Rate of Return) for the examined typical dwelling in Cyprus which has an area of 133 m^2. The results show that the PCM only case is not considered to be a very attractive solution, in monetary terms, due to the combination of the high initial cost and the annual money saving which result to a very long payback time which is estimated to be 14.5 years and a rather low IRR (4%). This is changing when the PCM is used together with insulation where the payback period is reduced to 7.5 years and the IRR is 13-14% respectively.

CONCLUSION

From the results of the previous discussed works it can be concluded that the application of PCM on the envelope of dwellings in Mediterranean countries is considered to be a very attractive solution in terms of energy saving and comfort conditions. Nevertheless, a major disadvantage is still due to the high initial cost of PCM but this could change if the cost is decreased in the near future.

As for the Lebanon case, Lebanon imports all its needed energy of which 70% is consumed by the residential buildings. Unfortunately, in most Lebanese houses, heating and AC are treated individually making every house and even every room have its own separate installation (Mourtada 2010). Lebanon has Mediterranean climate weather with about 300 sunny days per year, but its climate varies considerably from one region to another. So it is divided into four climatic zones: (1) Coastal plain (littoral), (2) Western Mid-Mountain, (3) Inland Plateau, and (4) High Mountain. In this context, it is possible to get a better thermal comfort in homes while saving energy consumption for heating and cooling. To achieve that, a number of measures have to be applied in the aim of improving the thermal and energetic qualities of the building. So, as concluded from the previous experiences in Cyprus, Greece, and Spain, the integration of PCM in buildings with an adequate insulation would improve their thermal capacity and reduce the energy consumption. Microencapsulated PCM in cement can be a good application for the Lebanese case, since most of the outer walls in buildings are of masonry with an air cavity in between. It would be interesting if a test experimental sample is conducted for a typical Lebanese house or at least a numerical simulation for a model is implemented to obtain comparable results with those of Cyprus. The increasing number of studies on PCM in many Mediterranean countries should therefore set the subject seriously into discussion in other neighboring countries like Lebanon, where energy saving is vital due to the increasing price of fuel.

ACKNOWLEDGMENTS

The authors wish to thank the national Research Agency (ANR: French Agency) for its continued financial support of this research under the research project Stock-E MICMCP.

REFERENCES

BP Statistical Review of World Energy 2013

Murat Ozdenefe, Phase Change Materials And Thermal Performance Of Buildings In Cyprus, Thesis 2012

Energy Efficiency Requirements In Building Codes, Energy Efficiency Policies For New Buildings, International energy agency agence internationale de l'energie 2008

Mandilaras I., Stamatiadou M., Katsourinis D., Zannis G., Founti M., Experimental thermal characterization of a Mediterranean residential building with PCM gypsum board walls, Building and Environment 61 (2013)

Castell A., Medrano M., Roca J., Vila A., Cabeza L.F., Experimental Study Of PCM In Mediterranean Buildings

Dutil Y., Rousse D., Lassue S., Zalewski L., Joulin A., Virgone J., Kuznik F., Johannes K., Dumas J.P., Bédécarrats J.P., Castell A., Cabeza L.F., Modeling phase change materials behaviour in building applications: selected comments, World Renewable Energy Congress 2011, Sweden

C. Castellón, M. Nogués, J. Roca, M. Medrano, L. F. Cabeza, Microencapsulated Phase Change Materials (PCM) For Building Applications

Joulin A., Zalewski L., Lassue S., Naji H., Experimental investigation of thermal characteristics of a mortar with or without a micro-encapsulated phase change material, Applied Thermal Engineering 66 (2014) 171e180

Gracia A.D., Rincón L., Castell A., Jiménez M., Boer D., Medrano M., Cabeza L.F., Life Cycle Assessment of the inclusion of phase change materials (PCM) in experimental buildings, Energy and Buildings 42 (2010) 1517–1523

Gregoris P. Panayiotou, Soteris A. Kalogirou and Savvas A. Tassou, Theoretical study of the application of Phase Change Materials (PCM) on the envelope of a typical dwelling in Cyprus, 2nd Conference on Power Options for the Eastern Mediterranean Region 07-08 October 2013, Nicosia, Cyprus (Paper No. POEM13/188)

Comair F.G., Mourtada A., Chehab S., Matar T., Sebrini C., Thermal Insulation Market in Lebanon Le Marché de l'Isolation Thermique au Liban

Hegger M., Fuchs M., Stark T., Zeumer M., Energy Manual, Sustainable architecture, Birkhauser edition, 51-52, 146-148

Mehling H., Cabeza L.F., Heat and cold storage with PCM : An Up to Date Introduction into Basics and Applications, Springer editions, 2008, pp. 6-9, 15-40, 218-250, 274-280

Kaltenbach F., PCM Latent Thermal Storage Media Heating and Cooling without Energy Consumption?, Detail magazine, 2005, 5: 544-549

Konstantinidou C.V., Integration of thermal energy storage in buildings, Master Thesis - Spring 2010

Cabeza, L. F., Catellon, C., Nogues, M., Medrano, M., Leppers, R., & Zubillaga, O. (2007), Use of Microencapsulation PCM in concrete walls for energy savings. Energy and Buildings 39 , 113-119.

Banu, D., Feldman, D., Haghighat, F., & Paris, J. H. (1998). Energy-storing Wallboard: flammability tests. J. Mater. Civ. Eng 10 , 98-105.

Telkes, M. (1975). Thermal Energy Storage for solar heating and cooling. Proceedings of the Workshop on Solar Energy Storage Subsystems for the Heatign and Cooling of Buildings. Charlottesville (Virginia, USA).

Mehling, H., Hielbler, S., Ziegler, F. (1999). Latent heat storage using PCM-graphite composite material: advantaged and potential applications. In Proceedings of the 4th Workshop of IEA ECEC IA Annex 10, Bediktbeuern, Germany.

Bentz, D. P., & Turpin, R. (2007). Potential Application of Phase Change Material in concrete Technology. Cement & Concrete Composites, 527-532.

Lane, G. A. (1983). Solar Storage: Latent Heat Materials. *Vol 1 CRC Press.*

Optimization of the Building Envelope in Hot and Humid Climates

Balsam A. Nehme **Ghina R. Annan**

ABSTRACT

This paper aims to promote building sustainable designs to reduce energy consumption by implementing optimized building envelope for high performance buildings. Several building envelope case studies for office, airport, retail and residential buildings, located in hot and humid climates, are analyzed in WUFI, IES VE and CadnaA software in order to assess, simulate and optimize building envelope performance.

A detailed building thermal energy performance simulation, assessing the impact of solar radiation and external shading devices, using IES VE, has been conducted to evaluate the effectiveness of the proposed building envelope designs and weigh the impact of sustainability measures on enhanced indoor environmental quality and building loads reduction. Also, a condensation analysis for external wall assemblies is conducted via WUFI software to study moisture influence on the thermal performance of the construction and long-term system behavior. Additionally, sound control software such as CadnaA assists in validating noise mitigation techniques from outdoor equipment to promote indoor occupant comfort.

The simulation results also suggest building envelope composition optimized for hygrothermal and acoustic performance, window to wall ratios and appropriate shading strategies that promote reductions in building annual cooling energy for the different case studies to a range of savings varying from 5 % to 35% with respect to ASHRAE standard 90.1-2007 for hot and humid climates.

INTRODUCTION

The envelope of a building represents a transition space that allows interaction between outdoor climatic conditions and the indoor environment. Accordingly, the selection of building envelope components can significantly impact its thermal performance which is a determining factor in the total energy consumption. At the early conceptual project design stage, the selection of building physical components, distribution and shape have great impact on the overall thermal performance. The objective of this paper is to analyze several case studies for the optimization of the building envelope for high performance buildings in hot and humid climates. Four different case studies are investigated in this work toward energy saving in high performance building design. All cases assess the optimization of building window to wall ratio (WWR) along with potential external wall and roof assemblies, further to the enhancement of the glass/frame assembly thermal performance. **Table 1** describes the four cases, each representative of a different building function and located in regions classified as extremely hot and humid.

Balsam A. Nehme is the head of the energy modeling, condensation and acoustics unit in the Department of Mechanical Engineering at Dar Al Handasah (Shair and partners), Beirut, Lebanon. Ghina R. Annan is a team member in the energy modeling, condensation and acoustics unit in the Department of Mechanical Engineering at Dar Al Handasah (Shair and partners), Beirut, Lebanon.

Table 1. Case Studies Description and Weather Files

Case study	Case Description	Case Type	Weather file
Case 1	Abraj Kudai	Retail & Hotels	Makkah- KSA
Case 2	Dubai Airport	Airport Concourse	Dubai – UAE
Case 3	King Abdallah District	Offices & Residential	Riyadh- KSA
Case 4	Al Madinah Convention	Convention Center	Al Madinah- KSA

METHODOLOGY

Weather Data

Based on ASHRAE 90.1-2007 standard Table B1-4, all climates characterized by more than 5000 cooling degree days (10°C) are classified as very hot and humid, where high temperatures and moisture represent significant problems. Hot and humid climates must be given special attention because they are more likely to have problems related to excess moisture than the drier climates. External relative humidity exceeds 80% throughout the year and reaches a maximum of 100%. As such, external wall and roof assemblies are further optimized to enhance overall building thermal performance while keeping moisture out and avoiding potential condensation likely to occur during exchange between very hot and humid external climate conditions and internal air-conditioned occupied spaces.

Cities located within this climatic zone are considered in this study and are Makkah (KSA), Dubai (UAE), Riyadh (KSA) and Al Madinah (KSA). All annual simulation weather files have been acquired from the Meteonorm software version 6.1.

Parameters under analysis

External opaque construction

The use of thermal insulation is not a common practice, however due to stringent requirements for sustainable high performance building design, thermal insulation has been added to the opaque assemblies of all building types considered in Cases 1 through 4, as described in **Table 2** below. All propositions of building envelope components are compared with respect to a base building wall and roof assembly as described by ASHRAE 90.1 standard, Table 5.5-1 and Appendix A. The baseline external wall is characterized by an insulation installed within the cavity of the steel stud framing; the total wall assembly R-value is equal to 1.42 $m^2.K/W$ (8.06 $hr.ft^2.°F/Btu$). The base roof assembly is a continuous insulation over a structural deck with an overall thermal resistance of R-2.78 $m^2.K/W$ (15.78 $hr.ft^2.°F/Btu$).

Window-to-wall ratio (WWR)

While recommendations for designing for high performance glazing assemblies in sustainable buildings still hold, it is also advised to limit the window to wall ratio thus minimizing large heat exchange between external hot and humid environment and internal cool conditions.

Based on ASHRAE 90.1-2007 section 5.2.1, the prescriptive building envelope option recommends that the vertical fenestration area does not exceed 40 % of the gross wall area as adopted in all four cases tabulated below.

External wall and roof Insulation

The types of external insulation used in cases 1 through 4 are: 80 mm (0.26 ft.) mineral fiber insulation, 75 mm (0.24 ft.) extruded polystyrene insulation, 100 mm (0.33 ft.) extruded polystyrene insulation, and 75 mm (0.24 ft.) extruded polystyrene insulation respectively. Also, the types of insulation installed in the roof for cases 1 through 4 are: 200 mm (0.65 ft.) expanded polystyrene, 240 mm (0.79 ft.) dual density Hardrock, 50 mm (0.16 ft.) expanded polystyrene, and 185 mm

(0.61 ft.) dual density Hardrock, respectively. Furthermore, a cool roof has been applied with highly reflective white coating on all roofs of all four cases.

Shading Coefficient (SC)

External solar shading devices can substantially reduce the cooling load of buildings and large energy savings can be achieved. Hence, Islamic arcades shading are added on all building exposures in case 1 and 4, similarly in case 2 a 1.2 meter extruded overhang is built on East and West facade and finally in case 3, multiple Vertical fins & overhangs are constructed on all building exposures. It is important to note that skylight properties and selections are similar to vertical glazing in their respective cases. In addition to the thermal resistance, a glazing assembly thermal performance is assessed by its shading coefficient. The shading coefficient (SC) is referred to as an indicator to how the glass is thermally insulating (shading) the interior when the window is exposed to direct sunlight, such that low SC is needed to lower the solar heat gain through the glass. Four levels of SC are chosen to be studied in this analysis: 0.28, 0.40 and 0.20 as described in **Table 2** below.

Table 2. Thermal assessment of case studies envelopes

Parameters	Baseline Case	Case 1	Case 2	Case 3	Case 4
WWR (%)	Similar to the case under study	27	20	30	17
U value of external Wall (W/m².k) [Btu/hr.ft².°F]	0.7 [0.123]	0.22 [0.039]	0.3 [0.053]	0.28 [0.049]	0.2 [0.035]
U value of roof (W/m².k) [Btu/hr.ft².°F]	0.36 [0.063]	0.17 [0.15]	0.15 [0.026]	0.5 [0.088]	0.2 [0.035]
U value of glass assembly (W/m².k) [Btu/hr.ft².°F]	6.81 [1.199]	1.98 [0.349]	1.98 [0.349]	2 [0.352]	1.56 [0.275]
SC-glass Assembly	0.28	0.28	0.4	0.4	0.2

RESULT ANALYSIS

Lower annual chiller energy savings are obtained in C4 and Al Madinah versus the other two cases due to the presence of unshaded skylight with glazing to roof ratio of 9 % and 4 %, respectively. Moreover, the shading coefficient of the skylight in Dubai Concourse 4 (Case 2) is higher than ASHRAE 90.1 standard requirement as it was selected with SC-0.4, thus explaining the drop in annual chiller energy savings to 5 %. **Figure 1** below represents the annual chiller energy load percent savings between the baseline and proposed models of cases 1 through 4. The cooling load and chiller energy savings computed by specifying an HVAC system allowing for 11°C (51.8 °F) temperature differential between cold air supply and room set point temperature. The load and energy were calculated by means of the loads and annual simulation modules in IES VE software.

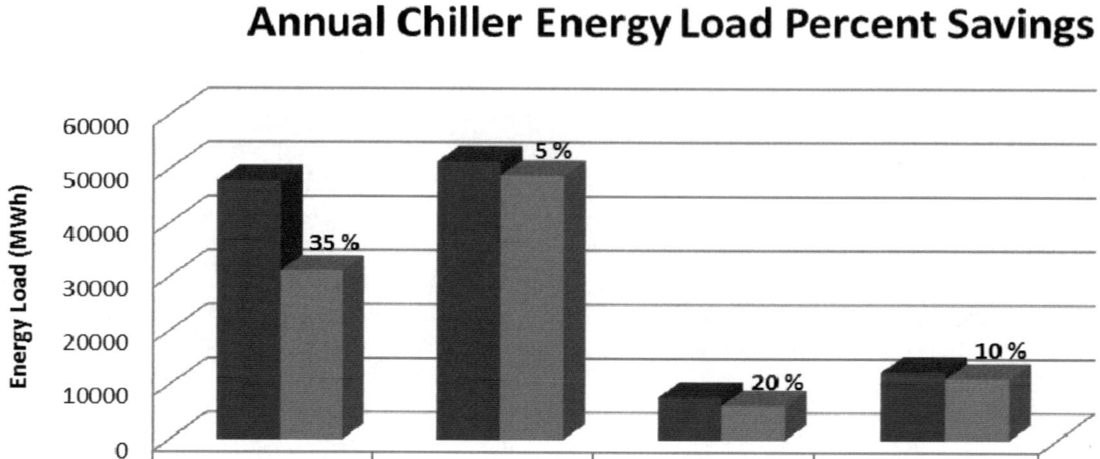

	case 1	case 2	case 3	case 4
■ Base line	47920	51455	7865	12580
■ Proposed	31150	48880	6290	11320

Figure 1 Annual chiller energy load percent savings

Hygrothermal Assessment

There are many considerations when designing a building to optimize heating and cooling efficiency, control moisture and provide occupant comfort especially in very hot and humid climates. A condensation analysis has been performed using WUFI software onto the walls separating external and internal environments of Kudai Hotel Towers & Retail podium and Dubai airport concourse 4. WUFI tool allows realistic calculation of the transient coupled one- and two dimensional heat and moisture transport in multi-layer building components exposed to natural weather. WUFI is a tool for estimating the likelihood for condensation to occur within walls, ceilings, and other assemblies. To reduce the building energy consumption in an effective manner, the interaction between the building, usage and system technology must be taken in account during the planning. Typically, a hygrothermal simulation is modeled for a period of three consecutive years. In order to eliminate the possibility for water vapor to condense within a building assembly, it is important to seal building assemblies to reduce air leakage into wall/ceiling cavities and coordinate the performance of the building's heating and cooling system so as to control the indoor air humidity and the pressure of the indoor air relative to the pressure of the exterior. Key factors that need to be considered in assessing and managing condensation risk include: environmental conditions, interior environment, and thermal vapor properties of wall structure. External walls are environmental separators that must exclude, air vapor and heat as preventative steps to avoid condensation and thus ensuring optimal indoor environmental quality. The temperature on each wall construction layer, referred to as the Interface Temperature, should always be greater than the total dew point temperature of the assembly. Additionally, in order to prevent any potential mold growth within the assembly, relative humidity, at wall layers exposed to internal conditioned spaces is maintained less than 75%. **Figures 2 and 3** show the variation of temperature in red and relative humidity in green at the insulation layer of both cases 1 and 2.

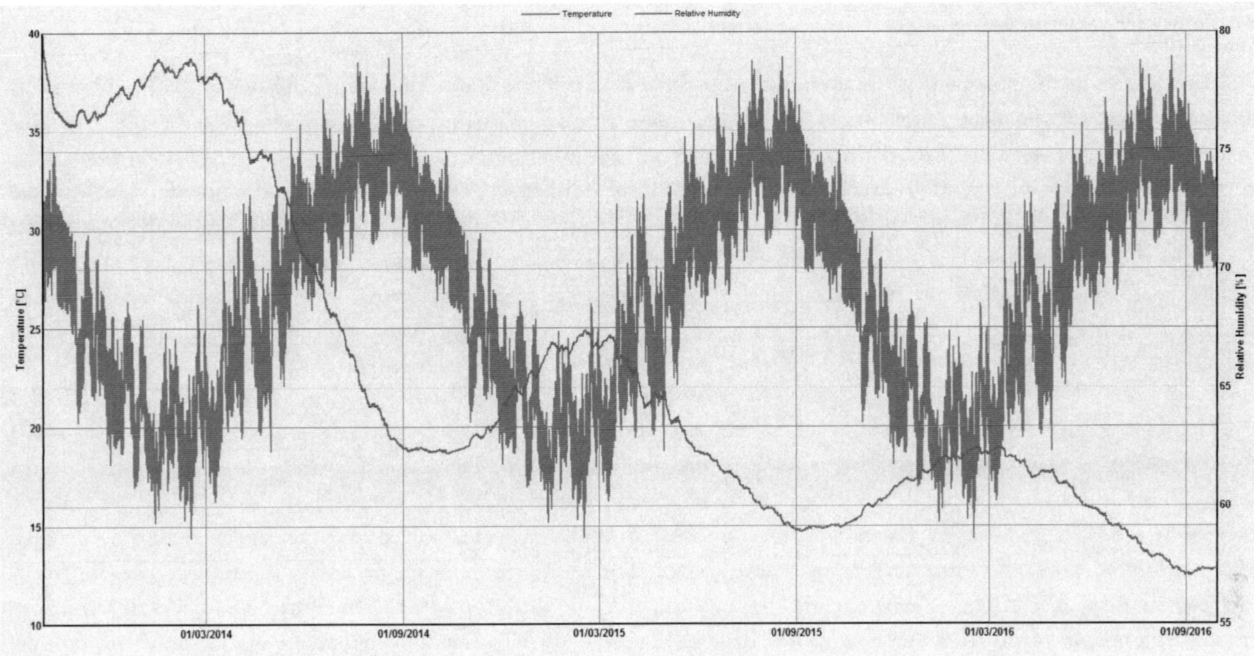

Figure 2 Variation of temperature and relative humidity at the insulation layer of external wall – Case 2

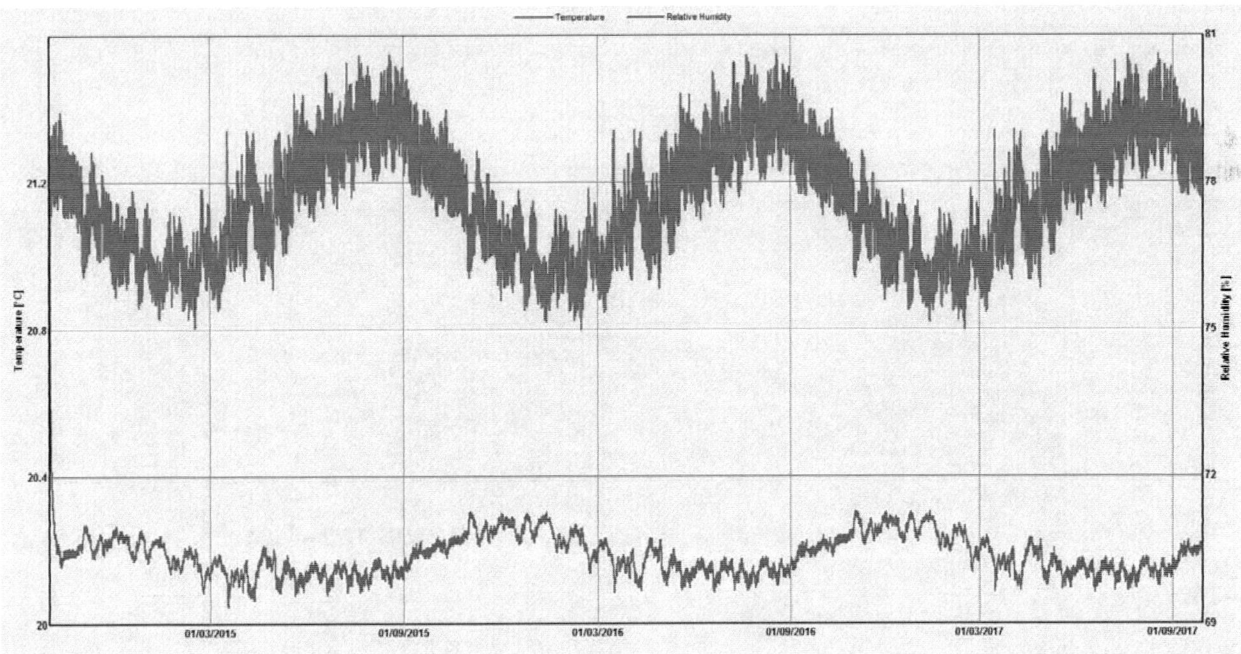

Figure 3 Variation of temperature and relative humidity at the insulation layer of roof – Case 2

Acoustic performance

In addition to the possibility of improving the building's solar, thermal and moisture performance, building envelope systems present supplementary advantages in reducing noise transmission to the occupied spaces within the building. Thermal comfort is not only limited to the thermal effect but the envelope should be optimized to account for noise mitigation from adjacent external sources. The objective of the present work was to develop an approach, which would allow the construction of a building envelope with optimal performance with respect to thermal and acoustic comfort conditions thereby taking properties of the building envelope such as function, position, dimensions and orientation with respect to potential disturbing external noise sources into account. A comfortable environment is created by controlling objective measures like decibel level (sound pressure), reverberation time, and the sound reflection and damping properties of materials. Also, creating barriers and sound breaks between sources of noise is important.

Sound analysis is performed, using CadnaA version 4.2.140, onto the cooling towers located at the roof of Hotel towers in Abraj Kudai project (Case 1). In order to properly account for the sound transmission effect from the three cooling towers placed on the roof of the tower, the sound spectrum at top surface as well as motor sides and tower ends are modeled based on a total sound power level (110 dB (A)). Several main areas of concern are considered; the occupied spaces inside the tower as well as facades of nearby towers. The results of the acoustical study as displayed in **Figures 4 and 5**, show that noise reaches a maximum of about 80 dB(A) at the tower on top of which the cooling towers are placed and a maximum of 84 dB(A) at the nearest hotel towers' facades, for all equipment operating at full speed without any sound attenuators in place and without any additional background noise source.

As a matter of fact, sound attenuation has been optimized at cooling towers to include a vertical sound barrier wall system providing sound reduction of no less than 12 dB from the noise emitted from all sides of the cooling towers, in addition to silencers with potential reduction of no less than 25 dB placed on top of the cooling towers. Upon applying noise attenuation to the cooling towers, the noise level perceived at the towers' facades will not record a value higher than 61 dB(A) as shown in **Figure 6** below.

While it is recommended that the sound level inside the hotel occupied rooms does not exceed 30 db(A) and for improved indoor thermal comfort, we have adopted an insulated roof with 80 mm (0.26 ft.) mineral fiber insulation allowing a noise reduction of 65 db(A). While for the glazed section of the elevations, a double glazed assembly with Rw-Ctr of 37 db allows for adequate sound transmission to the occupied spaces.

Figure 4 Sound Levels at Towers – Base Case

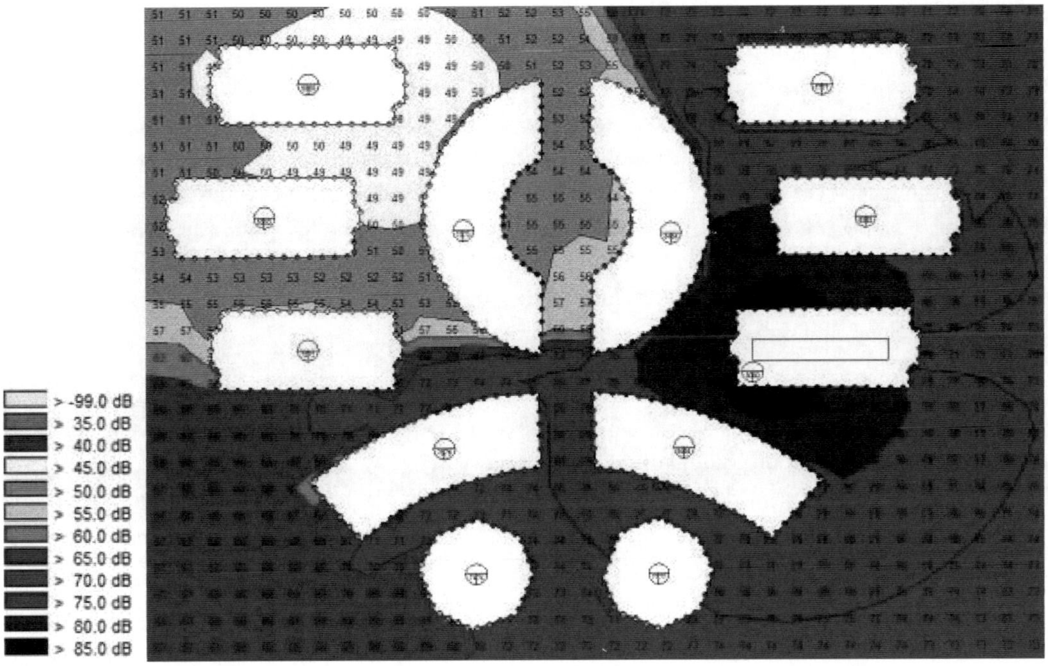

Figure 5 Sound Levels Contours at Towers – Base Case

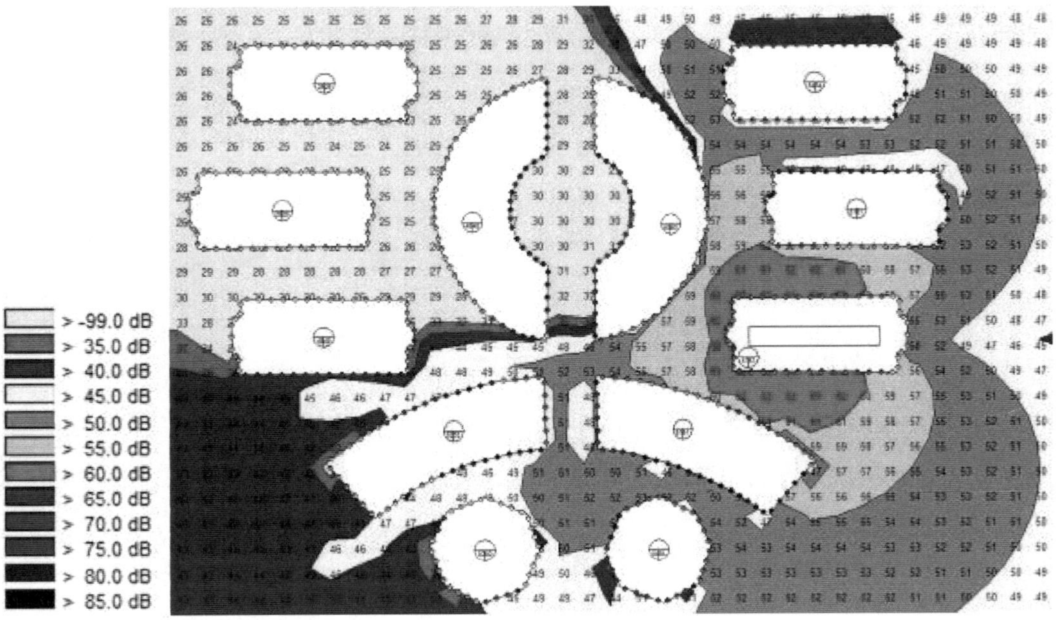

Figure 6 Sound Levels Contours at Towers – with Attenuation

CONCLUSION

The main objective of this study was to investigate building sustainable designs that can bring significant performance benefits in order to reduce energy consumption. The parameters under analysis focused on percent glazing window to wall ratio, thermal resistivity of external Wall, roof and glass / skylight assemblies along with their respective shading coefficients. The annual chiller energy load percent savings achieved in the four study cases, comparing ASHRAE 90.1 baseline model with respect to the proposed design models are achieved due to optimization of the above mentioned parameters. Sustainable high-performance design is further coupled with prevention of transmission of hot and humid air and high disturbing noise into the occupied spaces thus enhancing indoor air quality. The conclusions presented in this paper refer only to the building and climate characteristics described here.

ACKNOWLEDGEMENTS

We would like to express our gratitude and deep appreciation for the continuous support of our department directors Mr. Maroun Khoury, Mr. Naji Edde and our head of Department Mr. Jad Laham. We would like to further extend a special thanks to Mr. Sharbel El-Haber along with every member of the mechanical design teams for their cooperation, assistance and valuable contributions.

REFERENCES

Chapter 25, Heat, Air, and Moisture control in building assemblies - 2009 ASHRAE handbook – Fundamentals (SI).

The ASHRAE Guide for Buildings in Hot and Humid Climates, Harriman L. & Lstiburek J.

ASHRAE 90.1-2007 standard "Energy Standard for Buildings except Low-Rise Residential Buildings".

ASHRAE 198.1-2011 standard "Standard for the Design of High-Performance Green Buildings except Low-Rise Residential Buildings".

Mantha, P., Arena, L. B., & Winter Associates Inc.. A Systematic Approach to Hygrothermal Modeling and Compliance with Failure Criteria using WUFI. Fifth National Conference of IBPSA-USA, 414-422.

Acoustics and Noise Control, BJ Smith & RJ Peters & S Owen.

"The Framework of an Optimization Model for the Thermal Design OF building Envelopes", S. Al-Homoud &O. Degelman & L. Boyer.

"Thermal Insulation of Building Envelope Toward Zero Energy Design In Hot-Humid Climate", S. Westphal & A. Yamakawa, T. de Castro.

www.iesve.com
www.meteonorm.com
www.datakustik.com
www.wufi-pro.com

Use of Solar Regenerated Membrane-Desiccant System to Control Humidity and Decrease Energy Consumption in Office Buildings

Khoudor Keniar

Kamel Abou Ghali, PhD
Member ASHRAE

Nesreen Ghaddar, PhD
Member ASHRAE

ABSTRACT

The proposed system is a hybrid air-conditioning system that uses a novel liquid desiccant cycle with porous membrane material. The cycle is integrated with a conventional HVAC system to allow for independent control of indoor humidity and temperature. Conventional liquid desiccant cycles dehumidify the air before supplying it to the indoor space, while the proposed cycle absorbs the humidity directly from indoor space using a permeable cylindrical membrane acting as a dehumidifier. The air leaving the indoor space is cooled using an evaporative cooler. The cooled air is then used to decrease the temperature of the desiccant entering the indoor space in order to reduce the sensible load that might be added to the space air and to increase the potential of the desiccant to absorb moisture. The size of the proposed cycle is significantly reduced as compared to the conventional cycles that rely on evaporative cooling towers or regenerators or even desiccant wheels. The regeneration of the desiccant is also done through the use of permeable membrane tubes with the aid of parabolic panels to collect solar energy. An integrated mathematical model of the permeable membrane with the space modal was developed to study the feasibility and efficiency of the proposed cycle. It was found that the system can achieve a 32-35 % reduction in electric energy consumption compared to conventional air conditioning systems operating on a typical office of size 4 m×5 m×6 m.

INTRODUCTION

The control of indoor humidity is of fundamental importance for building material sustainability, energy consumption and thermal comfort of indoor space occupants. Thermal comfort of the occupants deals with regulating the thermal environment around them, which includes controlling temperature, air velocity and in particular humidity. Extreme levels of humidity, whether high or low, can irritate people's comfort (Morse, et al.) and affect the building structure (Ghali, et al., 2011).

In order to eliminate the negative impacts of elevated moisture levels, indoor humidity should be controlled, either by conventional or non-conventional techniques. The conventional way of moisture removal is using HVAC systems that cool the supply air to temperatures below its dew point to condense the excess moisture, followed by reheating to the adequate supply air temperature. The process of sub cool-reheat requires additional electric energy supply to the HVAC system due to the addition of an extra heating coil and oversizing the cooling system to meet the extra reheat and cooling loads

Khoudor Keniar is a Master's student at the American University of Beirut. Dr. Kamel Abou Ghali and Dr. Nesreen Ghaddar are professors of mechanical engineering at the American University of Beirut.

respectively. This increase of electric energy consumption can have negative economic and environmental outcomes.

Recent sustainable dehumidification techniques that substitute conventional vapor compression systems use desiccant technology. When employing a liquid desiccant system, the installment of dehumidification and regeneration tower beds is required (Mohammad, et al., 2013). Two issues rise from using such technology; the first is the direct contact of air with the liquid desiccant in the towers which might cause health issues and corrosion problems (Studak, et al., 1988) especially with the entrainment of hazardous salts to the ventilation system. The second is that in dehumidifying the air outside the indoor space, the humidity ratio of the supply air must be further lowered to below indoor space humidity ratio and this requires a strong desiccant solution. Recent novel developments, such as the HAMP, have shown that direct indoor cooling and dehumidification is a very efficient technique as it responds directly to changes in indoor latent and sensible loads and that a strong liquid desiccant solution is not needed. The HAMP, Heat and Moisture Transfer Panel (HAMP) (Fauchoux, et al., 2009), is able to control simultaneously the temperature and indoor relative humidity (Fauchoux, et al., 2010) using liquid desiccant as the working fluid. The panel absorbs moisture directly from the indoor room through a permeable membrane. The HAMP was tested in an open cycle system where the liquid desiccant was conditioned before it enters to the indoor space. Although the HAMP has shown good results in performing indoor cooling and dehumidification, one of the major concerns of using it is the energy consumption it might take to regenerate the liquid desiccant or to cool it.

In order to investigate the energy consumption of a permeable membrane performing indoor dehumidification, a similar heat and mass exchanger setup that is operating in a closed loop cycle is presented. The dehumidifier and regenerator being used are made from a set of cylindrical pipes formed from a permeable material. The dehumidification of the air will take place directly in the indoor room (similar to the HAMP), which insures better response to changes in latent loads. The decoupling of humidity and temperature control is important to save energy, as it will diminish the use of subcool-reheat process in conventional HVAC systems.

In this study, the dehumidification/regeneration ability of the membrane material will be modeled. The whole solar-regenerated-desiccant cycle with both membranes (dehumidification and regeneration) will be integrated with a typical office space in Beirut City to independently control the humidity and supply air temperature. The performance and efficiency of the membrane desiccant system will be compared with the conventional air conditioning system. The cycle will totally operate on sustainable energy, i.e. it will use solar energy as a heat source and evaporative cooling to meet the cooling loads of the cycle.

SYSTEM DESCRIPTION

The proposed dehumidification system is composed of the dehumidification permeable pipes, desiccant regeneration permeable pipes, sensible heat exchangers, indirect evaporative cooling heat exchanger, and parabolic solar collectors as shown in Fig.1. The piping system is made of solid piping lines except in the regenerator and dehumidifier parts. In those sections the pipe is made of porous material (polypropylene) that is permeable to water vapor but not liquid. The permeable parts will not be subjected to direct sunlight-especially UV rays- and they will be put in shades in case they are exposed to ambient air. The cool strong liquid desiccant enters the vapor moisture membrane at (7) picking the moisture from the indoor space (1) to leave at (2) where it is pumped through a solid piping system and being heated from two sources: heat exchange with the high temperature desiccant leaving the regenerative membrane (exchanger A) and solar energy stored in the tank (exchanger B). The heated desiccant then enters the regenerator, at (4), i.e. the permeable pipes in contact with ambient air where it will be regenerated and expel absorbed moisture to the ambient environment. To ensure that the desiccant will not impose an additional indoor heat source and to increase its ability in picking the indoor moisture, the desiccant will exchange heat with ambient air through the auxiliary heater then with the solution leaving indoor room, through exchanger (A). The desiccant will then be cooled through the cross flow heat exchanger with the evaporative-cooled indoor space air. The exit air of the indoor space undergoes evaporative cooling by a commercial evaporative cooler. The cool air then enters a cross flow heat exchanger where it cools the desiccant that is going to enter to the indoor space.

After passing through the cross flow heat exchanger the desiccant enters again the dehumidifier at (7). The liquid desiccant needs to be heated (regenerated) in order to expel the moisture it absorbed from the indoor environment. The energy source required for regeneration will be solar energy. The solar irradiance will be collected through parabolic plate collectors which will heat a working fluid, which in turn will heat the liquid desiccant through a heat exchanger (B). Solar energy will be stored in a storage tank. The liquid desiccant will leave the heat exchanger to enter the regeneration membrane at an elevated temperature to expel the space collected moisture to the ambient external air.

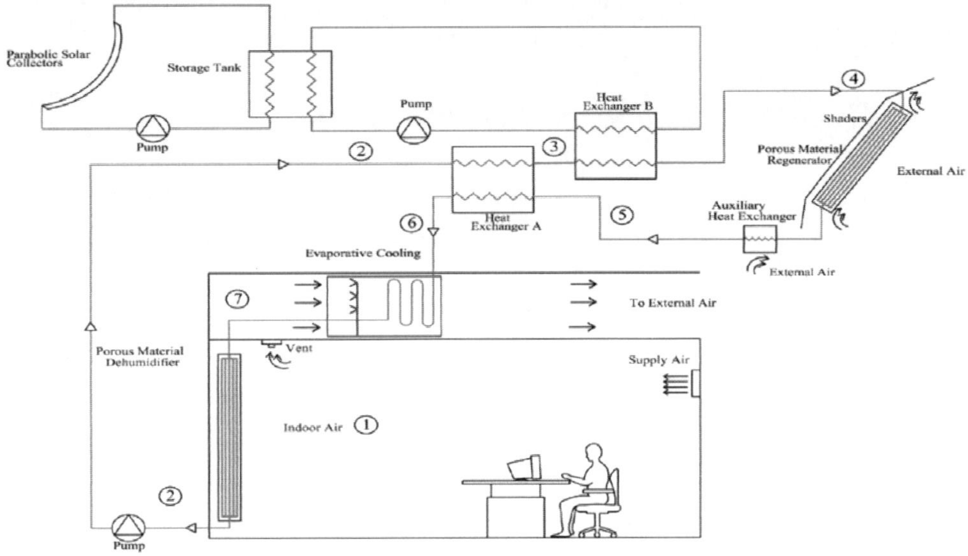

Figure 1: Schematic of the system

MODELING OF DEHUMIDIFIER/REGENERATOR

The dehumidifier/regenerator consists of the liquid desiccant in the permeable pipes of the cycle. The models used by (Bergero, et al.), (Fan, et al., 2011) and (Fauchoux, et al., 2010) will be adopted. The mass flow on dry basis, i.e. the desiccant, $CaCl_2$, flow rate per second is assumed constant in the pipes. Using the models in literature, the energy and mass balance for the liquid desiccant are shown in equations (1) and (2).

$$0 = -m_d \frac{d}{dy}\{h_{sol}(1+c_{sol})\} + U_c(T_o - T_{sol}) + h_{fg}\rho_a U_m(\omega_0 - \omega^*_{sol}) \tag{1}$$

$$0 = -m_d \frac{dc}{dy} + \rho_a U_m(\omega_0 - \omega^*_{sol}) \tag{2}$$

"y" represents the length coordinate along the direction of desiccant flow in the pipes and " ω^*_{sol} "represents the humidity ratio at the surface of the liquid desiccant in the pipe at the temperature and concentration of the desiccant. The heat and mass transfer coefficients per unit length are given by equations (3) and (4) respectively. The thermal resistance modal is shown in Fig. 2.

$$U_c = \left\{\frac{1}{2\pi r_o h_{c,o}} + \frac{\ln(r_o/r_i)}{2\pi k} + \frac{1}{2\pi r_i h_{c,i}}\right\}^{-1} \tag{3}$$

$$U_m = \left\{ \frac{1}{2\pi r_o h_{m,o}} + \frac{\ln(r_o/r_i)}{2\pi D} + \frac{1}{2\pi r_i h_{m,i}} \right\}^{-1} \tag{4}$$

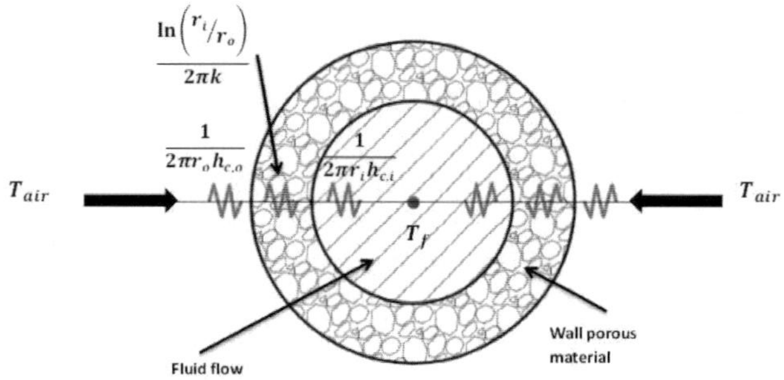

Figure 2: Thermal resistance modal

MODELING OF INDOOR SPACE

In this study an indoor space model is developed under the assumptions of steady state lumped approach and the space loads are mostly internal with no external loads (Yassine, et al., 2012). The space energy equation can be written as:

$$m_a c_{pa}(T_i - T_o) + m_a c_{pv}(T_i \omega_i - T_o \omega_o) + Q = \sum_{i=1}^{n} \int_0^{l_i} U_c (T_o - T_{sol}) dy \tag{5}$$

The term on the right side of the equation represent the sensible energy transfer with the desiccant membrane system. Similarly, the space moisture balance equation can be represented as:

$$m_a(\omega_i - \omega_o) + m_g = \sum_{i=1}^{n} \int_0^{l_i} U_m \rho_a (\omega_o - \omega_{sol}^*) dy \tag{6}$$

The right side of the equation represents the moisture exchange with the desiccant membrane piping system.

MODELING OF THE SOLAR COLLECTORS, EVAPORATIVE COOLER AND HEAT EXCHANGERS

Solar collectors will be used to harvest the solar energy and supply the required heat for the cycle. The model developed by (Duffie, et al.) will be used in this study. The evaporative cooling is used to cool the air exiting the indoor space which is used to further cool the liquid desiccant entering the indoor room. The evaporative cooler model done by (Kulkarni, et al., 2013) will be adopted in this study. A performance factor of 0.8 will be used in this study, i.e. the evaporative cooler can decrease the dry bulb temperature of the air by 80% of the difference between the wet and dry bulb temperatures. The sensible heat exchangers used in the cycle are modeled using the effectiveness method found in (Incropera, et al.). An effectiveness of 0.85 will be used for all the heat exchangers in the cycle.

NUMERICAL SOLUTION

To simulate the whole cycle of the integrated model the following input information is required: the space air supply conditions (temperature, humidity and flow rate), the internal space loads (sensible and latent) and the physical properties of the microporous membrane material (thermal conductivity and vapor diffusivity), the physical dimensions of both the membrane dehumidifier and regenerator and the flow rate of the desiccant system and inlet regeneration temperature. By performing iterations and making sure that the moisture content of the desiccant exiting the dehumidifier matches the moisture content of the desiccant at inlet of the regenerator, the model will output the steady state space temperature and humidity conditions as well as the concentration of the liquid desiccant solution. The solar energy input for regeneration can be determined and accordingly the flat plate collector system can be sized. The solution assumes quasi-equilibrium conditions, i.e. steady state with variable internal and external loads every hour.

CASE STUDY

Designing for maximum load operation, the cycle pump mass flow rate per pipe is set to be m_p= 1.6 Kg/hour (3.52 lb/hour), the number of pipes in the room, n=225, the regeneration temperature of the liquid desiccant in the cycle is T=45°C. The input supply air temperature is at 17°C (62.6°F) and 100 % relative humidity. The air flow rate is m_a= 0.19 Kg/s (25.13 lb/min). The ambient conditions are taken for a typical day in August for Beirut with an average temperature of 28°C (82.4°F) and relative humidity of 75 %. The prosed dehumidification cycle is applied to a typical internal office room. The internal latent and sensible loads and the number of occupants are shown in table 1.

Table 1: Sensible and latent loads of the occupants

Time (hr)	8-9	9-10	10-11	11-12	12-13	13-14	14-15	15-16	16-17
Number of occupants	2	2	3	4	5	5	5	4	3
Sensible load (W)	906	906	1149	1392	1635	1635	1635	1392	1149
Latent load (W)	110	110	165	220	275	275	275	220	165

The temperature, relative humidity and humidity ratio of the indoor space, with and without the membrane system operating, are shown in figures 3, 4 and 5. The thermal energy consumption of the cycle and the sub-cool/reheat energy required for a conventional vapor compression system to reach the same temperature and humidity ratio of the indoor space with the cycle existing is shown in table 2. The vapor compression cycle used for the reheat and subcool process has a coefficient of performance (COP) of 3.2.

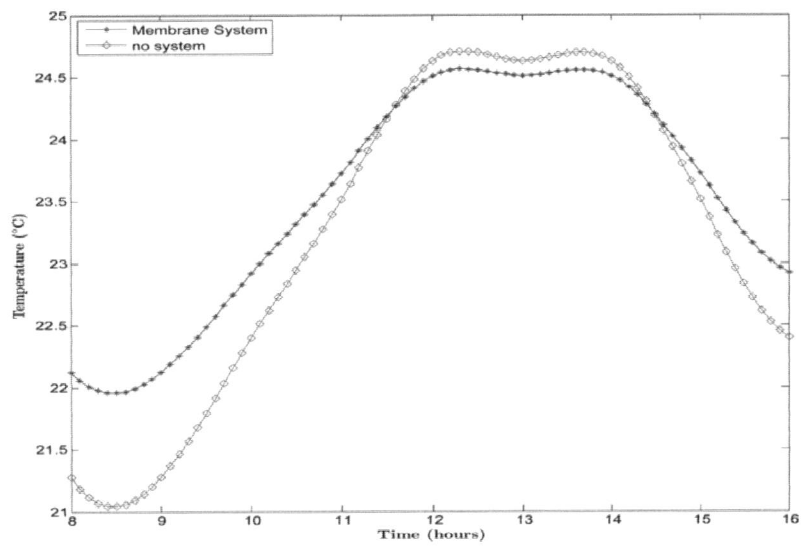

Figure 3: Comparison of the temperature variation of indoor space with the membrane system operating (blue) and without it (green)

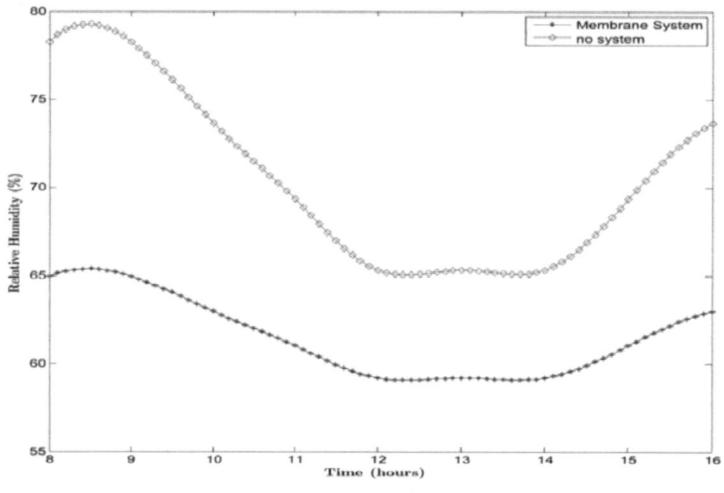

Figure 4: Comparison of the relative humidity variation of indoor space with the membrane system operating (blue) and without it (green)

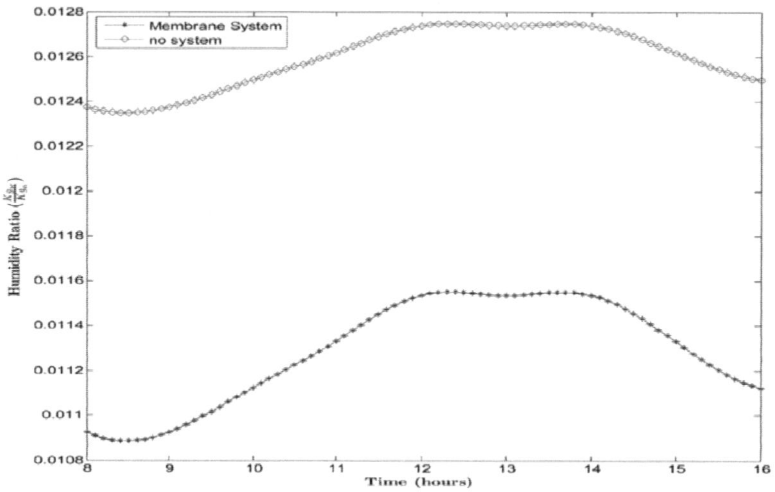

Figure 5: Comparison of the humidity ratio variation of indoor space with the membrane system operating (blue) and without it (green)

Table 2: Energy requirements of the cycle and corresponding subcool/reheat loads.

Time (hr)	8-9	9-10	10-11	11-12	12-13	13-14	14-15	15-16	16-17	Total Energy
Thermal Energy (kW)	4.35	4.35	4.33	4.31	4.29	4.29	4.29	4.31	4.33	38.8
Subcool energy (kW)	1.15	1.15	1.09	1.02	1.05	1.00	1.05	1.00	1.02	9.53
Subcool Electric energy (kW)	0.36	0.36	0.341	0.319	0.328	0.313	0.328	0.313	0.319	2.981
Reheat energy (kW)	0.54	0.54	0.46	0.372	0.29	0.29	0.29	0.372	0.46	3.614

DISCUSSION

The indoor room temperature without the desiccant system operating was on average 23.5°C (74.3°F), while with it the temperature slightly increased to 23.7°C (74.66°F). The effect of the membrane system was to increase the indoor temperature by 0.9°C (1.8°F) during the morning and by 0.5°C (0.9°F) at end of the day while decreasing during the noon by 0.1°C (0.18°F), as shown in Fig. 3. Although the sensible load increases at noon, due to the lower temperature of the desiccant entering through the permeable tubes, the desiccant pipes removed a small portion of the sensible load. This shows that the desiccant pipes act as a buffer for the indoor temperature, i.e. raising it when it gets low and lowering it when it increases. As for the humidity ratio, in the absence of the cycle it was 12.6 g_w/Kg_a (0.201 oz_w/lb_a) on average, while with the dehumidification cycle operating the humidity ratio decreased to 11.3 g_w/Kg_a (0.1809 oz_w/lb_a) on average, as shown in Fig. 4. The humidity ratio of the liquid desiccant, ω^*_{sol}, is equal to 7.7 g_w/Kg_a (0.123 oz_w/lb_a) is much lower than the room humidity ratio thus forming a driving potential to absorb the moisture from the indoor space. The effect of absorbing moisture from the indoor air has increased the temperature of the desiccant flowing in the pipes, due to the absorption energy (condensation), which in turn has increased the room temperature in the morning and at the end of the day. The reduction in the indoor air humidity ratio has in turn reduced the indoor relative humidity from an average value of 71% to 61%, as shown in Fig. 5. The dehumidification cycle has reduced the room relative humidity in the morning by 14 %, in the noon by 8 % and in the afternoon by 11 %. The thermal energy need for the cycle operation is shown in table 2. The thermal energy consumption varies from 4.29 kW (4.03 BTU/s) to 4.35 kW (4.12 BTU/s). This energy is to be supplied through available low grade thermal energy sources such as solar collectors. A solar collector of width one meter and total length of 12 meters is capable of supplying the required energy in a typical day in August. On the other hand, the energy required for the vapor compression system varies from 1.8 kW (1.7 BTU/s) to 1.5 kW (1.42 BTU/s) every hour. Although this energy is less than the energy required by the cycle nonetheless it is mainly of the form of electric energy (AC system is oversized to meet the sub cooling loads and heating coils are added for the reheat process). The electric energy associated with the cooling of the external ambient air to the supply condition is 5.26 kW (4.98 BTU/s). Using the vapor compression cycle of COP 3.2, the required electric energy is 1.6 W (1.51 BTU/s). Thus by using the membrane system the sub cool-reheat consumption part will no longer be used and hence the electric energy savings range from 35.6% in the morning to 32.7 % at the end of the day. The COP of the desiccant system cycle was found to be 0.21 on average. The low value is due to the use of the auxiliary heat exchanger, which lowers the liquid desiccant temperature.

CONCLUSION

The installation of the cycle inside the indoor space has decreased the relative humidity in the indoor space by 10 % on average. This reduces the electric energy consumption that a conventional sub cool-reheat system would use by 32-35%. The system has minimal effects on the indoor room temperature which makes it adequate for moisture removal in indoor spaces. The system proves to be efficient over conventional sub cool-reheat systems as it consumes no electric energy.

NOMENCLATURE

T = temperature (k)

ρ = density (Kg/m^3)

c = concentration of water per desiccant (Kg_{H_2O}/Kg_{CaCl_2})

c_p = specific heat $(J/Kg \cdot k)$

h = enthalpy (J/Kg)

h_c = heat convection coefficient $(W/m^2 \cdot k)$

h_m = mass convection coefficient (m/s)

r_o: = external radius of the pipe (m)

r_i = internal radius of the pipe (m)
ω: = humidity ratio (Kg_{H_2O}/Kg_{dryair})
ω^* = the equilibrium humidity ratio at surface of solution at temperature and concentration of solution (Kg_{H_2O}/Kg_{dryair})
D = diffusion constant of vapor in the pipe wall material (m^2/s)
k = thermal conductivity $(W/m \cdot k)$
m = mass flow rate (Kg/s)
n = number of dehumidification/regeneration pipes
l = length of the pipe (m)
h_{fg} = latent heat of vaporization of the water (J/Kg)
U_c = heat resistance coefficient per unit length $(W/m \cdot k)$
U_m = mass resistance coefficient per unit length (m^2/s)
Q = internal heat generation (W)

Subscripts

sol = $CaCl_2$ and H_2O solution
o = outlet conditions
i: = inlet conditions
a: = air
v: = water vapor
w: = liquid water
d: = desiccant
g: = vapor generation

REFERENCES

Bergero S, Chiari A. Experimental and theoretical analysis of air, humidification/dehumidification process using hydrophobic capillary contractors. Applied Thermal Engineering 2001; 21: 1119-1135.
Conde MR. Properties of aqueous solutions of lithium and calcium chloride: formulations for use in air conditioning equipment design. International Journal of Thermal sciences 2004;43: 367- 382.
Duffie JA, Beckman W. Solar Engineering of Thermal Processes. 2nd ed. USA: JOHN WILEY & SONS, INC., 1980.
Fan H, Besant RW, Simonson Cj, Shang W. Performance of a Run-Around system for HVAC Heat and Moisture Transfer Applications Using Cross-Flow Plate Exchangers Coupled with Aqueous Lithium Bromide. HVAC & R Research 2006; 12: 313-336.
Fauchoux MT, Simonson CJ, Torvi D. Tests of a Novel Ceiling Panel for Maintaining Space Relative Humidity by Moisture Transfer from an Aqueous Salt Solution. ASTM International 2009; 6.
Fauchoux M, Bansal M, Talukdar P, Simonson CJ, Torvi D. Testing and modeling of a novel ceiling panel for maintaining space relative humidity by moisture transfer. International Journal of Heat and Mass Transfer 2010; 53: 3961-3968.
Ghali K, Katanani O, Al-Hindi M. Modeling of the effect of hygroscopic curtains on relative humidity for spaces air conditioned by DX split air conditioning system. Energy and Buildings 2011; 43: 2093-2100.
Incropera FP, Dewitt DP, Bergman TL, Lavine AS. Fundamentals of Heat and Mass Transfer. 6th ed. USA: John Wiley & Sons, 2007.
Larson M D. The performance of membranes in a newly proposed runaround heat and moisture exchanger. 2006.
Mohammad AT, Bin Mat S, Sulaiman MY, Sopian K, Al-abidi AA. Historical review of liquid desiccant evaporation cooling technology. Energy and Buildings 2013; 67: 22-33.
Morse RG, Haas PE, Zehnter DA. Causes and Effects of High Humidity in South Florida Schools. ASHRAE Building X International Conference, 2007.
Studak JW, Peterson JL. A Preliminary Evaluation of Alternative Liquid Desiccants for a Hybrid Desiccant Air Conditioner. Proceedings of the Fifth Symposium on Improving Building Systems in Hot and Humid Climates 1988.
Yassine B, Ghali K, Ghaddar N, Srour I, Chehab G. A numerical modeling approach to evaluate energy-efficient mechanical. Energy and Buildings 2012; 55: 618-30.

A Review on HVAC Heat Recovery Systems

Mohamad Ramadan, PhD.

Mostafa GAD EL RAB, PhD.

Farouk Hachem , PhD.

Mahmoud Khaled, PhD.

ABSTRACT

The energy crisis that has occurred during the last 15 years has pushed the researchers to study new sources of energy. The focus is generally concentrated on renewable energy, however another solution has recently been considered, that is energy recovery. Its main idea is to use the waste energy of any engineering system in another application. Energy recovery concept covers a wide range of domains among them heat recovery. The works existing in the literature concerning HVAC (heating, ventilation, and air-condition) heat recovery are few. However in this paper an investigation on heat recovery in HVAC applications is presented. The three main heat recovery strategies applied to HVAC applications are explained, "immersed condenser heat recovery system", "air to water heat recovery system" and "dual heat recovery system". Advantages and drawbacks are briefly discussed.

INTRODUCTION

Energy recovery [1-5] is one of the solutions to reduce the energy cost without debasing the comfort of our daily life. During the last 10 years, this area has undergone a magnificent growth along with the domain of renewable energy[6-8]. This evolution in these two domains is due to two main reasons that are the increase of fuel cost and the level of pollution that has reached dangerous limits. The principle of energy recovery (see Figure 1) may be applied to almost any system dissipating waste energy. For instance a system to recover the kinematic energy of vehicles is proposed in [3].

HVAC (heating, ventilation, and air-condition) refers to the control of temperature, moisture content, cleanliness, air quality and air circulation as required by occupants. In the last 20 years HVAC systems have become widely utilized in such a way that in a region like middle east in every house there is at least one HVAC system. Knowing that an HVAC system dissipate a high amount of heat during the process of condensation, the amount of waste heat that is dissipated by HVAC systems of an entire building is huge. That is why it is interesting to investigate on the different strategies of heat recovery that may be applied on HVAC system. The number of work in the literature that are dedicated to study HAVC heat recovery system are limited. In this paper different heat recovery concept applied to HVAC systems are briefly presented.

Author A, B, C and D are Assistant Professors at the Department of Mechanical Engineering of the Lebanese International University, Beirut, Lebanon. Author B is also an Assistant Professor in the Department of Mechanical and Power Engineering of Minoufyia University, Shebin El-Kom, Egypt.

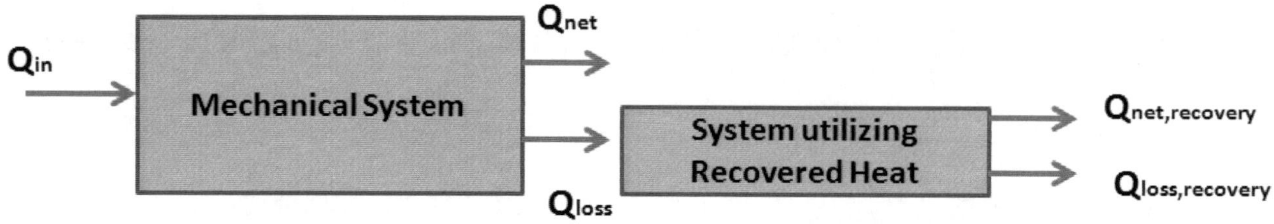

Figure 1 Heat recovery concept

The remaining of this paper is organized as follows: next section is devoted to present the refrigeration cycle, then the heat recovery concept as well as the different systems utilized for HVAC heat recovery are presented. Finally the main ideas of the paper and concluding remarks are summarized in the conclusion.

REFRIGERATION CYCLE

The refrigeration cycle applies to most of air conditioning and refrigeration systems [9]. The refrigerant circulates through the system in order to absorb heat from a medium and dissipate it into another one. When the refrigerant absorbs heat, the refrigerant's temperature increases so that it changes from liquid to gas, which represents the evaporation process. The refrigerant then undergo the condensation process during which it releases heat and changes its states to liquid. Six main components are needed to perform the refrigeration cycle as shown in Figure 2.

- Evaporator : In the evaporator the liquid refrigerant boil at a low temperature, which causes the refrigerant to absorb heat and therefore the inside space is cooled.
- Suction line: It represents the link between the evaporator and the compressor. After absorbing the heat, the suction line carries the superheated refrigerant to the compressor.
- Compressor: It is considered the heart of the system, it makes the refrigerant flow in the system. It separates the two parts of the system that are the low pressure and the high pressure zones. The refrigerant enters the compressor at low pressure and leaves the compressor at high pressure
- Hot gas discharge line: This tube connects the compressor to the condenser.
- Condenser : Contrarily to the evaporator, in the condenser, the refrigerant passes from the gas state to the liquid state by releasing heat.
- Expansion valve : It partially restricts the flow of the refrigerant producing the needed decrease in pressure, which allows the refrigerant to start evaporating.

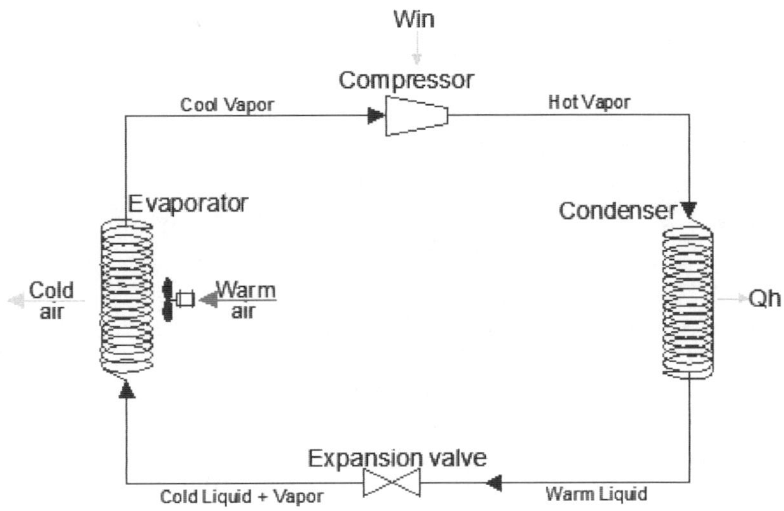

Figure 2 Refrigeration cycle

HVAC HEAT RECOVERY

In our daily life a huge amount of heat is dissipated as waste heat. HVAC systems are among those systems that are sources of waste heat. In the literature, the number of works that concerns the HVAC heat recovery domain is very limited. The existing systems may be classified within three different categories: "Immersed condenser Heat Recovery" system, "Air to water heat recovery system" and "Dual heat recovery system".

Immersed condenser Heat Recovery system

The concept of the "Immersed condenser Heat Recovery system" consists in modifying the air condition system by replacing the air cooled condenser by a water cooled condenser (as shown in Figure 3). The condenser coil is immersed into a water tank so that the heat dissipated by the condenser is absorbed by the water. The experimental set-up of this type of heat recovery system [10 to 14] is composed mainly of the three components, an air conditioning unit, a supply cold water tank , and a heat exchanger in the form of an insulated tank of different sizes where the condenser coils is immersed. When the HVAC system operates the condenser heats the water that can be then supplied as hot water. Another strategy may be utilized, [15] by installing a circulating pump and an additional water tank as a hot reservoir to collect and store the hot water supplied from the heat exchanger (see Figure 4). The role of the circulation pump is to circulate the water between the heat exchanger and the hot reservoir. The advantage of this heat recovery system is that it provides a higher quantity of hot water.

The average water temperature increase that may be obtained using this strategy of heat recovery, on an HVAC system utilized for domestic cooling, is 25°C [10-14]. However the main drawback of this type of system is that the system is modified and the coil may be damaged because of the permanent contact with water.

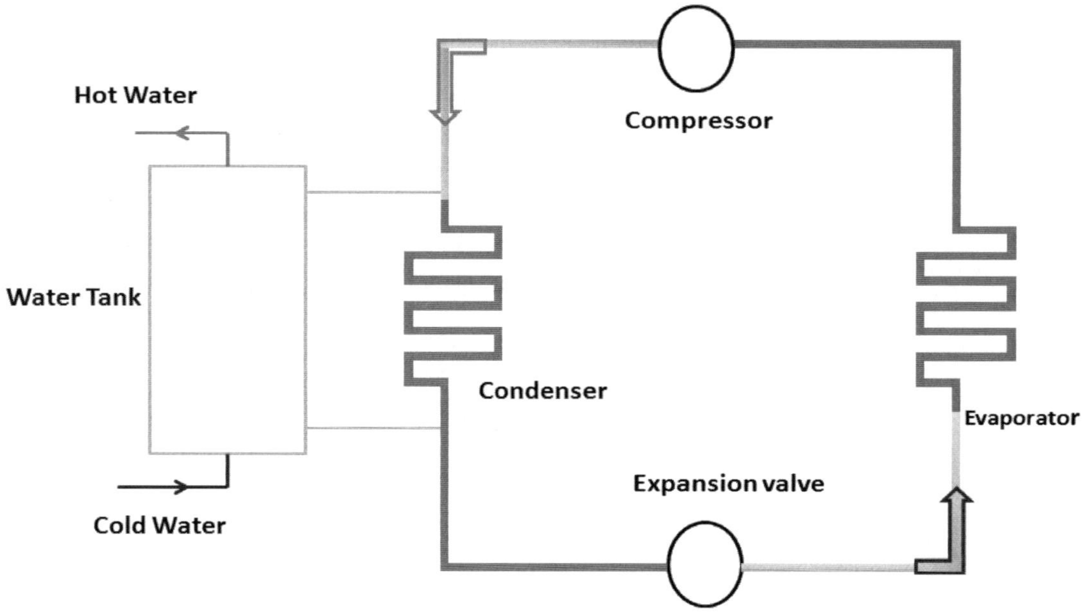

Figure 3 Heat recovery using the immersed condenser concept

Air to Water Heat Recovery System

In this strategy of heat recovery [16-17] a heat exchanger such as fin and tube heat exchanger or concentric tube is utilized. The heat exchanger is placed at the outdoor unit of the air conditioning system so that the hot air of the condenser fan is enters the heat exchanger as the hot fluid that heats the water which is the cold fluid. The hot water can be utilized directly or stored in a hot water reservoir. Figure 5 shows a schematic of the system. In [17] a numerical code is presented it allows to simulate the air to water heat recovery system. The water temperature increase is obviously lower than in the"Immersed condenser heat recovery system". The increase of water temperature varies depending on the characteristics of the system mainly the number of ton refrigerant. The average temperature increase is equal to 10°C. The main advantage of this system is that it does not require a modification in the HVAC system. Indeed the heat exchanger, piping system and other tools can be considered as accessories, that is to say it does not have any effect on the original system.

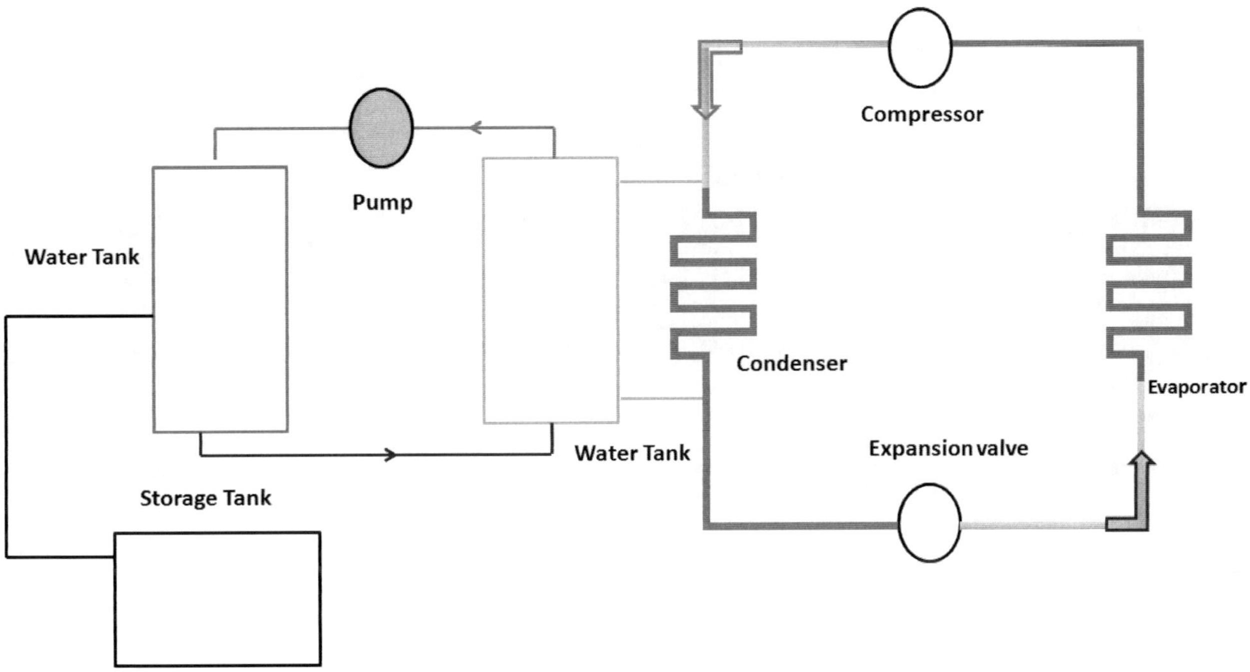

Figure 4 Immersed condenser with storage tank

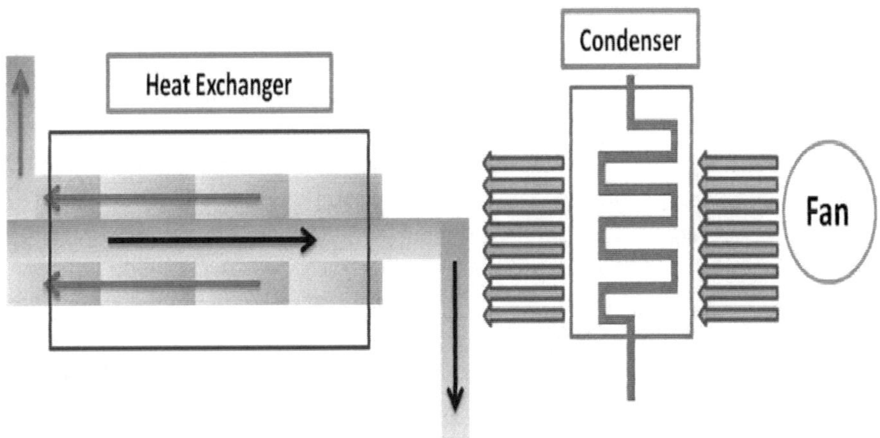

Figure 5 Direct heat recovery using the air to water concept

Dual heat recovery system

The dual scheme [18] is another way to recover the heat of a refrigeration cycle. A heat pump utilizes the refrigeration condenser output as its heat source. The result is a two-vapor compression system (see Figure 6). The two cycles are interconnected by a refrigerant-to-refrigerant heat exchanger. The drawback of this scheme is that it requires to have two systems which means it is not a general solution to recover heat and it can be applied only in some specific applications.

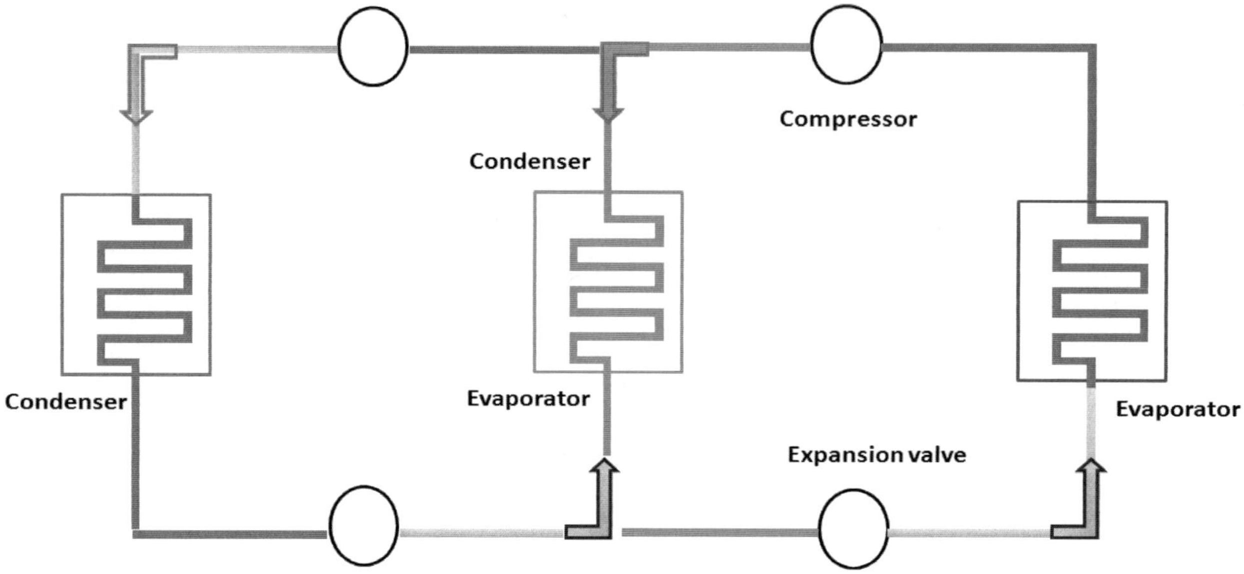

Figure 6 Dual heat recovery system

CONCLUSION

Heat recovery is one of the solution that may be adopted to face the energy crisis. HVAC systems dissipates considerable amount of heat that can be recovered. Three main strategies may be considered to recover the heat of HVAC systems. Immersed condenser heat recovery system can be utilized to heat water, it allows to raise the water temperature of an average value of 20°C. However it requires to change the operational mode of the system which may damage the system due to the contact with water. The air to water system offers lower temperature gain with the respect to the immersed condenser system, with an average gain of 10°C. Its advantage is that it does not require any modification of the system, the recovery part is added as an accessory tool. Dual heat recovery concept consists in combining two systems by using the condenser of the first system as evaporator for the second system. Since it requires to use two systems, this strategy can be utilized only in some specific applications and hence it cannot be generalized.

REFERENCES

[1] Westerlund, L., Hermansson, R., and Fagerstrom, J. 2012. Flue gas purification and heat recovery: A biomass fired boiler supplied with an open absorption system. Applied Energy, 96: 444-450.
[2] Xie, H., and Yang, Y. 2013. Dynamic behavior of Rankine cycle system for waste heat recovery of heavy duty diesel engines under driving cycle. Applied Energy, 112: 130-141.
[3] Ramadan, M., Khaled, M., El Kady, M., and El Hage, H. 2014. Generating power from Speed-Bump. Prototype development and experimental study. IEEE-MELECON, Lebanon
[4] Sprouse, C., and Depcik, C. 2013. Review of organic Rankine cycles for internal combustion engine exhaust waste heat recovery. Applied Thermal Engineering, 51: 711-722.
[5] Zhang, H., Wang, H., Zhu, X., Qiu, Y. J., Li, K., Chen, R., and Liao, Q. 2013. A review of waste heat recovery technologies towards molten slag in steel industry. Applied Energy, 112: 956-966.
[6] Solangi, K. H., Islam, M .R., Saidur R, Rahim N. A., and Fayaz, H. 2011. A review on global solar energy policy. Renew Sust Energ Rev. 15: 2149-2163.

[7] Joselin Herbert G.M. Iniyan S. Sreevalsan E, Rajapandian S. 2007. A review of wind energy technologies. Renew Sust Energ Rev. 11: 1117- 1145.

[8] Lopez, I., Andreu, J., Ceballos, S., Martinez de Alegria, I., and Kortabarria, I. 2013. Review of wave energy technologies and the necessary power-equipment. Renew Sust Energ Rev. 27: 413-434.

[9] McQuiston, Parker and Spitler. 2005. Heating, Ventilating and Air Conditioning: Analysis and design. Handbook Sixth Edition, John Wiley and Sons, Inc.

[10] Kongre, U.V., Salunkhe, M.B., Pohekar, A.A., Shendekar, S.S. 2013. Design Methodologies of air-conditioner cum water dispenser. International Journal of scientific research and management, 1(1): 18-22.

[11] Kongre, U. V., Salunkhe M, A. R., Dhumatkar, P. C., Aris, A. B. 2013. Testing and Performance Analysis on Air Conditioner cum Water Dispenser. International Journal of Engineering Trends and Technology , 4(4): 772-775.

[12] Mei, V. C., Chen, F. C., Domitrovic, R. E., Kilpatrick, J.K., Carter, J. A. 2003. A Study of a Natural Convection Immersed Condenser Heat Pump Water Heater. ASHRE Transactions, 109(2): 4628.

[13] Ji, J., Chow, T., Dong, J., He, W. 2003. Domestic air-conditioner and integrated water heater for subtropical climate. Applied Thermal Engineering, 23: 581-592.

[14] Keith Rice, C., Murphy, RichardW., Baxter, Van D. 2008. Design Approach and Performance Analysis of a Small Integrated Heat Pump (IHP) for Net Zero Energy Homes (NEH). International Refrigeration and Air Conditioning Conference.

[15] Stalin, M. J., Krishnan, S. M., Kumar, G. V. July 2012. Efficient Usage of Waste Heat from Air Conditioner. International Journal of Advances in Engineering and Technology, 4(1): 414-423.

[16] Rahman, M.M., Rahman, H. Y. 2011. Performance of Newly Developed Integrated Space Conditioning and Domestic Water Heating. Journal of Energy and Environment, 3 (1): 23-27.

[17] Ramadan, M., Khaled, M., Hachem, F., Al Shaer, A., Chahine, K., and Assi. 2013. Design and Analysis of an HVAC-Based Heat Recovery System, IEEE ICM, Beirut.

[18] Herbas, T. B.; Dalvi, E. A., and Parise, J. A. R. 1990. Heat Recovery from Refrigeration Plants Meeting Load and Temperature Requirements. International Refrigeration and Air Conditioning Conference.

Fresh Air Cooling Plant in Hot and Humid Climate Comparative Study

Charbel Seif
Senior Mechanical Engineer

Mazen Hassanieh
Project Mechanical Engineer

Rani Garzidin
Mechanical Engineer

ABSTRACT

In the design of air conditioning systems, introducing outdoor air is essential in maintaining a good indoor environmental air quality that helps building occupants feel stimulated and comfortable. In hot and humid weathers the fresh air cooling load may reach up to 50% of the building cooling load. Since air-conditioning equipment are the major energy consumers in buildings, reducing the Fresh air cooling energy demand forms a major challenge for engineers designing green and energy efficient buildings. The aim of this paper is to provide a comparative study for the energy performance of four different central fresh air treatment plant arrangements. The first arrangement considers the widely used tradition system of using a cooling coil to cool and dehumidify outside air to a low temperature followed by a heating coil to maintain the required supply temperature at the plant outlet. In the second arrangement air is divided in two streams, one that passes through a cooling coil while the other bypasses the coil; the two streams are then mixed to achieve dehumidification requirements. Combined air stream passes through a heating coil to maintain a fixed supply temperature at the plant outlet. A horseshoe heat pipe configuration is used in the third arrangement as a mean for pre-cooling air entering as well as reheating air leaving the plant; in this arrangement a cooling coil is used to dehumidify outside air. In the fourth arrangement similar to arrangement 2, air is divided into two streams the first passes through the cooling coil and the second through an enthalpy wheel, then the streams are mixed and reheat is achieved through a heating coil. Doha, Qatar, weather condition is considered the basis for this study since it is classified as having hot and humid summer weather. All plant arrangements are simulated to have a leaving condition of 18°C (64.4 °F) and 9.5g/kg (0.0095 lb/lb) of air moisture content. The moisture content level is selected to be equivalent to a comfort indoor condition of 24°C (75.2 °F), 55% relative humidity. The fourth arrangement showed the best result in terms of the Peak cooling load, Peak heating load and total cooling energy when compared to the other three arrangements. Utilizing arrangement four in designing fresh air treatment plant may reduce the plant cooling energy load by 22% and the reheating energy load by 80% compared to traditional systems.

INTRODUCTION

The overall aim of this paper is to provide a comparative study for different arrangements of fresh air treatment plants that are used in hot and humid climates. A total of four different arrangements and strategies are studied by calculating the peak cooling and reheat load and compared to each other to identify the arrangement that results in maximum energy saving.

The first arrangement consists of a cooling coil to cool and dehumidify the outside air to a low temperature followed by a heating coil to regulate the supply temperature.

In the second arrangement outside air is divided into two streams; the first passes through the cooling coil and the

Charbel Seif is a senior mechanical engineer at Dar Al Handassah, Shair and partners, Beirut, Lebanon.
Mazen Hassanieh is a project mechanical engineer at Dar Al Handassah, Shair and partners, Beirut, Lebanon.
Rani Garzidin is a design mechanical engineer at Dar Al Handassah, Shair and partners, Beirut, Lebanon.

second stream bypasses the coil and mixes with the cooled air stream to achieve the required dehumidification and supply air temperature.

In the third arrangement a hose-shoe heat pipe system is used to precool the air before entering the cooling coil, cooling coil to dehumidify the stream and a heating coil to reheat the air stream to the required supply temperature.

In the fourth arrangement pre-cooled fresh air is divided in two streams one that passes through an enthalpy wheel to reclaim sensible and latent heat form building exhaust air. The other stream is cooled through a cooling coil. The two streams mix and a reheat coil is used to heat the supply air to the design temperature.

This paper is divided into three sections with the first being the outdoor conditions used which are the peak conditions in Doha city, Qatar. Section two will discuss the different arrangements in detail and with schematic drawings and Section three summarizes the results and provides an energy comparison to identify the most energy efficient arrangement.

Load calculations were performed using "Hourly Analysis Program v. 4.51" from carrier and psychrometric charts were drawn using "ASHRAE Psychrometric Analysis"[1]. As for the energy wheel calculations were made using "Energy Conservation Wheel Selection Program".

OUTDOOR CLIMATIC CONDITIONS

The outdoor conditions of Doha, Qatar are considered in this study since it is a hot and humid climate. The peak conditions for July are used with maximum dry bulb temperature 46 °C (114.8 °F), wet bulb temperature 31 °C (87.8 °F) and a daily range of 11°C (19.8 °F) as shown in figure 1 below [2].

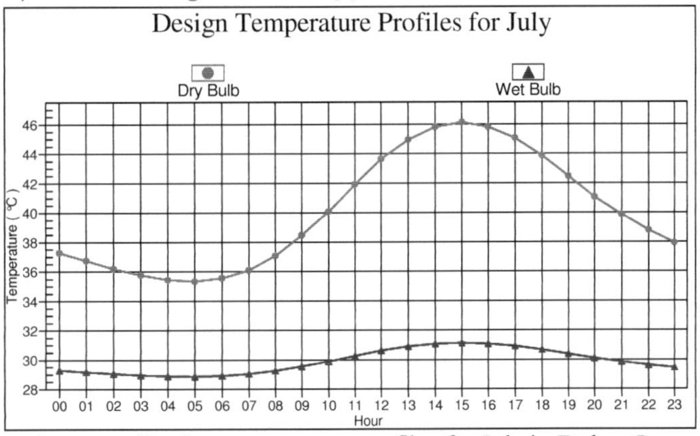

Figure 1 Design temperature profiles for July in Doha, Qatar.

Leaving air conditions from the treatment plant are: dry bulb temperature of 18°C (64.4 °F) and a water content of 9.5 g/kg (0.0095 lb/lb). These values are used to ensure indoor conditions where dry bulb temperature is 26 °C (78.8 °F), relative humidity of 50% +/-5% and an indoor air water content of 10.5g/kg (0.0105 lb/lb) of air [3].

FRESH AIR COOLING PLANT ARRANGEMENTS

The four different arrangements are explained below with a schematic drawing that shows the different components used and direction of air flow.

Arrangement 1

This arrangement consists of a cooling and a reheat coil. The cooling coil cools the outside air to a temperature of 14.4°C (57.9 °F) in order to achieve a water content of 9.5g/kg (0.0095 lb/lb) air. Then the heating coil heats the air back to 18 °C (64.4 °F) as shown below.

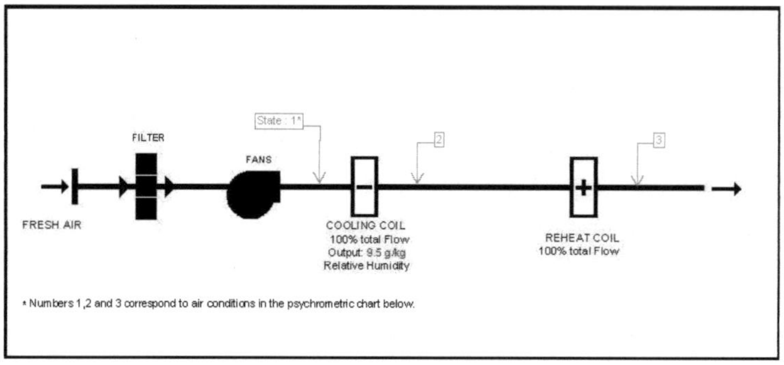

Figure 2 Schematic showing the components of arrangement 1.

Arrangement 2

In arrangement 2 outdoor air is divided into two streams the first passes through the cooling coil while the other bypasses the coil. The 2 streams are then mixed together and enter the reheat coil to achieve a leaving temperature of 18 °C (64.4 °F). In this arrangement there are two variables that would control the water content of the air, the first is the percentage of air passing through the coil and the second is the set point temperature of the cooling coil[4]. After several iterations and optimization the best combination was to cool the air to 10°C (50 °F) and a percentage of cooled air of 86.5%, as shown below.

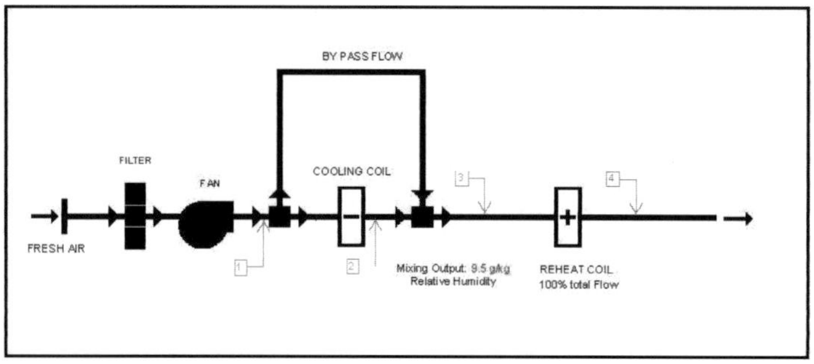

Figure 3 Schematic showing the components of arrangement 2.

Arrangement 3

In arrangement 3 a heat pipe is used to reheat the air leaving the cooling coil and at the same time to precool the air before entering the cooling coil. The cooling coil setpoint temperature is 14.4°C (57.9 °F) to achieve a water content of 9.5g/kg (0.0095 lb/lb).

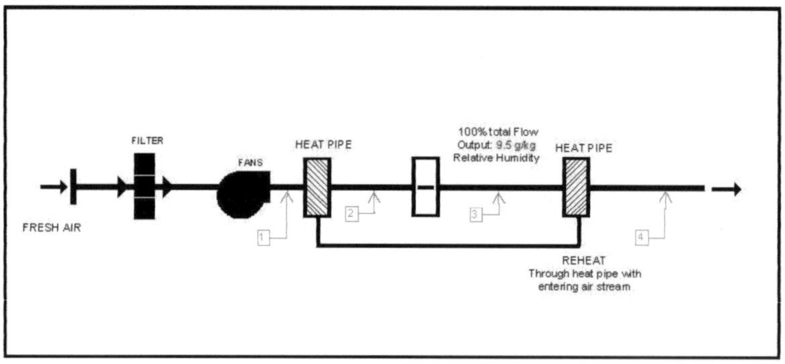

Figure 4 Schematic showing the components of arrangement 3.

Arrangement 4

In arrangement 4 bypass air and an enthalpy wheel are considered. The outdoor air is divided into two streams one goes to the cooling coil set at 10.5°C (50.9 °F) and the other bypasses the coil and passes through an enthalpy wheel. Then the two streams are mixed and reheated by a reheat coil. The percentage of flow passing through the cooling coil is optimized at 68% to achieve the desired water content of 9.5g/kg (0.0095 lb/lb). The enthalpy wheel allows for cooling of the outdoor air by exchanging heat with recovered exhaust air that is at 26 °C (78.8 °F) and 50% relative humidity (RH)[5].

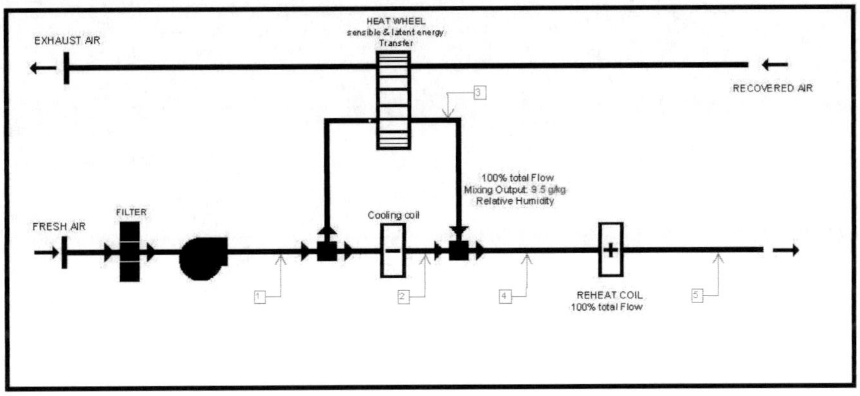

Figure 5 Schematic showing the components of arrangement 4.

COMPARISON OF RESULTS

For the simulation we used a flow of 165000 L/s (349615 cfm) for outdoor air to be cooled and 70000 L/s (148320 cfm) for the exhaust passing through the enthalpy wheel for arrangement 4. The below charts show the psychrometric analysis for each arrangement:

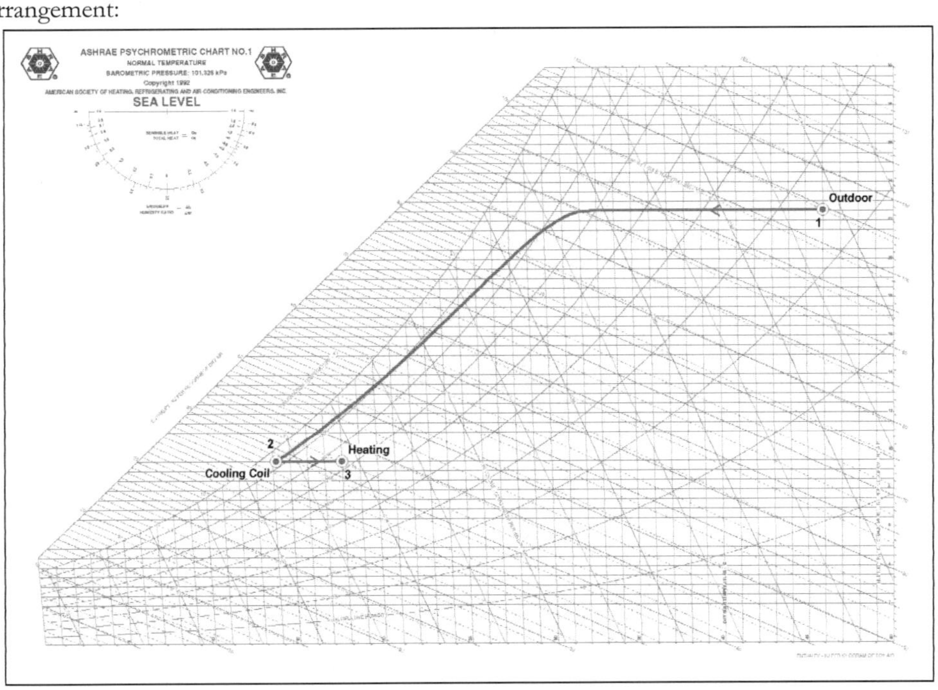

Figure 6 Psychrometric chart for arrangement 1.

Figure 6 shows the psychrometric chart for arrangement 1, where outdoor air is cooled by the cooling coil to reach the desired water content of 9.5 g/kg (0.0095 lb/lb), and then heated to 18 °C (64.4 °F). And in figure 7 below outdoor air is divided into 2 streams, the first goes through the cooling coil and the second bypasses the coil and then the 2 streams are mixed. At the mixing point the required water content is achieved and then the air is heated to the design supply temperature [8]

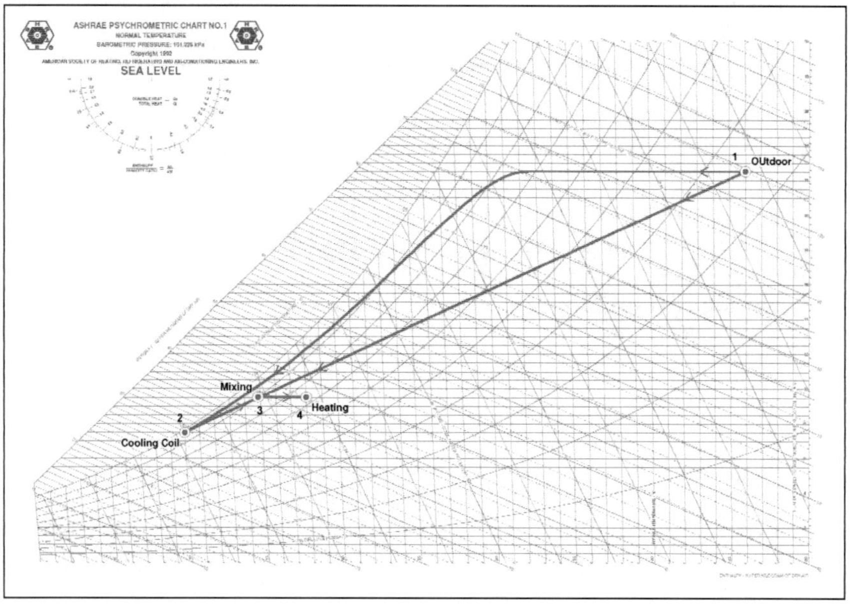

Figure 7 Psychrometric chart for arrangement 2.

In figure 8 below outdoor air is precooled through the heat pipe and then it passes through the cooling coil where it is cooled to reach the desired water content. Reheat is then achieved through the heat pipe and if needed a reheat coil. In the simulation the heat pipe was designed to cover the entire reheat load needed, so no reheat coil was required.

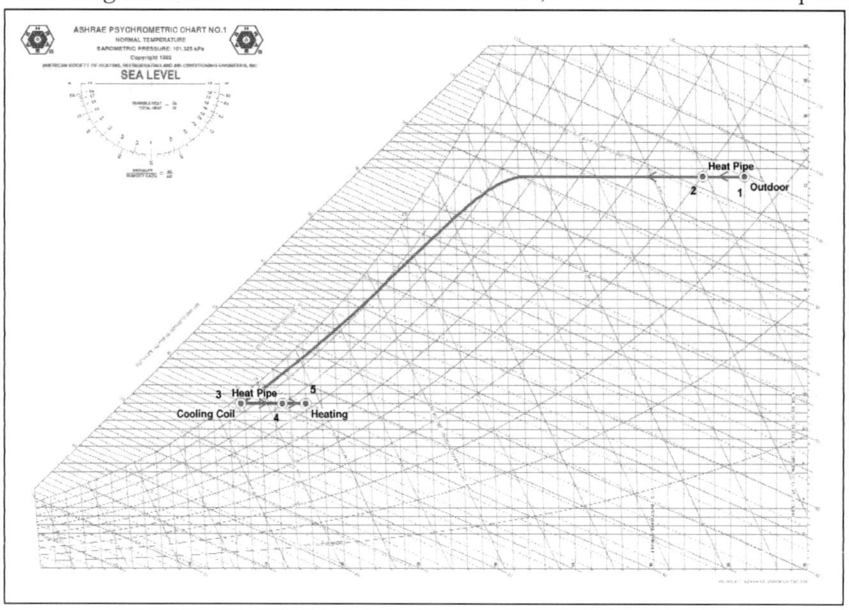

Figure 8 Psychrometric chart for arrangement 3.

And finally for arrangement 4 in the below figure outdoor air is split to 2 streams, the first passes through the cooling coil while the second through an enthalpy wheel that exchanges heat with recovered exhaust at 26 °C (78.8 °F) and 50% RH. The two streams are then mixed together and enter a reheat coil to have desired output temperature [7].

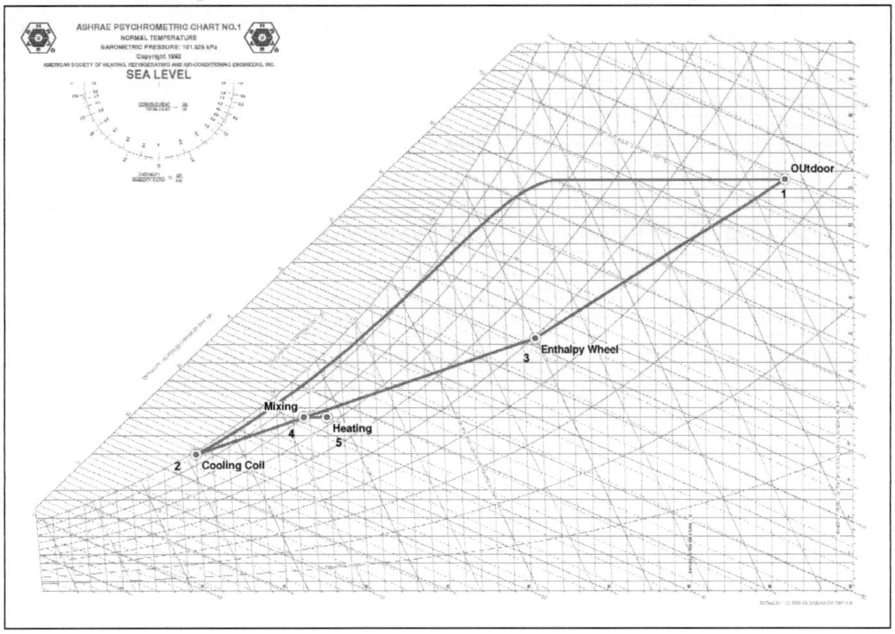

Figure 9 Psychrometric chart for arrangement 4.

The cooling and reheat load at peak condition was calculated for each case using HAP v4.51 and the results are tabulated below.

Table1 Comparison of Cooling and Reheat Coils Load

	Cooling coil Load KW (BTU)	Reheat Coil Load KW (BTU)
Arrangement 1: Direct Flow	12891 (3670)	713 (203)
Arrangement 2: Bypass Flow	12773 (3636)	871 (248)
Arrangement 3: Direct Flow with Heat Pipe	12062 (3434)	0 (0)
Arrangement 4: Bypass Flow with Heat Pipe and Enthalpy Wheel	9965 (2837)	138 (40)

Taking arrangement 1 as the basis for comparison, arrangement 2 shows a slight decrease in cooling load but coupled with an increase in reheat load. Arrangement 3 on the other hand provided a 6% decrease in cooling load and 100% decrease in reheat load, as for arrangement 4 the cooling load decreased by 22% and the reheat load by 80%.

CONCLUSION

With the growing energy challenges and the need to design sustainable and energy efficient buildings, it is vital to realize the huge savings that can be made in the ventilation systems for cooling fresh air especially in hot and humid climates as seen in the results above. A 22% decrease in cooling load and 80% reduction in reheat load were obtained in arrangement 4 by utilizing bypass flow coupled with a heat pipe and enthalpy wheel, proving to be the best between the available 4 arrangements. Even more efficient systems might be made possible with the advancement in technology and this is vital and important step in designing more sustainable buildings in the future.

ACKNOWLEDGMENTS

We would like to give special thanks to Mr. Maroun Khoury, director of Mechanical Department at Dar al Handasah Shair and partners, Beirut, Lebanon, for his support on developing this paper. Also we would like to acknowledge Dar al Handasah for their dedication and continuous support.

REFERENCES

[1] ASHRAE. 2009. ASHRAE Handbook-Fundamentals. Atlanta: American Society of Heating Refrigeration and Air Conditioning Engineers, Inc.

[2] Long Period Means& Extremes of climatologically elements Doha International Airport period (1962-1999), state of Qatar ministry of Comm& Transport Dept. of civil Aviation& Meteorology

[3] ASHRAE. 2011. ASHRAE Handbook- HVAC Applications. Atlanta: American Society of Heating Refrigeration and Air Conditioning Engineers, Inc.

[4] ASHRAE. 2012. ASHRAE Handbook- HVAC Systems and Equipment Handbook. Atlanta: American Society of Heating Refrigeration and Air Conditioning Engineers, Inc.

[5] Gommed, K. and G. Grossman. 2004. A liquid dessicant system for solar cooling and dehumidification. ASME Journal of Solar Energy Engineering. 126(3), 879-885 (Jul 19, 2004)

[6] Daou, K., Wang, R.Z. and Z.Z. Xia. 2006. Desiccant cooling air conditioning: a review. ELSEVIER Renewable and sustainable Energy Reviews, Volume 10, Issue 2, April 2006, Pages 55-77.

[7] Kolokotroni, M. and A. Aronis. 1999. Cooling-energy reduction in air-conditioned offices by using night ventilation. ELSEVIER Applied Energy, Volume 63, Issue 4, August 1999, Pages 241-253.

[8] Zhang, L.Z. and J.L. Niu. 2003. Indoor humidity behaviors associated with decoupled cooling in hot and humid climates. ELSEVIER Building and Environment, Volume 38, Issue 1, January 2003, Pages 99-107.

Heating Water Using the Recovered Waste Heat from Boilers in HVAC Applications – Thermal Modeling and Parametric Analysis

Mahmoud Khaled, PhD **Mohamad Ramadan, PhD** **Charbel Habchi, PhD**

Mehdi Chouman, PhD **Farouk Hachem, PhD**

ABSTRACT

The current tendency in the energy sector is to reduce fossil fuel-based energy investments in favor of an energy approach focusing on renewable energy concepts. On the other hand, Heating, Ventilating and Air Conditioning HVAC has changed from being luxury to essential need for people and comprises many energy components that require to be well managed. The present work concerns a coupling between these two energy domains and particularly aims to apply heat recovery concepts to HVAC applications. It particularly uses the energy waste from the exhaust gases of boilers to heat/preheat water. The heat exchanger considered is concentric tube heat exchanger. A thermal modeling of the heat recovery system applied to boilers as well as a corresponding iterative code are developed and presented. Calculations with the code are performed and give magnitude orders of energy saving and management in HVAC applications. A parametric analysis based on several geometrical configurations and boiler heating loads is carried out. It was shown that water can be heated from 20 °C (68 F) to up to 100 °C (212 F) depending on the mass flow rate. The most efficient configuration is for water flowing in the inner tube with an inner to outer diameter ratio of 0.75.

INTRODUCTION

The main concern of science, nowadays, is to find new sources of energy and apply them in life equipments. Recent researches ensure that after 30 year whole world will suffer from the lack of fuel which is the main actual source of energy [1]. Wind, River, Solar, Sea waves, even nuclear are different forms of renewable energy which scientists is trying to benefit from [2-4], in many sectors, by generating electricity and other forms of energy. These renewable energies as well as heat recovery concepts [5-6] are actually the main means towards the reduction of fuel consumption and carbon dioxide emissions.

Heat recovery is defined as the capture of heat contained in fluids or gases of systems and its use in production of other forms of energy and that otherwise would be lost. Depending on the system in which hot fluids or gases are contained, heat recovery has different applications: internal combustion engines, heat pumps, chillers, power generators, chimneys and shower water [6]. On the other hand and depending on the final form of energy captured, the heat recovery is of different types: heat exchanger systems when thermal energy output is required and thermodynamic heat recovery system when electrical power output is desired. Heat exchanger systems are systems based on transferring heat from high temperature stream (exhaust in the present case) into low temperature stream (water pipe in this case) [6].

Authors of this paper are faculty members in the Mechanical Department of the School of Engineering of the Lebanese International University

On the other hand, Heating, Ventilating and Air Conditioning HVAC systems has changed from being luxury to essential need for people and thus involve many energy components which need to be more and more managed. In this context, the present work suggest a new design that permits to couple between the two energy domains described above: energy recovery and HVAC. It particularly uses the energy waste from the exhaust gases of boilers to heat/preheat water. For this purpose, a thermal modeling of the recovery system is presented and iterative procedure is described. The heat exchanger considered is concentric tube heat exchanger. A parametric analysis testing the effect of the exchanger geometry (inlet and outlet diameters) is also performed.

The rest of the manuscript is organized as follows. In the next section, the thermal modeling of the complete system as well as the operational mode of the code are exposed. The following section is then devoted to the results and parametric analysis. Finally, the main conclusions of the work are drawn.

THERMAL MODELING AND ITERATIVE PROCEDURE

The heat exchanger considered in the present study is counter flow concentric tube heat exchanger with inner diameter D_i, outer diameter D_o and length L. In the thermal modeling presented below and for presentation clarity, the exhaust gases of the boiler are considered flowing in the inner tube of the exchanger and water flowing in the annulus of the exchanger. Later in the calculations, the configurations can be inversed and the governing equations inversed accordingly. In this type of exchanger and neglecting the heat transfer between water in the annulus and the ambient air, the energy balance can be written as [7]:

$$\dot{m}_w C_{p,w}\left(T_{w,out}-T_{w,in}\right)=\dot{m}_g C_{p,g}\left(T_{g,in}-T_{g,out}\right)=U.A.\Delta T_{\ln} \tag{1}$$

Where:

$\dot{m}_w$ and $\dot{m}_g$ are respectively the water and gas mass flow rates, $C_{p,w}$ and $C_{p,g}$ are respectively the water and gas specific heats, $T_{w,in}$ and $T_{g,in}$ are respectively the water and gas inlet temperatures, $T_{w,out}$ and $T_{g,out}$ are respectively the water and gas outlet temperatures, U the overall heat transfer coefficient between the two fluids streams, A the area of heat transfer between the two fluids, and $\Delta T_{\ln}$ the logarithmic mean temperature difference between the two fluids calculated from:

$$\Delta T_{\ln}=\frac{\left(T_{g,out}-T_{w,in}\right)-\left(T_{g,in}-T_{w,out}\right)}{\ln\left[\frac{\left(T_{g,out}-T_{w,in}\right)}{\left(T_{g,in}-T_{w,out}\right)}\right]} \tag{2}$$

In most heat exchanger applications, a thin wall of large conductivity is generally used between the two fluid streams and then the overall heat transfer coefficient and the heat transfer area can be calculated from:

$$U=\frac{1}{\frac{1}{h_w}+\frac{1}{h_g}} \tag{3}$$

$$A=\pi D_i L \tag{4}$$

Where h_w and h_g are the convective heat transfer coefficients of respectively the water and gas flows calculated from heat transfer correlations found in [7].

To calculate the heat transfer rate that can be recovered from the exhaust gas, one of the outlet temperatures should be calculated. To proceed, Equations (1) to (4) are rearranged in order to obtain one equation with only one unknown the water outlet temperature $T_{w,out}$:

$$\dot{m}_w C_{p,w}\left(T_{w,out}-T_{w,in}\right)=U.A.\frac{\left(T_{g,in}-\frac{\dot{m}_w C_{p,w}}{\dot{m}_g C_{p,g}}\left(T_{w,out}-T_{w,in}\right)-T_{w,in}\right)-\left(T_{g,in}-T_{w,out}\right)}{\ln\left[\frac{\left(T_{g,in}-\frac{\dot{m}_w C_{p,w}}{\dot{m}_g C_{p,g}}\left(T_{w,out}-T_{w,in}\right)-T_{w,in}\right)}{\left(T_{g,in}-T_{w,out}\right)}\right]} \tag{5}$$

Equation (5) is solved by iteration to obtain the water outlet temperature $T_{out,w}$ for given flow rates, inlet temperatures and specific heats of water and gas. Then, the heat rate recovered from exhaust gases can be determined:

$$P_{exchanger}=\dot{m}_w C_{p,w}\left(T_{w,out}-T_{w,in}\right) \tag{6}$$

Now, to calculate the gas flow rate $\dot{m}_g$ and inlet temperature $T_{g,in}$ that are input to the heat exchanger calculations presented above, the boiler side is considered. The efficiency of a boiler is defined as the ratio of the heating load provided to the space $\dot{Q}_{heating}$ (and calculated in HVAC designs) and the heat of combustion $\dot{Q}_{combustio}$ [8]:

$$\eta_b=\frac{\dot{Q}_{heating}}{\dot{Q}_{combustion}} \tag{7}$$

The difference between the heating load $\dot{Q}_{heating}$ and $\dot{Q}_{combustio}$ is lost in the exhaust gases and then the energy rate contained in the exhaust gases is:

$$\dot{Q}_{ex}=\dot{Q}_{combustion}-\dot{Q}_{heating} \tag{8}$$

Combining equations (7) and (8) yields to:

$$\dot{Q}_{ex}=\left(\frac{1-\eta_b}{\eta_b}\right)\dot{Q}_{heating} \tag{9}$$

At the same time, the energy rate of the exhaust gases can be written as:

$$\dot{Q}_{ex}=\dot{m}_g C_{p,g}\left(T_{in,g}-T_{amb}\right) \tag{10}$$

Where T_{amb} is the ambient temperature.

The fuel (Diesel) flow rate can be calculated from:

$$\dot{m}_f = \frac{\dot{Q}_{combustion}}{q_c} = \frac{\dot{Q}_{heating}}{\eta_b q_c} \tag{11}$$

Where q_c is the heat of combustion of the fuel (Diesel).

Noting A_{af} the air to fuel ratio, the gas mass flow rate can be written as:

$$\dot{m}_g = (A_{af} + 1)\dot{m}_f = \frac{(A_{af} + 1)\dot{Q}_{heating}}{\eta_b q_c} \tag{12}$$

Combining equations (9), (10) and (12), one can write for the gases inlet temperature:

$$T_{g,in} = T_{amb} + \frac{(1-\eta_b)q_c}{(A_{af} + 1)C_{p,g}} \tag{13}$$

The operational mode of the code developed for the present calculations consists on an iterative procedure manipulating the different equations presented in the thermal modelling presented above in order to have a set of input parameters and calculating the main output: the heat exchanger performance. Figure 1 shows a schematic of the overall scheme of the code.

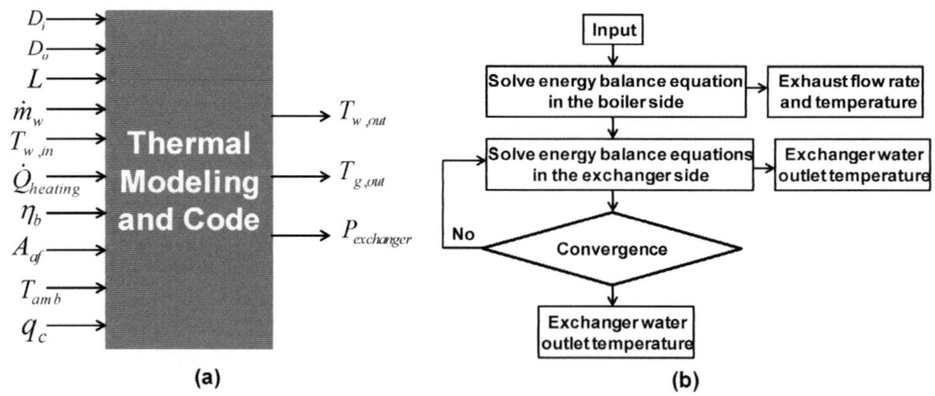

Figure 1 Schematic of the (a) overall scheme and (b) operational mode of the code developed.

The code solves the problem iteratively using a partitioned algorithm allowing coupling the two physical problems: cooling the condenser by the air and heating the water by the heated air. The inputs of the developed code are: the operational parameters of the HVAC system, the characteristics of the heat exchanger, and the physical properties of air and water.

Figure 1-b shows schematically the operational mode of the code. In the first part of the solver the energy balance equation applied to the boiler side is solved, which gives the value of the exhaust gases mass flow rate and temperature. In the second part the energy balance equation applied to the heat exchanger is solved. The temperature and mass flow rate obtained from the first part are input for the heat exchanger calculation part, whereas the temperature of the water is the inlet cold temperature. The solver iterates until reaching a convergence.

RESULTS AND ANALYSIS

The thermal modeling presented in the previous section is employed to perform parametric analysis on the thermal performance of the boiler heat recovery concept proposed. The performance of the heat exchanger of the system is explored when the geometry of the exchanger is varied. A counter-flow concentric tube heat exchanger of three meters length is used to heat water by the exhaust gases is considered. The water inlet temperature is fixed at 20 °C (68 F). The heating load is considered equal to 100 kW (341 200 Btu/hr) (magnitude orders for an apartment of 5 to 8 people) and the boiler efficiency is considered equal to 0.8. The ambient temperature is equal to 20 ºC (68 F) and the air-to-fuel ratio is equal to 15. The six geometrical configurations tested are detailed in Table 1.

Table 1. Details of the different geometrical configurations tested.

Configuration	Geometry	Diameters
1	Geometry 1 – Figure 2-a	Outlet to inlet diameter ratio = 1.25 Inner diameter = 0.1 m
2	Geometry 1 – Figure 2-a	Outlet to inlet diameter ratio = 1. 5 Inner diameter = 0.1 m
3	Geometry 1 – Figure 2-a	Outlet to inlet diameter ratio = 1.75 Inner diameter = 0.1 m
4	Geometry 2 – Figure 2-b	Inlet to outlet diameter ratio = 0.25 Outlet diameter = 0.1 m
5	Geometry 2 – Figure 2-b	Inlet to outlet diameter ratio = 0. 5 Outlet diameter = 0.1 m
6	Geometry 2 – Figure 2-b	Inlet to outlet diameter ratio = 0.75 Outlet diameter = 0.1 m

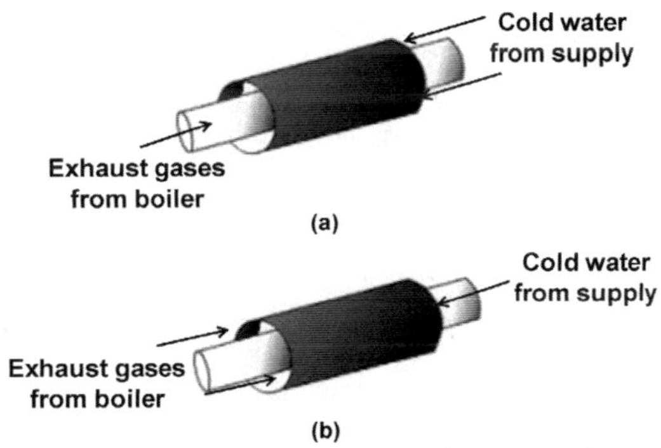

Figure 2 (a) first tested geometry and (b) second tested geometry.

A first set of calculations is performed by varying the water flow rate from 0.02 kg/s (158 lb/hr) to 0.5 kg/s (3968 lb/hr) with a step of 0.02 kg/s (158 lb/hr) and for the three first configurations 1 to 3. Figure 3 shows the variation of the water outlet temperature and waste power (heat rate) recovered in terms of the water flow rate and for configurations 1 to 3. From Figure 3-a, it can be shown that the water outlet temperature decreases exponentially with the water flow rate and decreases with the outer diameter for a fixed inner diameter. As illustration for an outer to inner diameter ratio of 1.25, by increasing the water flow rate from 0.02 kg/s (158 lb/hr) to 0.5 kg/s (3968 lb/hr) the water outlet temperature decreases from 131.6 °C (268.9 F) to 25.7 °C (78.3 F). For outer/inner diameter ratios of 1.5 and 1.75, the water outlet temperature decreases from 116.1 °C (241 F) to 25.3 °C (77.5 F) and from 105.4 °C (221.7 F) to 23.6 °C (74.5 F) respectively. It should be noted that actually water temperature greater than 100 °C (212 F) does not exist and this mean that water will evaporate before travelling the entire length of the exchanger. Values of water temperature higher than 100 °C (212 F) will be kept in the calculations only to give magnitude orders on waste heats that can be recovered.

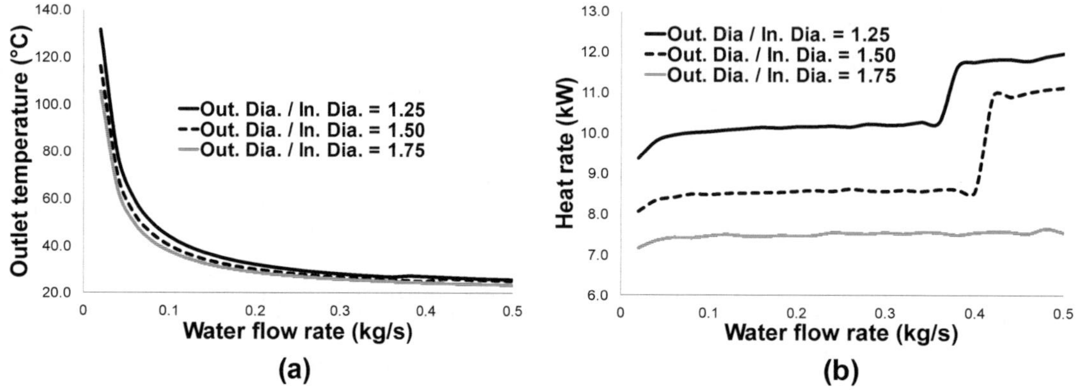

Figure 3 Variation of (a) the water outlet temperature and (b) waste heat recovered in function of the water mass flow rate for different configurations.

From Figure 3-b, it can be shown that for an outer/inner diameter ratio of 1.75 the heat rate is almost invariant equal to 7.5 kW (25 590 Btu/hr). For the diameter ratios of 1.25 and 1.5, sudden increases in the heat rate exchanged occur at given water flow rates. As illustration for an outer/inner diameter ratio of 1.25, the heat rate is almost constant around 10 kW (34 120 Btu/hr) up to a water flow rate of 0.36 kg/s (2857 lb/hr) from which it increases suddenly to 11.7 kW (39 920 Btu/hr) and continue its increase with an exponential manner. The same trends are noted for the diameter ratio of 1.5 but with different magnitudes and transition for a water flow rate of 0.4 kg/s (3175 Btu/hr). These sudden increases are due to the transition from laminar to turbulent regimes that occur at constant water flow rate and because of the change of the geometry. To illustrate Figure 4 shows the variation of the Reynolds number of water in terms of the water flow rate for the three configurations tested above. On this figure, the limit of Reynolds number of 2300 for transition to turbulent regime is shown also.

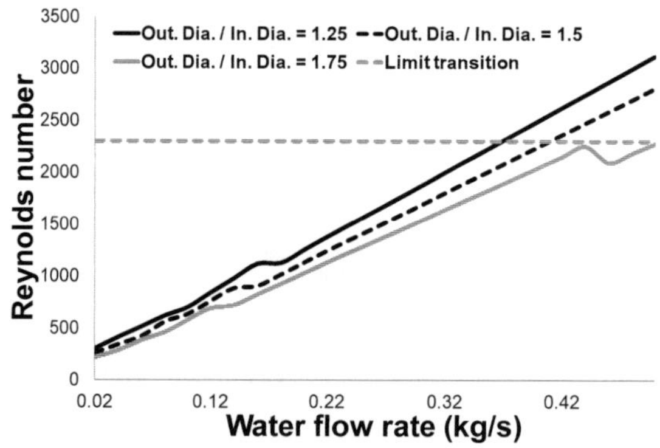

Figure 4 Variation of the Water Reynolds number in terms of the water flow rate for the first three configurations tested.

It can be shown that by varying the water flow rate from 0.02 kg/s (158 lb/hr) to 0.5 kg/s (3968 lb/hr) for an outer/inner diameter ratio of 1.75, the flow remains laminar with the Reynolds number varying from 212 to 2271. For an outer/inner diameter ratio of 1.5, the Reynolds number varies from 252 to 2808 with 2300 reached at a water flow rate around 0.4 kg/s (3175 Btu/hr). For an outer/inner diameter ratio of 1.25, the Reynolds number increases from 300 to 3120 with 2300 reached at a water flow rate around 0.36 kg/s (2857 lb/hr). It should be noted that the inner Reynolds number of gases is invariant between the three configurations and with the water flow rate and equals 17 628. This high Reynolds number induces high convective coefficient of the gases flow. In the three configurations and before transition to turbulence, when the water flow rate increases the water convective coefficient increases with magnitude orders remaining negligible compared to the gases convective coefficient yielding to a negligible effect on the overall heat transfer coefficient of equation (3) and then on the rate of waste heat recovered. For the outer/inner diameter ratios of 1.25 and 1.5, when transition occurs the water convective coefficient increases to a limit that it becomes no longer negligible compared to the

gases convective coefficient and its increase induces an increase in the overall heat transfer coefficient and then in the rate of waste heat recovered.
From a practical point of view, the results exposed above can be shown in the form of recommendations. One should use an outer diameter of outer flow the most closer to the inner diameter in order to enhance the heat recovery occurring. By decreasing the outer diameter for a fixed inner diameter, one is fastening the transition to turbulence and thus enhancing the heat transfer rate for the same quantity of water flowing. At the same time for each geometrical configuration, the mass flow rate corresponding to the transition should be calculated preliminary and flow rates higher than this value are recommended.

A second set of calculations is performed by varying the water flow rate from 0.02 kg/s (158 lb/hr) to 0.5 kg/s (3968 lb/hr) with a step of 0.02 kg/s (158 lb/hr)and for the three last configurations 4 to 6. Figure 5 shows the variation of the water outlet temperature and waste power recovered in terms of the water flow rate and for configurations 4 to 6.

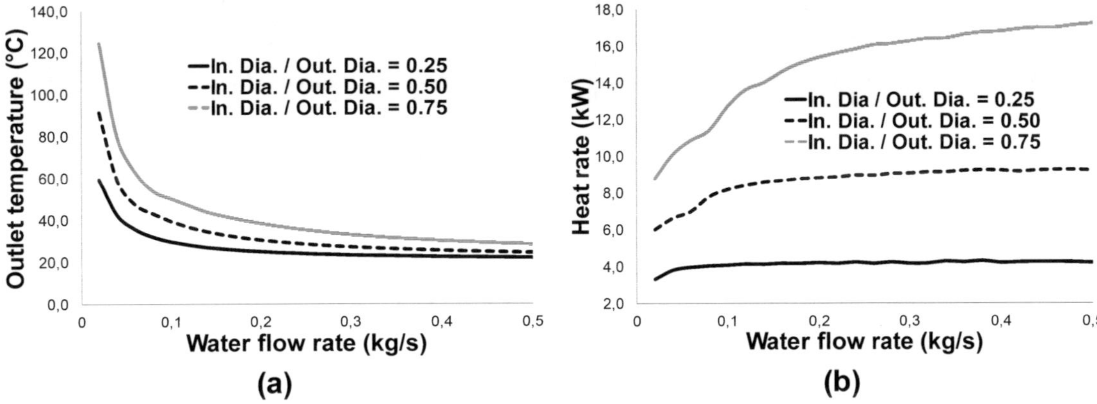

Figure 5 Variation of (a) the water outlet temperature and (b) waste heat recovered in function of the water mass flow rate for the three last configurations.

From Figure 5-a, it can be shown that the water outlet temperature decreases exponentially with the water flow rate and increases with the inner diameter for a fixed outer diameter. As illustration for an inner/outer diameter ratio of 0.25, by increasing the water flow rate from 0.02 kg/s (158 lb/hr) to 0.5 kg/s (3968 lb/hr)the water outlet temperature decreases from 59.3 °C (138.7 F)to 22 °C (71.6 F). For inner/outer diameter ratios of 0.5 and 0.75, the water outlet temperature decreases from 91.6 °C (196.9 F) to 24.4 °C (75.9 F) and from 124.7 °C (256.5 F) to 28.2 °C (82.8 F) respectively.

From Figure 5-b and differently from the case of water flowing in the outer annulus, the heat rate is all the time increasing exponentially with the water flow rate. As illustration for an inner/outer diameter ratio of 0.25, the heat rate increases from 3.3 kW (11 260 Btu/hr) to 4.2 kW (14 330 Btu/hr). The increase in the heat rate is more sensible at higher inner diameter: the heat rate increases from 6.0 kW (20 472 Btu/hr) to 9.2 kW (31 390 Btu/hr) for an inner/outer ratio of 0.5 and from 8.8 kW (30 026 Btu/hr) to 17.2 kW (58 686 Btu/hr) for an inner/outer ratio of 0.75. To explain the features of Figure 5-b the Inner and outer Reynolds number are considered in this case since the outer Reynolds number is varying between the three configurations considered. The outer Reynolds number in each case is invariant with the water mass flow rate and equals to 13 499, 11 249 and 10 073 for the cases of inner/outer diameter ratios equal respectively to 0.25, 0.5 and 0.75. Figure 6 shows the variation of the water Reynolds number in terms of the water flow rate for the three configurations considered. Compared to the three first configurations (configurations 1 to 3 in Table 1), the transition for the three last configurations (configurations 4 to 6) occur at smaller water flow rate due to the decrease in the inner diameter. It can be noticed that starting from a water flow rate of around 0.1 kg/s (794 lb/hr)the water flow becomes turbulent in the three configurations. At fixed inner/outer diameter ratio and by increasing the water flow rate, the outer Reynolds number remains constant and the inner Reynolds number increases. This yields to an increase of the water convective coefficient for a constant gases convective coefficient. With both flows turbulent, the water convective coefficient is non-negligible compared to the gases convective coefficient and its increase can be traduced with an increase in the overall heat transfer coefficient of equation (3).

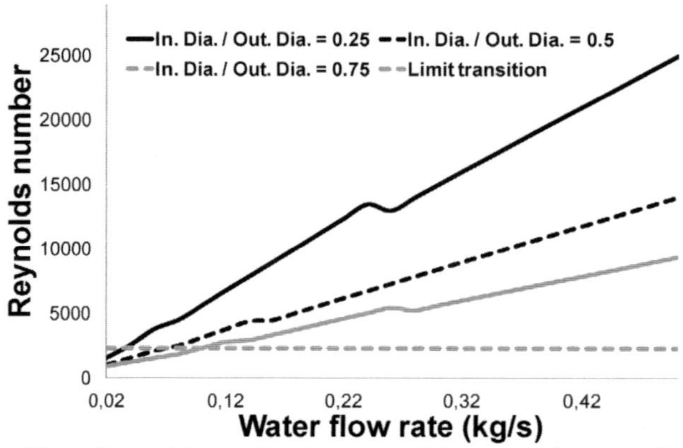

Figure 6 Variation of the Water Reynolds number in terms of the water flow rate for the first three configurations tested.

Compare now between the three cases of diameter ratios, it can be shown that the water Reynolds number (water convective coefficient) decreases when the inner/outer diameter ratio increases. On the other hand, the gases Reynolds number (gases convective coefficient) increases when the inner/outer diameter ratio increases. The increase in the gases convective coefficient dominates on the decrease of the water convective coefficient and results on an increase in the overall heat transfer coefficient of equation (3) and then the heat rate of Figure 5-b.

CONCLUSION

The present work concerns a parametric analysis of recovering heat from boiler exhaust through the use of counter flow concentric tube heat exchanger. Different geometrical configurations are tested and comprise mainly two geometries: a first one where exhaust gases flow in the inner tube and water in the annulus and a second one where water flows in the inner tube and gases in the annulus. A case of a 100 kW (341 200 Btu/hr) boiler is considered.

It was shown that water temperatures up to 100 °C (212 F) can be obtained depending on the mass flow rate. Configurations of geometry 2 show higher heat rates than configurations of geometry 1. When geometry 1 is implemented, it is more recommended to have the outer diameter the most closely possible to the constant inner diameter. When geometry 2 is under consideration, the inner diameter should be the most closely possible to the constant outer diameter.

REFERENCES

Shi, L.S., Yit Lin Chew, M. 2012. A review on sustainable design of renewable energy systems. Renewable and Sustainable Energy Review, 16:192-207.

Solangi, K.H., Islam, M.R, Saidur, R., Rahim, N.A., and Fayaz, H. 2011. A review on global solar energy policy. Renewable and Sustainable Energy Review, 15:2149-2163.

Joselin Herbert, G.M., Iniyan, S., Sreevalsan, E., and Rajapandian, S. 2007. A review of wind energy technologies. Renewable and Sustainable Energy Review, 11:1117-1145.

Lopez, I., Andreu, J., Ceballos, S., Martinez de Alegria, I., and Kortabarria, I. 2013. Review of wave energy technologies and the necessary power-equipment. Renewable and Sustainable Energy Review, 27:413-434.

Panwar, N.L., Kaushik, S.C., and Kothari, S. 2011. Role of renewable energy sources in environmental protection: A review. Renewable and Sustainable Energy Review, 15:1753-1766.

Mardiana-Idayu, A., and Riffat, S.B. 2012. Review on heat recovery technologies for building applications. Renewable and Sustainable Energy Review, 16:1241-1255.

Incorpera, F.P. and DeWitt, D.P. 2007. Fundamentals of heat and mass transfer. Sixth Edition. John Wiley & Sons.

Bujak, J. 2008. Mathematical modeling of a steam boiler room to research thermal efficiency. Energy. 33 (12): 1779-1787.

Towards Sustainable Design with Cutting Edge CFD Simulations

Adnan Akhdar **Youssef Ghoussoub**

ABSTRACT

With the ever increasing demands for sustainable buildings, engineers are coming up with ever more complex and diversified designs to find ways to reduce loads, boost efficiency, and utilize renewable resources. CFD simulations have become an instrumental tool at Dar Al-Handasah, a multi-national EA consulting firm, to provide state-of-the-art designs for high performance buildings and meet the ongoing challenge of developing models that are energy efficient, sustainable, and LEED compliant. At an early phase in the design process, CFD simulations are extensively used as an optimization and validation tool in supporting the Company's design process through accurate prediction of thermal comfort conditions, design validation and optimization, and most importantly energy reduction for the HVAC building systems. Specifically, the simulations investigate insightful strategies for design and ingenious air conditioning systems and other energy saving measures that are geared towards decreasing the overall building energy cost, thus, qualifying the design for more points under LEED EA Credit 1, Optimize Energy Performance. Some of the innovative air-conditioning systems include radiant cooling coupled with dedicated outdoor air system (DOAS) in dry climates, under floor air distribution (UFAD) in high assembly spaces and optimized air flow distribution system for open stadia and high ceilinged spaces such as mosques.

INTRODUCTION

The built environment has significant impacts on the environment and natural resources. In fact, buildings account for a significant fraction of the primary energy usage and electricity consumption. Consequently, poor design of buildings and mechanical systems not only misuses energy, wastes resources and causes adverse impacts to the environment, but also creates uncomfortable and unhealthy indoor environments. With the ever-increasing demands for sustainable and energy-efficient buildings and as expectations of health standards and comfort levels rise, building engineers are developing more complex and diversified designs to reduce loads, boost efficiency and utilize renewable resources. In recent years, computational fluid dynamics (CFD) has been playing an increasingly important role in building design, following its continuing development for over a quarter of a century. An overview of applications in ventilation [Chen, 2009] has indicated that CFD simulations were a most popular tool used to predict ventilation performance of buildings and that CFD simulations have been widely used to study airflow behavior in the indoor environment, as well as the outdoor environment [Awbi, 1989; Chen & Srebric, 2000; Chow, 1996]. For most applications of ventilation and turbulent flows, CFD simulations can simultaneously predict airflow, heat transfer and contaminant transportation in and around buildings with reasonable accuracy; and this detailed information can be used to quantify indoor environment quality and gain invaluable insight into the underlying physical phenomena.

While the conventional design approach for buildings typically relies on hand or simple electronic calculations, the use

Adnan Akhdar, M.E., LEED Green Associate is the leader of the CFD Unit in the Mechanical Department at Dar Al-Handasah. Youssef Ghoussoub, B.E., is a mechanical engineer and a member of the CFD Unit at Dar Al-Handasah.

of charts provided by the manufacturers and formulas and correlations that were developed from laboratory data, the use of such empirical data can result in significant uncertainties when they are applied to large spaces (such as atria, concert halls and sports facilities) or applications that are dissimilar from those upon which the laboratory data were developed. When a novel HVAC system is used, there are predictably no data available for the engineering design which leads to deliberate over-design and the use of conservative safety and correction factors to account for the uncertainity [Zhai, 2006; Fletcher et al., 2001]. At Dar Al Handasah, Computational Fluid Dynamics (CFD) simulations proved to be an inexpensive yet effective design tool to meet the challenge of increasingly complex and demanding projects and assist the engineering design process to achieve the desired occupant comfort as well as sustainable development in terms of capital cost and energy consumption. Typical projects range from simple fluid flow simulations, to the modeling of wind effects, the assessment of HVAC systems, and the investigation of fire scenarios for a variety of building designs ranging from parking, to hospitals, hotels, malls, stadia and tunnels.

This paper will illustrate the integration of CFD in the building design process at Dar Al-Handasah through the presentation of two case studies:

1. Case study 1: Under floor air distribution (UFAD) in the convention center at Princess Noura Bint AbdulRahman University for Women in Riyadh, Saudi Arabia.
2. Case study 2: Optimized outdoor air cooling system for Al Khor International Stadium in Doha, Qatar.

CASE STUDY 1: THE CONVENTION CENTER AT PRINCESS NOURA BINT ABDULRAHMAN UNIVERSITY FOR WOMEN IN RIYADH, SAUDI ARABIA.

The convention center at Princess Noura Bint AbdulRahman University for Women is a 31,000-square-meter (333681.2 square-feet) construction, where challenging comfort and efficiency design goals were established. The Main Assembly Hall consists of four levels (basement, ground, mezzanine and first floor) interconnected via one air continuum including the stage and seating area, with all levels occupied. A full-scale three-dimensional model of the assembly hall, representing the geometry of the building, was built as **shown in Figure 1** below in order to analyze the performance of the proposed air conditioning system.

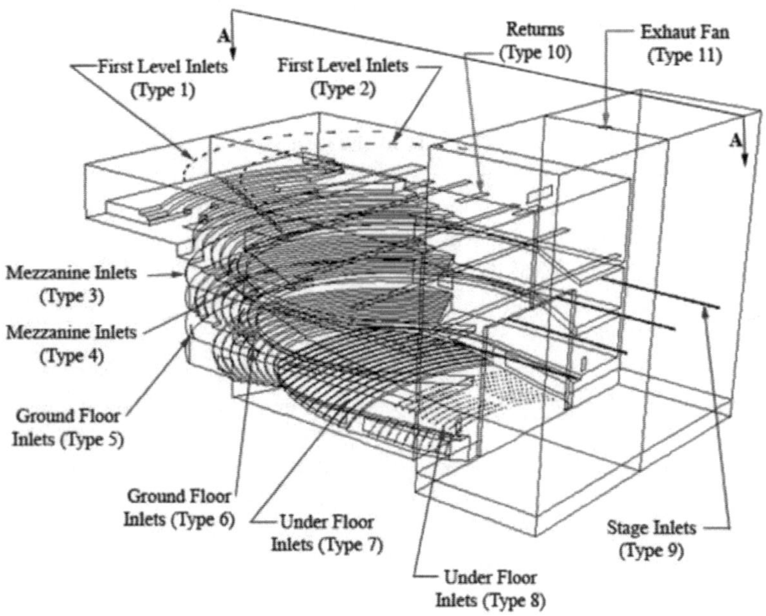

Type	Size (m)	# diffusers	l/s each
1	1 x 0.1	14	259
2	1 x 0.1	18	202
3	10.6 x 0.1	4	986
4	6.8 x 0.1	8	492
5	4.6 x 0.1	6	635
6	5.1 x 0.1	8	474
7	0.45 x 0.15	522	16.2
8	Φ 15	380	16.2
9	15 x 0.1	4	3405
10	10 x 0.1	4	10990
11	1 x 1	1	7000

Figure 1 Full-scale CFD model of the convention center assembly hall

A hybrid cooling system coupling under-floor air distribution (UFAD) and conventional ceiling air supply was designed using CFD simulations at the conceptual design phase. Several design iterations were investigated to maximize energy savings and reduce energy costs. Specifically, the total supply airflow of 50,960 l/s (107978.1 cfm) should be able to uniformly cool the occupied zones but only temper the upper regions where maintenance catwalks are located. Moreover, the distribution of air supply outlets needed to be optimized to ensure proper air delivery in the 24-meter-high (78.7 feet) assembly hall while avoiding disturbance of the thermal stratification of the air, a key energy-saving measure.

The computational grid consists of approximately 2 million tetrahedral cells whereas special attention was paid to the high grid resolution of the supply, return and exhaust air diffusers, as well as the occupant region up to 2m from floor level. Indeed, the 902 air diffusers beneath the audience seats in the hall and the 62 ceiling flow bars distributed throughout the different levels were accurately modeled. The minimum and maximum cell sizes used were 0.1 x 0.1 x 0.1 m (0.328 x 0.328 x 0.328 ft) near the diffusers and in the occupant zone and 1m x 1m x 1m (3.28 x 3.28 x 3.28 ft) in the core volume, respectively. The design criteria that the CFD simulations have been based on are listed below:

- UFAD diffusers supply at 18^{0}C (64.4^{0}F) whereas all other diffusers supply at 12^{0}C (53.6^{0}F).
- The supply diffusers for the stage area at 14m AFFL (45.9 ft) were modeled as a series of long and narrow slots with high velocity.
- An exhaust fan was added at the highest elevation to relief the space contaminants. It was a modeled as a single 1 x 1 m (3.28 x 3.28 ft) opening with a flow of 7000 l/s (14830 cfm) to maintain a slight space positive pressure.
- Four 10 x 1 m (32.8 x 3.28 ft) return outlets, also modeled as long and narrow slots, provide a path for return air to the air handlers.
- The simulation is based on peak occupancy with 1950 occupant (1 person/m^2). People sensible and latent loads are respectively 70 W/person (239 BTU/hr per person) and 35 W/person (119.5 BTU/hr per person).
- The occupants were modeled as porous zones where energy source terms were specified. The limit of the occupant zone is defined at 1m (3.28 ft) above finished floor level in the seating area and at 2m (6.56 ft) above finished floor level in the stage area.
- The sensible load for lighting in the stage and seating areas is 15.3 W/m^2 (4.85 BTU/hr.ft^2); 70% of this load is assumed in the occupied zone.
- The sensible load for equipments in the stage area is 165 W/m^2 (52.3 BTU/hr.ft^2); 70% of this load is assumed in the occupied zone.
- The external loads affecting the main hall are summarized in Table 1 below.

Table 1. External Loads

Exposed walls	Loads (W/m^2) [BTU/hr.ft^2]
Stage Walls	12.3 [3.9]
Mezzanine Walls	12.5 [3.96]
First Level Wals	10.8 [3.42]
Stage Roof	127 [40.26]
First Level Roof	9.02 [2.86]
Catwalks	1025 [325]

Steady-state simulations were then carried out to assess and optimize the performance of the proposed hybrid air conditioning system. In particular, velocity and temperature distributions were monitored and airflow delivery was optimized so that no disturbances occurred in the hot stratified region. The initial simulation, **as shown in Figure 2a** below, indicated a potential for improving the temperature distribution at various levels. In fact, high temperatures were observed at the first-floor-level seating area (around 31°C (87.8⁰F)) and at the stage area (around 28°C (82.4⁰F)), implying insufficient supply air flow to these zones. Moreover very high temperatures (around 40°C (104⁰F)) were reported in the core volume mainly due to high heat loads from the equipment on the catwalk. In addition, the mezzanine-level seating area was overcooled, with average temperature around 21°C (69.8⁰F), indicating that the supply air flow delivered to that zone could be reduced.

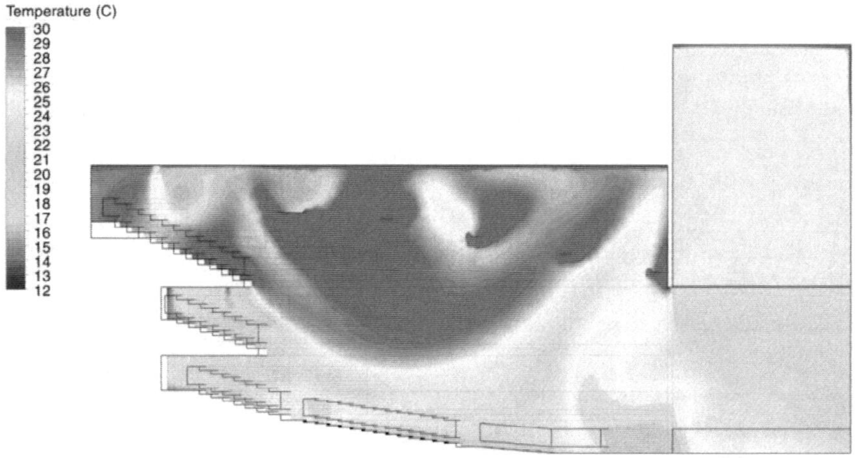

Figure 2a Temperature distribution across the Main Assembly hall for the base supply configuration

Using the results from the initial flow simulation, we were able to clearly visualize the airflow behavior inside the high-ceilinged hall **as shown in Figure 2b below** and identify measures to improve airflow delivery.

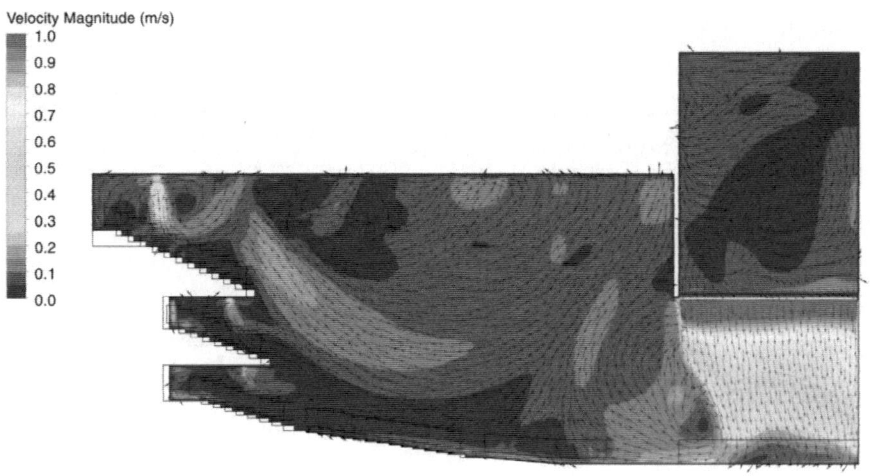

Figure 2b Flow field across the Convention center's Main Assembly Hall for the base supply configuration

Those measures were further investigated and validated in a second simulation **as shown in Figures 3a and 3b**

below. Specifically, the air flow serving the first-floor seating area was increased by 68% and directed at predefined angles (also optimized via simulations), to improve air delivery toward the occupant zone. These changes, **as shown in Figure 3a**, resulted in lower and better distributed temperatures in the first-floor seating area where the average temperature in the occupant zone dropped to be around 26^{0}C (78.8^{0}F), a 6^{0}C (42.8^{0}F) decrease compared to the base design. Similarly, **as depicted in Figure 3b**, the flow serving the stage area was directed at 30° from the vertical and away from the wall. This helped deliver more flow toward the center of the stage where improved thermal conditions were observed. However, as a result higher temperatures were reported at the back of the stage compared to the base design. Additionally, the base simulation results indicated that the side flowbar diffusers at the balconies were inducing significant disturbances to the hot stratified air layer and, thus, those diffusers were removed in the optimized design to prevent unnecessary mixing with the stratified layer. Spot cooling was used to balance the high heat emitted by the catwalk's equipment so disturbances would not be introduced into the upper hot stratified air layer. Finally, to capture and promote the rising movement of buoyant hot air from the occupied zones, air returns were added above the catwalks with an exhaust fan installed at the highest elevation of the stage. This fan also assisted in exhausting any contaminants.

As a result of the CFD study, the realized energy savings for the Convention Center HVAC system ranged between 20 to 25% which greatly facilitated achieving the required 14% energy cost reduction of LEED EAc1, while ensuring that thermal comfort conditions are met in the stage and seating areas **as shown in Table 2 below.**

Table 2. Average temperatures in Stage and Seating areas for the base and optimized models

Model	Stage-adjacent Seating area	Ground Floor Seating Area	Mezzanine Seating area	First-Floor Seating area	Stage area
Base	~23.6^{0}C [~74.5^{0}F]	~22.7^{0}C [~72.9^{0}F]	~21^{0}C [~69.8^{0}F]	~31^{0}C [~87.8^{0}F]	~25.7^{0}C [~78.3^{0}F]
Optimized	~23^{0}C [~73.4^{0}F]	~22.7^{0}C [~72.9^{0}F]	~22.5^{0}C [~72.5^{0}F]	~25.8^{0}C [~78.4^{0}F]	~26.6^{0}C [~79.9^{0}F]

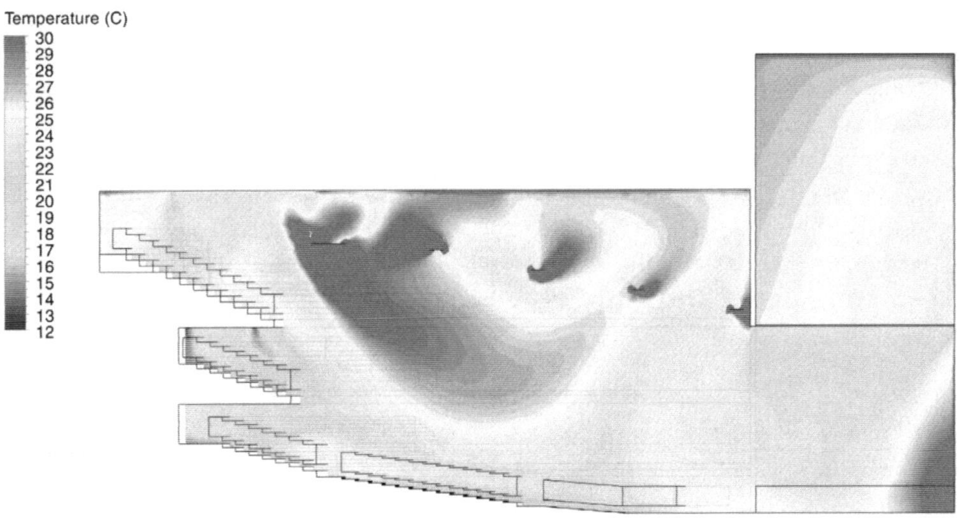

Figure 3a Temperature distribution across the Main Assembly hall for the optimized supply configuration

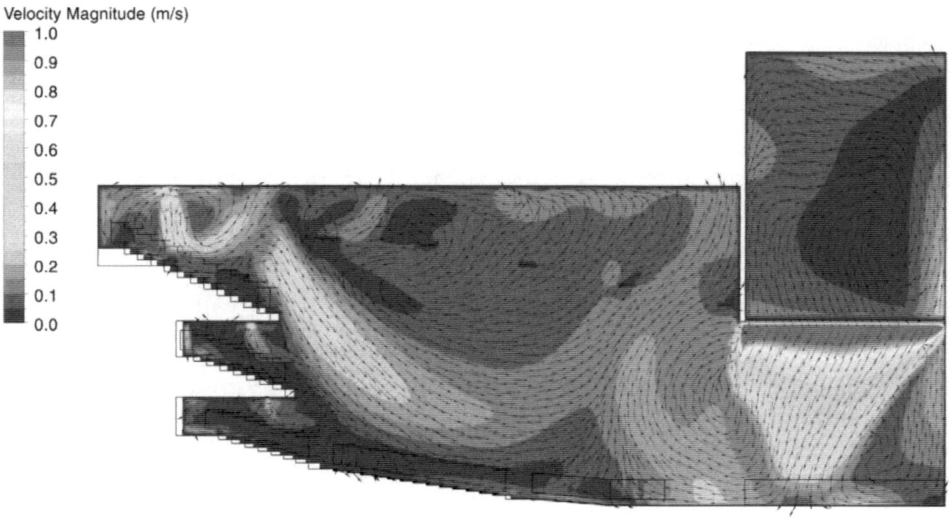

Figure 3b Flow field across the Convention center's Main Assembly Hall for the optimized supply configuration

CASE STUDY 2: OPTIMIZED OUTDOOR AIR COOLING SYSTEM FOR AL KHOR STADIUM IN DOHA, QATAR.

The 70,000+ seat Al Khor Stadium (Figure 1) designed to host Qatar 2022 FIFA World Cup is a state-of-the-art facility with a 120m x 93m (393.7 ft x 305.1 ft) retractable roof system to provide the natural grass playing surface with adequate sunlight for proper growth. Our challenge, from a mechanical design point of view, was to provide comfortable environmental conditions for spectators and players during games when the retractable roof is open, with the least energy expenditure. We also had to devise a cooling strategy to mitigate the infiltration of hot ambient air due to critical winds - a task exacerbated by the high summer ambient temperatures often reaching 47°C (116.6^{0}F) during the scheduled football tournaments in June and July.

An extensive number of simulations have been conducted early in the engineering design process to design and optimize the stadium's mechanical ventilation systems while considering external weather parameters such as temperature, humidity and wind speed and detailed internal heat sources such as body heat from occupancy, solar radiation, and heat gains from lighting and equipments.

In fact, the weather parameters that the analysis has been based on are chosen to present severe climatic conditions in Doha and consist of three scenarios as listed below:

- Scenario 1 (peak DB): 47^{0}C DB (116.6^{0}F), 24^{0}C MCWB (75.2^{0}F), 5.9 m/s MCWS (19.36 ft/s)
- Scenario 2 (peak WB): 36.8^{0}C MCDB (98.2^{0}F), 33.4^{0}C WB (92.1^{0}F), 3.5 m/s MCWS (11.48 ft/s)
- Scenario 3 (peak WS): 40.7^{0}C DB (105.3^{0}F), 22.3^{0}C MCWB (72.1^{0}F), 12.3 m/s MCWS (40.35 ft/s)

Several models have been simulated for the critical wearther scenarios for three critical wind directions: NNW, North and East. These wind directions were determined by inspecting Doha's long term meteorological data (1983 to 2011) and by considering the shape of the stadium's outer tent and the positioning of the opening. Scenario 3 was found to present the most severe climatic conditions in all the simulated models and it was thus used to optimize the capacity of the cooling system.

Initially a complete floor radiant system with displacement ventilation was considered, however based on the client's request that an all-air system be adopted for cooling the bleachers and field of play areas, alternative cooling strategies were investigated: from an all-air system consisting of overhead ductwork and jet diffusers to the adopted all-air system consisting of jet cooling via high throw jets to mitigate wind-driven hot air infiltration. A full-scale three-dimensional model of the stadium was built inside a wind tunnel of dimensions 4000 m (13120 ft) length x 1750 m (5740 ft) width x 600 m (1970 ft) height and the 2056 high throw air diffusers serving the field of play and bleachers areas were modeled in detail **as shown in Figure 4** below.

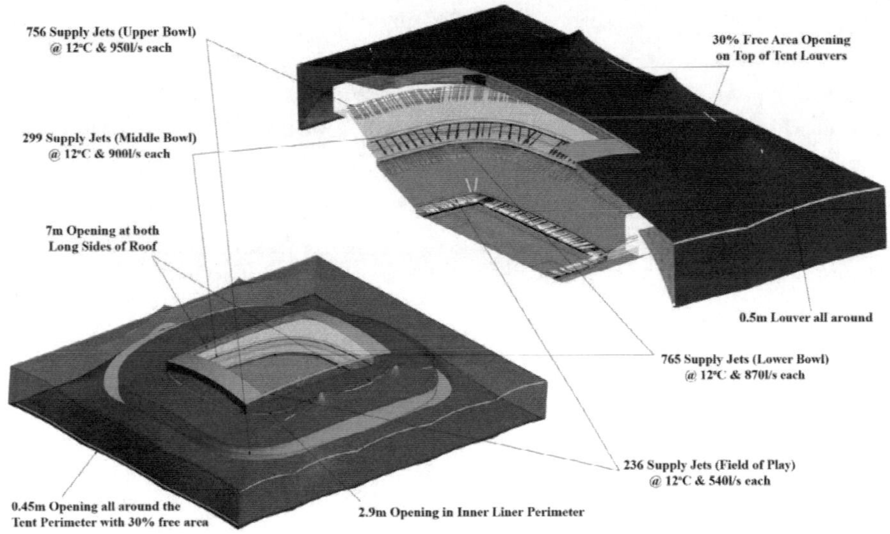

Figure 4 Full-scale 3-D model of Al Khor Stadium

The computational grid consists of approximately 7.6 million polyhedral cells whereas special attention was given to the precise modeling and high grid resolution of the supply jets, as well as the occupant region in the bleachers and field of play areas, up to 2m (6.56 ft) from floor level. A partial view of the conformal mesh **is shown in Figure 5** below. The minimum cell size used is 0.1 x 0.1 x 0.1 m (0.328 x 0.328 x 0.328 ft) near the diffusers and the maximum size is 12 x 12 x 12 m (39.37 x 39.37 x 39.37 ft) in the uppermost region of the virtual wind tunnel.

Figure 5 Simulated Wind Tunnel setup (left) and computational mesh (right) for Al Khor Stadium

The design criteria that the CFD simulations have been based on are listed below:

- The total capacity of the outoddor cooling system in 1405 m^3/s (2,977,000 cfm) in tournament mode and the supply temperature is 12°C (53.6°F) at 80% relative humidity.
- Inner liner emissivity of 0.58 and thermal conductivity of 0.285 W/m.K (0.1647 BTU/ft.hr.°F)
- Outer tent emissivity of 0.1 and thermal conductivity of 0.285 W/m.K (0.1647 BTU/ft.hr.°F)
- Peak occupancy of 70,000 spectators, whereas the people sensible and latent loads are respectively 70 W/person (239 BTU/hr per person) and 45 W/person (153.7 BTU/hr per person).
- 26 players and referees are considered in the field of play. Their sensible and latent loads are 210 W/person (717.2 BTU/hr per person) and 310 W/person (1058.7 BTU/hr per person), respectively.
- 184 occupants (coaches and cameramen) are considered on the sidelines of the field of play (75 W/person (256.1 BTU/hr per person) sensible load and 55 W/person (187.8 BTU/hr per person) latent load)
- The sensible gain from lighting and equipment loads in the bleachers and field of play areas is 905 kW (3,090,737 BTU/hr) and 388kW (1,325,089 BTU/hr), respectively.
- The bleachers are shaded in tournament mode while the field of play is subjected to diffuse solar flux of 40 W/m^2 (12.7 $BTU/hr.ft^2$).

Concurrent studies were carried out to assess the impact of wind effects and optimize the ventilation strategy for the double-tent cavity. In one hand, the HVAC systems serving the bowl and field of play were simulated to achieve the targeted comfort conditions and several design parameters, such as the number and size of supply diffusers, the supply air temperature and terminal velocity, were optimized to establish a stratified cold layer in the occupied zones and avoid unnecessary mixing **as shown in Figure 6** below.

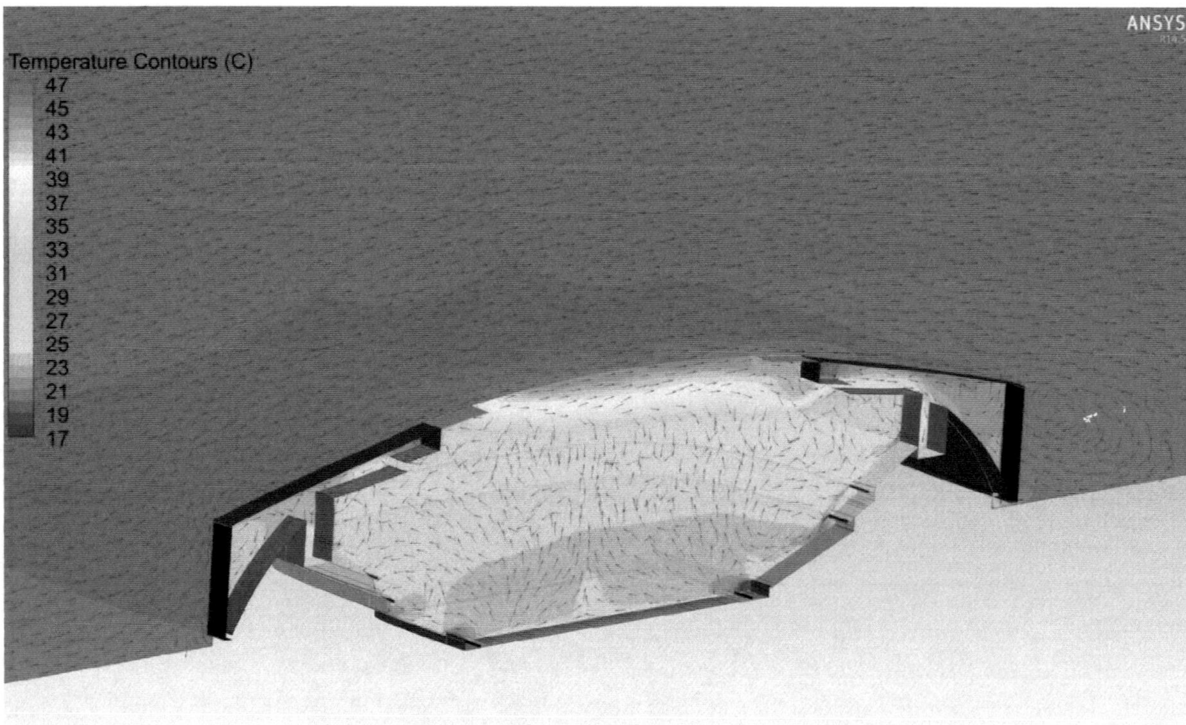

Figure 6 Contours of Temperature magnitude and overlaid velocity vectors at central Noth-South Section

In another hand, the simulations allowed us to gain invaluable insight on the established flow pattern inside the cavity and helped us improve the ventilation strategy to effectively remove and prevent any heat build-up within the tent cavity from the transmitted and absorbed solar radiation on the outer tent surface **as shown in Figure** 7 below.

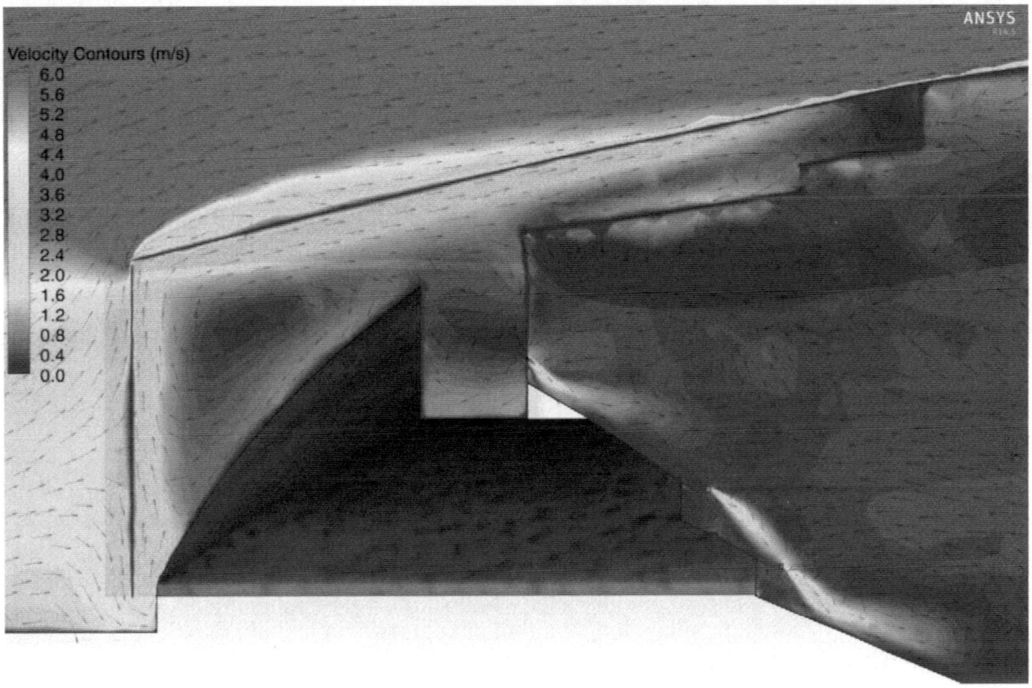

Figure 7 Upstream close up view of velocity contours through central North-South section of the Stadium

Coincidently, the simulations helped quantify the volume flow of hot air infiltrating into the stadium through the roof opening and assess its impact on the bowl and field of play temperature **as depicted in Figure 8, Figure 9 and Table 3** below.

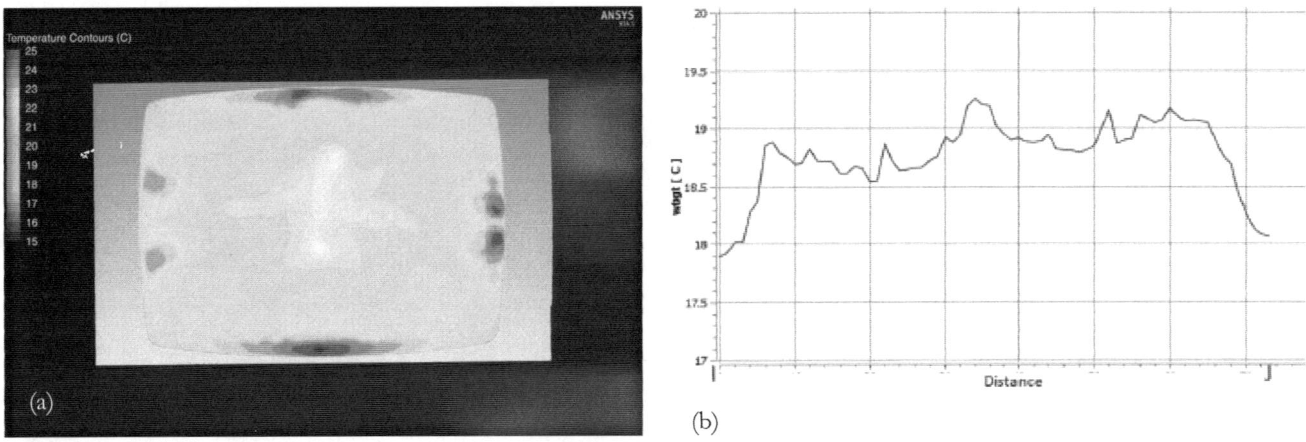

Figure 8 (a) Temperature contours and (b) variation of Wet Bulb Glob Temperature (WBGT) at 1 m (3.28 ft) above Field of Play FFL

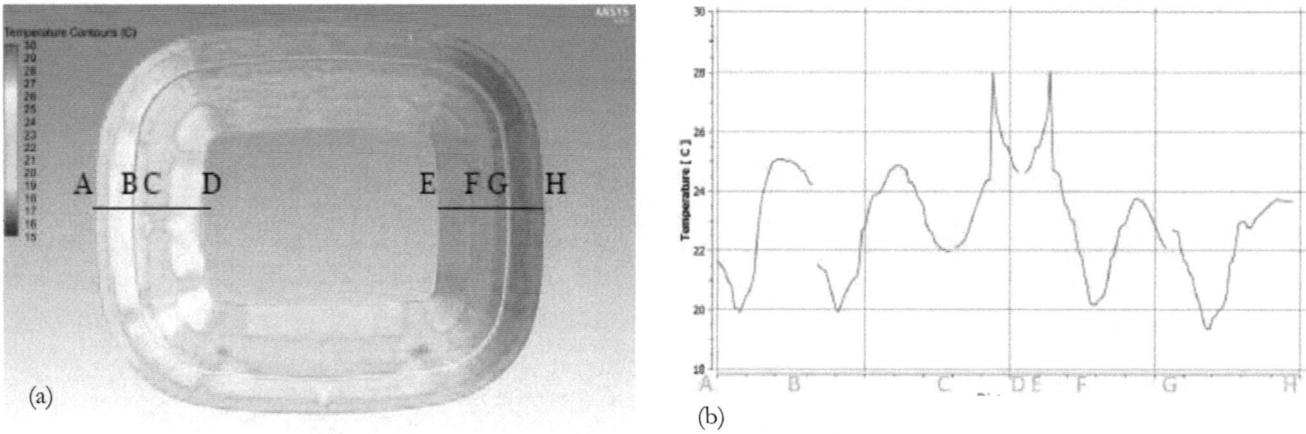

Figure 9 (a) Temperature contours and (b) Temperature variation at 0.5 m (1.64 ft) above bleachers FFL

Table 3 Average temperature, air speed and RH in bleachers and field of play areas

Zone	Scenario 3			
	DB (^{0}C) [^{0}F]	RH (%)	Air Speed (m/s) [ft/s]	SET or WBGT(^{0}C) [^{0}F]
Playfield	~21.3 [~70.3]	~41	~0.7 [2.3]	~17.5 [63.5]
Lower Bleachers	~21.8 [~71.2]	~42	~0.3 [0.98]	~23.4 [~74.1]
Middle Bleachers - Lower	~23.2 [~73.8]	~39	~0.3 [0.98]	~24.2 [~75.6]
Middle Bleachers - Upper	~25.4 [~77.7]	~36	~0.2 [0.66]	~26.5 [~79.7]
Upper Bleachers	~22.9 [~73.2]	~39	~0.3 [0.98]	~24.5 [~76.1]

CONCLUSION

Computational Fluid Dynamics (CFD) simulations can provide detailed flow distributions, temperatures and pollutant concentrations with excellent accuracy and at minimal cost, permitting engineering decisions to be made with considerable precision to optimize HVAC designs and meet the ongoing challenge of developing efficient and sustainable designs. They have become an instrumental tool in supporting the company's design process and have allowed us to tackle a wide range of complex and challenging design problems, such as those presented in the current paper, while exceeding client expectations.

ACKNOWLEDGMENTS

We would like to dedicate a special thank you to Mr. Sharbel El-Haber who has founded the team and thrived for ensuring continuous development, progress and support at all levels. We would also like to thank all design teams for their collaboration, assistance, and valuable contribution.

NOMENCLATURE

CFD	=	Computational Fluid Dynamics
UFAD	=	Underfloor air distribution
AFFL	=	Above Finished Floor Level
FFL	=	Finished Floor Level
DB	=	Dry bulb temprature
WB	=	Wet bulb temprature
WS	=	Wind speed
RH	=	Relative Humidity
MCDB	=	Mean coincident dry bulb temperature
MCWB	=	Mean coincident wet bulb temperature
MCWS	=	Mean coincident wind speed
SET	=	Standard effective temperature
WBGT	=	Wet bulb globe temperature

REFERENCES

Chen, Q. (2009). Ventilation performance prediction for buildings: A method overview and recent applications. Building and Environment, Vol. 44, pp. 848-858.

Awbi, H.B. (1989). Application of computational fluid dynamics in room ventilation. Building and Environment, Vol. 24, No. 1, pp. 73-84.

Chen, Q. & Srebric, J. (2000). Application of CFD tools for indoor and outdoor environment designs. International Journal on Architectural Science, Vol. 1, pp. 14-29

Chow, W.K. (1996). Application of computational fluid dynamics in building services engineering. Building and Environment, Vol. 31, pp. 425-36.

Chow, W.K. (1999). A discussion on using computational fluid dynamics in building services engineering. Architectural Science Review, Vol. 42, No. 2, pp. 121-128 (1999).

Zhai, Z. (2006). Application of Computational Fluid Dynamics in Building Design: Aspects and Trends. Indoor and Built Environment, Vol. 15, pp. 305.

Fletcher*, C.A.J. & Mayer, I.F. & Eghlimi, A. (2001). CFD AS A BUILDING SERVICES ENGINEERING TOOL. International Journal on Architectural Science, Volume 2, Number 3, p.67-82.

Haber, S. & Nehme, B. & Akhdar, A. (2011). Meeting Green Building Design Goals with Engineering Simulation. ANSYS Advantage, Vol. 5, Issue 1.

Earthen HVAC System

Tanmay Jain[1]
[Student Member ASHRAE]

Harish Mahadhasu[2]
[Student Member ASHRAE]

ABSTRACT

Cooling of air by evaporative cooling using earthen materials is proposed, the widespread use of water cooling using earthen pots is taken a step ahead by cooling air using the cold water. This cool air can be supplied directly into a space or into the intake of an air conditioner. The advantage over the existing desert coolers. No addition of moisture to the air. This system requires a part of the space to be conditioned to be made of a composite wall, which contains earthen material at one end facing the surroundings and the other surface to be made of metallic material, facing the conditioned space. The working of the setup initiates when air is supplied through the blower in the duct. The air first comes in contact with the metallic surface corresponding to a wall just ahead of the blower inlet. Then the air subsequently touches the other metallic surfaces before finally entering into the conditioned space from the recess. The metallic surface would absorb heat from the flowing air via two modes of heat transfer viz. Convection and conduction. The former mode can be greatly enhanced by increasing the surface area of the metallic surface area in contact with the air. If the amount of heat that is transferred to the water from the air is less than the heat lost due to evaporation at the outer surface. It can be said that the room will be cooled further. There can be several modifications to the metallic and evaporation surface, such increasing the surface area so as to improve heat transfer at both ends of the wall. The modifications are also mentioned. The system has various advantages in comparison to the existing systems. The electricity consumption will be minimal. Also no addition of moisture is present in the cooled air. The outlet air of the system can be fed to an air-conditioner as well.

INTRODUCTION

Evaporative cooling, as the name suggests, is a passive cooling process in which an object or a space is cooled by the evaporation of water from its surface. It is a popular technique that has been in use for several centuries in the Middle East and Africa. Evaporative cooling is a promising method that is used in the design of storage facilities for perishable food products like fruits and vegetables and also for air conditioning in rural areas of developing countries, especially when the local weather conditions are hot and dry, owing to the fact that only ambient energy is utilized for the evaporation. Evaporative cooling should be especially relevant at present when fossil fuels are depleting at an alarming rate and it should help in mitigating the environmental problems caused by the rapid use of such fuels which are of mounting global concern.

Evaporative cooling is a heat and mass transfer process in which some sensible heat from the surroundings of a wetted surface is used to evaporate moisture into the air and thus it is converted to latent heat of the moisture[1]. For a adiabatic saturation process the sensible heat from the air that is flowing past the waster surface is taken and temperature of the air approaches the wet-bulb temperature because the cooling of the air proceeds along the constant wet-bulb temperature. The water present in the wall(as in a pot) forms a part of "surroundings" of the outer wetted surface of the wall as in the case of porous earthen surface. Hence, a part of latent heat of evaporation is procured from it.

Appreciable evaporative cooling can be enjoyed in localities where high dry-bulb temperature (≥35°C) occurs simultaneously with low wet-bulb temperature (≤24°C).[2]

The long-term temperature of water in a porous ceramic pot is found to be 10.5–15 °C below the ambient temperature and it is is observed to have a smaller daily variation than the external temperature[Aimiuwu (1992, 1993)].[3] An experiment was conducted by Taha et al. (1994) in which a box shaped Evaporative cooled chamber (ECC) was constructed of zinc whose external surface was covered by continuous wetted charcoal layers.[4] In an ECC fruits and vegetables like promo, bananas, tomatoes and potatoes exhibited good appearance along with better psychological properties and also their shelf lives were increased (by a factor of 1.2-2.7) as compared to storage under ambient conditions.[5]

1 Tanmay Jain is a Student in the Department of Mechanical Engineering, Indian Institute of Technology Roorkee, Roorkee, India. .
2 Harish Mahadhasu is a student in the Department of Mechanical Engineering, Indian Institute of Technology Roorkee, Roorkee, India.

A transient response to the changes in the relative humidity (RH) and temperature on an ECC was obtained by Anyanwu (2004) in dry and wet seasons.[6] The shelf lives of tomatoes and pumpkins went up by factors of 3 and 5 above thier ambient-air storage values. A mathematical model was made to examine the effect of structural and operational parameters, which was affected by the rate of evaporation on the external surfaces of the ECC and also infiltration/exfiltration rates from the chamber, by Dash and Chandra (2001).[7]

Ibrahim et al.(2003) observed a drop of 6-8 °C in the dry bulb temperature along with the increase in the RH of the air at the inlet by 30%[8]. Because of the high porosity of the evaporator, increase in the pressure of water supply, and a single (as compared with a twin) row of evaporators in the air duct, the effect of cooling was improved a lot.

THEORETICAL ANALYSIS

The flow of water through the exterior of the earthen wall is driven by the gravitational pressure head of water stored in the wall. The seepage of water from the porous earthen wall is assumed to follow the Darcy's law. As the rate of seepage of water is greater than mass transfer from the external surface of the wall, the heat and mass transfer problem at the external surface of the wall is uncoupled from the flow through the porous wall. Thus, heat and mass transfer from the external surface of the wall will only be considered here.

Heat and mass transfer processes are unsteady due to diurnal variations in dry-bulb temperature and relative humidity of the ambient air. However, the heat and mass transfer can be assumed to be quasi-steady since they vary slowly with time; hence steady state equations can be used to describe them at any instant.

The sensible heat transfer rate through the external surface into the surrounding from the pot is driven by :-

$$Q_S = h_c*A*(T_W - T_A) \quad ......(1)$$

h_c is the transfer coefficient of heat convective and A denotes the area of the external surface of the pot(wall). T_W is the temperature of external surface of the pot(wall) whereas T_A signify the temperature of the ambient.

Also the latent heat transfer rate through the surface of the wall can be expressed as :-

$$Q_L = h_d*A*(W_W - W_a)*h_{FG} \quad ....(2)$$

Where, h_d is the proportionality constant of mass transfer. W_w represents the humidity ratio for the saturated air at the surface of the wall whereas Wa denotes the humidity ratio of the air in ambient. The latent heat of vaporization of water at T_W is represented by h_{FG}.

Therefore, the total heat transfer rate through the wetted surface is governed by :

$$Q_T = Q_S + Q_L \quad ....(3)$$

The above mentioned is the heat that can be transferred through the outer surface of the wall, whereas the inner metallic surface which is subjected two modes of heat transfer, namely convection and conduction.

The rate of heat transfer due to convection is given by:

$$Q_{convection} = h_{convection}*A*(T_{room} - T_{water}) \quad ....(4)$$

Where hconvection can be calculated from Nusselt number: Nu = (hconvection t)/kconduction Here, t is the thickness of the metallic surface and kconduction is described below.

The rate of heat transfer due to conduction is given by:

$$Q_{conduction} = k_{conduction}*A*(T_{room} - T_{water}) \quad ....(5)$$

In the limiting case, the heat transferred by the room to water is equal to the heat transfer from the external

$$Q_{conduction} + Q_{convection} = Q \quad ....(6)$$

EXPERIMENTAL SETUP

A cubic wooden container of one meter side is used to imitate a space to be cooled. A cuboidal composite wall of 80 cm *80 cm *8 cm is placed concentric to the wooden wall. CAD models of the test setup are shown below:

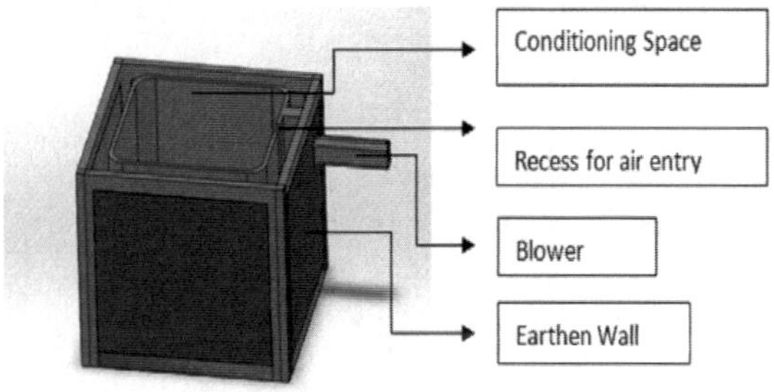

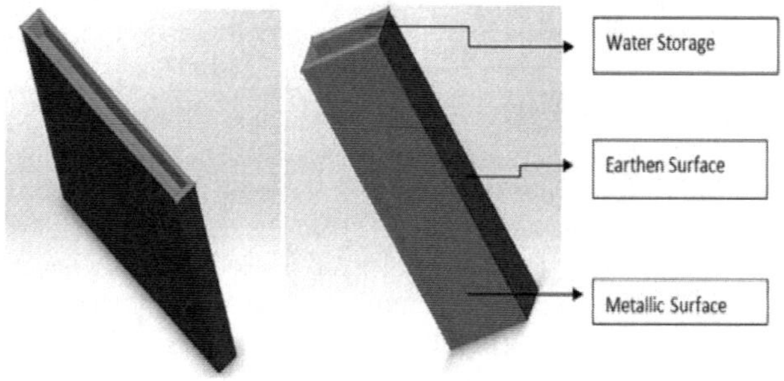

Figure 1 : Complete Setup and Cross Sectional View

Figure 2 : Earthen side(red) and Metallic side (silver)

The setup shown above works on the simple principle of evaporative cooling. The outer surfaces of the wall being earthen will lose heat and mass through evaporation, this heat will be lost from the water contained within the wall. This water is in constant contact with the metallic surface. Thus, as the temperature of the water within the wall decreases so does the temperature of the wall. The wall is in contact with the air that is being constantly supplied by the blower, which through forced convection would facilitate heat transfer between the air flowing and the metallic surface. This cooled air can either be directly supplied to the conditioned space or could be fed to an air conditioner. Either way it could save high grade electrical energy.

The wall is part earthen part metallic and will be filled nearly to capacity with water. The wall will also be attached with a drip system (not shown in fig), which will keep the wall full with water, as the amount that evaporates from the surface need to be recuperated. The setup will then be placed on the rooftop of the laboratory, in order to imitate the conditions of a high rise building in summers. Small resistance heaters will be placed within the conditioning space to compensate for the heat produced by humans and electrical apparatus. The temperature of room air, temperature of water in the wall, and ambient conditions will be monitored using thermocouples and psychrometers. Finally, the readings of the thermocouples will be used along with the equations given above to calculate the heat transfer through different surfaces of the wall.

The working of the setup initiates when air is supplied through the blower in the duct as seen in the figure. The air first comes in contact with the metallic surface corresponding to a wall just ahead of the blower inlet. Then the air subsequently touches the other three metallic surfaces before finally entering into the conditioned space from the recess. The metallic surface would absorb heat from the flowing air via two modes of heat transfer viz. Convection and conduction. The former mode can be greatly enhanced by increasing the surface area of the metallic surface area in contact with the air. The convection heat transfer coefficient can be controlled by adjusting factors like surface area, velocity of air passing and the convection heat transfer coefficient can be calculated using the Nusselt number as shown above. Since, the air flow rate required by the conditioned space can be calculated using the number of occupants. It can be determined what would be the air flow rate through the blower, further the convection heat transfer can be calculated using the velocity of air. Thus, the heat transfer through each wall can be calculated using the Newton's cooling law equation. In order to imitate a room resistance heaters will be placed within the cooling space, so that sensible heat can be generated within the room. The amount of heat generated by human body is known and can be approximated, given the number of occupants. Heating load of the room may be evaluated. If the amount of heat that is transferred to the water from the air is less than the heat lost due to evaporation at the outer surface. It can be said that the room will be cooled further.

Changes In Design

In order to achieve higher heat transfer rates, it was essential to either design a very big setup or a grid of smaller ones. Now, the system has to rely on natural convection, so using cuboidal shape in grid is not beneficial due to high amount of drag forces on the surface. Thus the system design was modified to a cylindrical one, as pressure drop would be much lesser.

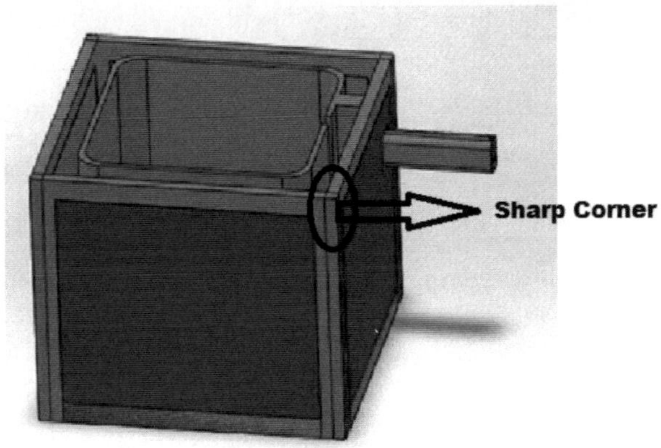

Figure 3: Failure in Cubical Tube

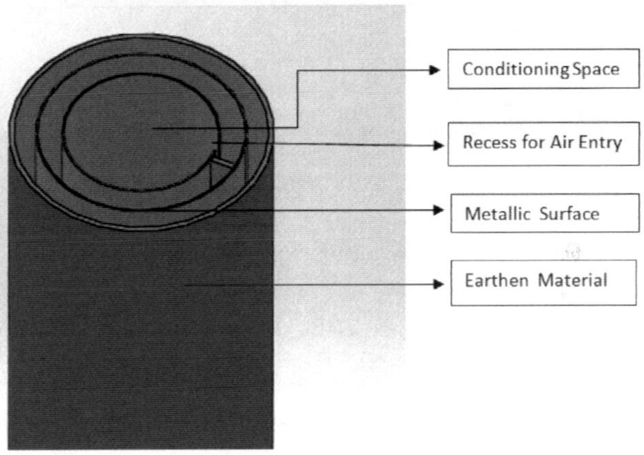

Figure 4: Cylindrical Tube

The ultimate design for a single setup is achieved, yet to generate large amount of cooling we need to design a grid that utilizes natural convection to rid the grid of moisture. To enhance the moisture removal from the vicinity of the cylinders, the inline formation of the cylinders in the bank was changed to a staggered configuration which incorporated a reduction of flow area from one cylinder to another, aiding in accelerating the air.

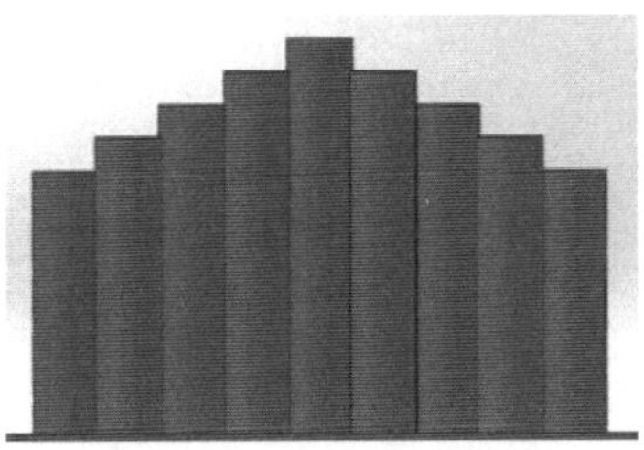

Figure 5: Front View Of Staggered Tube Bank Bank

Figure 6 : Isometric View Of Staggered Tube

THE FINAL GRID MODEL

Further modifications were incorporated to increase the air velocity and also to reduce warps and turbulent zones so as to eliminate moisture accumulation by making the system symmetric. The design of the system was then taken to a higher level. The Tube Bank was then transformed from rectangular staggered to circular staggered, so as to relieve the system of any dependencies on wind direction. Since the system is circular symmetric, no matter what the wind flow direction the performance would remain unchanged.

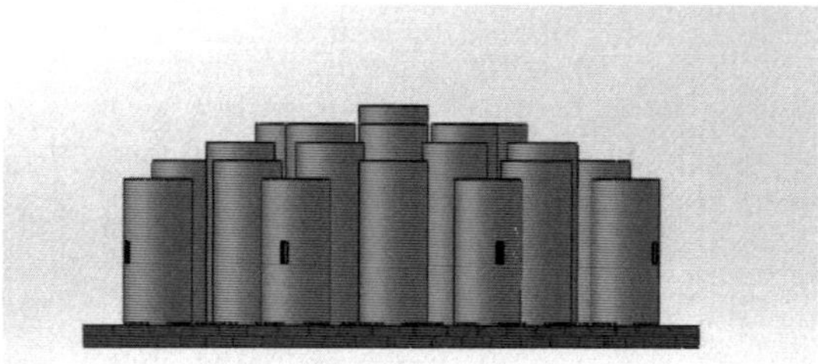

Figure 7 : Front view of Circular Staggered Tube Bank

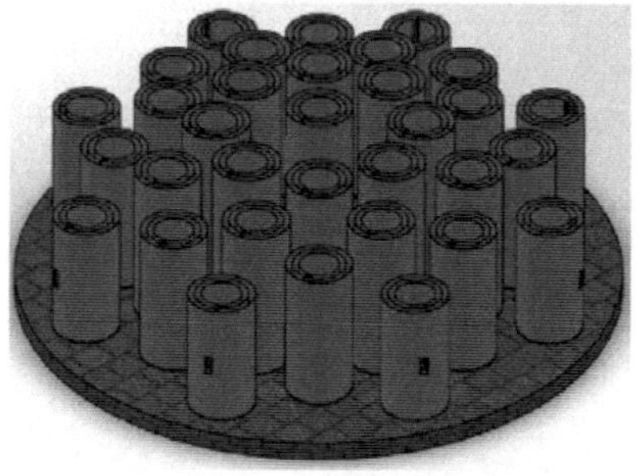

Figure 8 : Isometric view of Circular Staggered Tube Bank

THE EARTHEN POT ANOLOGY

In order to judge whether the system will be feasible, we can draw an analogy from the earthen pot. Several experiments have been performed in order to determine the heat transfer through the surface of an earthen pot. The experiment was performed in October leading to lower temperature. The results are expected to be much promising during summer.

Table 1: Variation in mass and temperature of water kept in earthen pot

Hour	Mass of water(kg)	Ambient Temp	Water temp	Relative Humidity
0	4.759	29.5	28.3	77.0
4	4.723	31.0	26.2	56.0
8	4.678	32.0	25.5	48.0
12	4.634	31.0	25.1	48.0
24	4.526	28.5	24.8	54.0
28	4.458	30.0	23.4	41.0
32	4.398	31.5	23.1	37.0
36	4.338	30.6	23.6	37.0

48	4.183	27.7	23.1	41.0
52	4.111	29.0	21.7	47.0
56	4.043	30.5	22.1	45.5
60	3.986	29.5	23.0	39.5

Key points of the experiments:

- The lowest temperatures were achieved between 2pm and 4pm.
- The pot gave the highest water depression during the highest temperature and lowest relative humidity.
- The temperature depression is further enhanced by increasing the surface area to volume ratio of the pot.
- The readings obtained for the experiments are shown in the table.
- Mass of water evaporated: 31% of the water initially put in the pot.
- The ideal temperature to be maintained in a conditioned space- 24 °C.(297K)

ENERGY SAVED

Several experiments have been performed in order to determine the heat transfer through the surface of an earthen pot.

Table 2: Calculating the air cooling in tons at noon.

Time	Day	Mass Of Water Evaporated (in kg)	Evaporative Cooling Rate (in W/m^2)	Water Cooling Rate (in W/m^2)	$T-T_1$	Relative Humidity
10:00 am to 12:00 pm	1.	.009	43.18	16.13	1.4	65.0-57.0
	2.	.020	65.98	10.82	1.1	33.0-32.0
	3.	.023	72.24	2.92	0.6	31.0-28.0
	4.	.018	61.94	0.0	0.0	29.0-32.0
	5.	.015	61.69	1.27	0.5	35.0-31.0

12:00 pm to 2:00 pm	1.	.011	48.02	8.18	0.9	57.0-53.0
	2.	.020	66.00	0.0	0.0	32.0-35.0
	3.	.025	76.37	0.0	0.0	28.0-26.0
	4.	.017	59.86	-6.47	-0.5	32.0-34.0
	5.	.011	47.45	-3.59	-0.3	31.0-32.0

Average water cooling rate: 7.804 W/ m2 (0.229 BTU/h/ft²)

CALCULATION OF ENERGY SAVED BY THE GRID PROPOSED

Under this we will calculate approximately the total energy saved by the proposed system in comparison to a 2-star AC equivalent.

The area of the grid system is supposed to be 798.84 m^2 (8598.64 ft^2)
Average Water cooling rate can be traced from Table 2:- 7.804 Watt/m^2 (0.229 BTU/h/ft^2).
The effectiveness of the heat exchanger was calculated by NTU method and was found to be: - 0.61
Energy Efficiency Ratio (EER) of a 2-star AC = 2.5

Step 1: Calculating Total Cooling rate

The total water cooling rate is in (Watt). To calculate the total water cooling rate we will take Average water cooling rate in (Watt/m^2) and multiply it with the total grid Area in (m^2):

Total Water Cooling rate = (7.804 Watt/m^2)*(798.84 m^2) = 6234.14 Watt (21270.88 BTU/h).

Step 2: Calculating Total Capacity of the setup

The total Capacity (sensible) of the setup is equal to total water cooling rate in (Watt) multiplied by the effectiveness of the setup:

Total Cooling Capacity of the setup = 6234.14*0.61 = **3802.82 Watt (12975.22 BTU/h)**

Step 3: Calculating Power consumed by a 2 Star-Rated Air Conditioner of similar cooling capacity

Energy efficiency Ratio (EER) for air conditioners is equal to the ratio of Cooling Capacity of the setup in (Watt) divided by Power consumption in (Watt).

Therefore, Power Consumption in (Watt) is equal to cooling capacity of setup in (Watt) divided by EER.

Power consumption of 2 Star-Rated Air Conditioner of similar cooling capacity = (6234.14*0.61)/2.5 = **1521.128 Watt (5190.08 BTU/h)**

Now, if the system is considered to be naturally operated without any electrical appliance, it does not consume any power, which is not possible for obvious reasons. To achieve required flow and circulation of air, the system needs a blower. To achieve flow rates mentioned in the above calculation one or more blowers of a cumulative power of 500-600 Watt (1706-2047.2 BTU/h are required.)

So by all means this system hopes to meet the cooling requirements of the consumer while consuming much lesser power than its contemporary air conditioners.

CONCLUSIONS

Certain conclusions regarding the setup discussed above can be drawn from a simple earthen pot. It has been noted that when earthen pots are kept in hot climates, and water can be stored at temperatures appreciably lower than the dry-bulb temperature of the quiescent air.

The depression in the temperature of water is highest when the relative humidity is minimum and temperature of room air is maximum. Another, peculiar observation is that the longest period of continuous cooling water occurs on the first day of filling the pot with water.

Heat transfer from both the surfaces of the wall, can be increased by increasing the surface area or increasing the air velocity touching the surfaces. On the outer surface (earthen-ambient), if air velocity is high the moisture content of the air adjacent to the surface would be kept almost equal to the ambient. Since, vapors are released from the earthen surface to the ambient, the relative humidity of the air adjacent to the earthen surface would be higher, and therefore a high velocity of air flow would be desirable to remove the excess moisture. Whereas on the inner surface (metallic), higher air velocity would result in higher Nusselt number, which in turn would cause a higher convection heat transfer coefficient.

NOMENCLATURE

A	=	Area of the external surface of the pot(wall), m2
hc	=	transfer coefficient of heat convective, W/m2.K
h_d	=	proportionality constant of mass transfer, kg/m2
h_{FG}	=	latent heat of vaporization of water, J/kg
k	=	thermal conductivity, W/m.K
t	=	thickness of the wall, m
T	=	Temperature, °C
W	=	humidity ratio.

Subscripts

W	=	Water in the wall
A	=	Ambient
L	=	Latent
S	=	Sensible
T	=	Total

REFERENCES

[1] J.A.Olorunmaiye, "Evaporative cooling of water in Earthen pots in Quiescent air".

[2] Harris, N.C. (1983), "Modern Air Conditioning Practice", McGraw-Hill Book Company, New York

[3]Aimiuwu, V. O. 1992. Evaporative cooling of water in hot arid regions. Energy Conversion and Management 33(1): 69 74.

[4] Taha, A. Z., A. A. A. Rahim and O. M. M. Eltom. 1994. Evaporative cooler using a porous material to be used for reservation of food. *Renewable Energy* 5(1): 474-476.

[5] Dhemre, J. K. and D. P. Waskar. 2003. Effect of post-harvest treatments on shelf-life and quality of mango in evaporative cool chamber and ambient conditions. *Journal of Food Science and Technology* 40(3): 316-318.

[6] Anyanwu, E. E. 2004. Design and measured performance of a porous evaporative cooler for preservation of fruits and vegetables. *Energy Conversion and Management* 45: 2187-2195.

[7] Dash, S. K. and P. Chandra. 2001. Effects of different structural and operational parameters on the thermal environment of an evaporatively cooled storage structure. *International Agricultural Engineering Journal* 10(3-4): 231-253.

[8] Ibrahim, E., L. Shao and S. B. Riffat. 2003. Performance of porous ceramic evaporators for building cooling application. *Energy and Buildings* 35: 941-949.

Geothermal Cooling Application for Fresh Air Treatment Plant in Doha Qatar, Case Study

Charbel Seif
Senior Mechanical Engineer

Nicolas Ibrahim
Mechanical Engineer

ABSTRACT

Geothermal energy is a cost effective, reliable, sustainable and environmentally friendly energy stored in the earth mass. Earth ground surface temperature is function of the ambient temperature. The soil temperature yearly fluctuations decrease as the soil is deeper. The deeper it is the more it converges to the mean yearly ambient temperature. Doha summer climate is considered to be hot and humid. Cooling humid fresh air contributes to around 50% of the building cooling load in that area. The purpose of this study is to investigate the effectiveness of using geothermal energy to pre-cool outside air in Doha Qatar. It evaluates two geothermal heat transfer systems to help reduce the overall building energy demand. The first geothermal system is based on closed loop principle. It consists of installing vertical loops that run perpendicular to the surface deep in the ground. The ground loop acts as a heat exchanger to cool the circulating water that will be used to pre-cool building fresh air through air coils. In this study the optimal bore depth is considered to be 300 ft. (91 m) with a heat exchange rate of 0.82 refrigeration tons per bore. The second geothermal system uses groundwater to pre-cool the outside air. This system consist on having a production well, an injection well and cooling air coils. Water extracted from the production well at ground temperature passes through air coils to pre-cool outside air. Heated water shall be injected into a second well called the injection well. In this system the minimum spacing between the production well and the injection well is estimated to be 170 ft. (51 m). Maximum ground water temperature increase is limited to 8°C to avoid inducing thermal disturbance in the aquifer. Each refrigeration ton would require 0.108 l/s (0.228 cfm) of ground water. To conclude vertical loops compared to the ground water open loops have a higher initial cost since it involves extensive earth works where large number of drilled holes as well as water loops needs to be installed; however, vertical loops are considered more appropriate for Doha area application since it involves lower running cost as well as it preservation of the Doha deep aquifer.

INTRODUCTION

This paper shall describe and emphasize on the geothermal System proposed in the New Doha International Stadium. In this project, two geothermal alternatives were investigated. One system based on the closed loop principle and the other one is based on the open loop principle.

Geothermal system is intended to reduce outdoor air temperature entering the air handling stations distributed in the stadium basements. **As shown in Figure 1** adjacent to each air handling station a parking space with a total area of 16,350 m2 is being used to dissipate heat captured from the entering air into the soil.

In this project geothermal analysis is being done based on "Ground source heat pumps - Design of geothermal systems for commercial and institutional buildings" by ASHRAE [1].

Charbel Seif is a senior mechanical engineer at Dar Al Handasah, Shair and partners, Beirut, Lebanon;
Nicolas Ibrahim is a design mechanical engineer at Dar Al Handasah, Shair and partners, Beirut, Lebanon.

Figure 1 Geothermal System

BASIS OF DESIGN

In the New Doha International Stadium, geothermal coils in each of the fan stations are used to pre-cool the outside air thus reduce the project chiller plant size and reduce the carbon foot print of the project.

To maximize geothermal cooling, fresh air passes through earth tubes before reaching the geothermal coils.

The Geothermal coils were selected to meet the following criteria.

1. Air flow rate per station: 195 000 l/s (413181 cfm).
2. Entering dry bulb air temperature: 44 °C (111.2 °F), (taking into consideration that Earth-tube has initially pre-cooled the outdoor air temperature from 46 °C (114.8 °F) to 44 °C (111.2 °F) by exchanging heat with the ground soil through the tube wall to)
3. Entering Wet Bulb air temperature: 30.56 °C (87.008 °F).
4. Leaving dry bulb air temperature: 38 °C (100.4 °F).
5. Leaving Wet bulb temperature: 29.32 °C (84.776 °F).
6. Total load on geothermal coil per station: 1400 KW (4781251 BTU/h).
7. Sensible load on geothermal coil per station: 1400 KW (4781251 BTU/h).

The ground temperature is chosen based on the state of Qatar Soil **as shown in Table 1**, The soil temperature fluctuations decrease in the earth's deeper layers. The deeper it is the more it converges to the yearly mean soil temperature. Therefore, the temperature has been assumed to be close to the mean Yearly soil temperature of 30 °C (86 °F).

Table 1 Qatar soil properties [2]

State of Qatar

Ministry of Communication & Transport

Department of Civil Aviation & Meteorology

Mean Ground Minimum Temperature (Deg. C)

Station : Doha International Airport

Lat: 25 15 N Long: 51 34 E

Year\Month	JAN	FEB	MAR	APR	MAY	JUN	JUL	AUG	SEP	OCT	NOV	DEC	YEAR
1982		11.1	14.7	18.0	23.7	25.8	26.3	27.0	25.1	22.6	17.2	11.1	20.2
1983	10.4	11.2	12.6	18.2	25.0	27.0	27.2	28.6	23.9	19.1	15.8	12.7	19.3
1984	9.8	12.1	16.1	20.5	23.4	26.3	26.9	27.2	24.4	20.9	17.6	13.7	19.9
1985	12.8	11.7	14.8	18.7	24.3	25.5	27.3	27.0	23.7	21.5	18.1	13.7	19.9
1986	10.2	12.1	14.7	20.1	24.9	27.4	27.5	28.9	26.1	22.8	18.6	12.5	20.5
1987	10.9	12.6	16.3	18.5	24.5	26.7	27.3	29.1	25.8	23.1	17.5	13.9	20.5
1988	10.6	12.8	15.8	19.9	25.1	26.8	28.2	28.2	27.0	22.4	17.7	14.0	20.7
1989	8.8	10.6	14.9	18.9	23.6	26.0	28.1	27.8	25.9	21.5	19.3	14.1	20.0
1990	11.1	13.2	14.9	19.8	24.7	26.0	28.1	26.9	24.8	21.9	17.7	13.9	20.3
1991	13.0	12.9	15.3	20.6	23.3	26.2	26.7	27.0	25.6	22.9	15.9	15.9	20.4
1992	9.8	11.4	12.6	17.2	24.1	26.7	27.8	28.1	25.4	21.2	18.0	15.1	19.8
1993	11.5	12.8	15.5	19.3	22.8	25.4	27.6	28.3	25.5	22.7	18.1	13.7	20.3
1994	12.6	11.6	15.1	19.6	25.2	27.2	29.5	28.0	26.2	23.7	21.4	13.2	21.1
1995	12.5	14.0	15.3	18.9	23.3	26.9	29.5	28.2	26.0	22.5	18.4	15.6	20.9
1996	13.1	14.1	17.3	20.1	23.3	27.1	28.3	29.1	26.8	21.5	18.1	12.9	21.0
1997	11.5	12.9	14.9	17.9	22.9	25.9	29.5	28.2	24.9	23.8	18.2	14.5	20.4
1998	11.8	13.4	15.8	19.3	24.2	27.2	29.7	29.7	27.4	23.2	17.9	15.1	21.2
1999	13.7	15.8	14.9	20.2	24.3	28.0	29.7	29.3	27.8	23.1	18.9	14.2	21.7
Mean	11.4	12.6	15.1	19.2	24.0	26.6	28.1	28.1	25.7	22.2	18.0	13.9	20.5

GROUND VERTICALCLOSEDLOOPES

Geothermal system based on Vertical loops that run perpendicular to the Surface is investigated. In order to maximize geothermal heat transfer ground loops has to run several hundred feet deep in the soils **as shown in Figure 2**.

Figure 2 Geothermal Schematic [2]

Ground loop installation details:

The ground closed loop acts as a heat exchanger, similar to a cooling coil or an evaporator in a chiller. Ground act as a mean to transfer energy form and to the loop. **Figure 3** illustrates a typical installation of a ground.

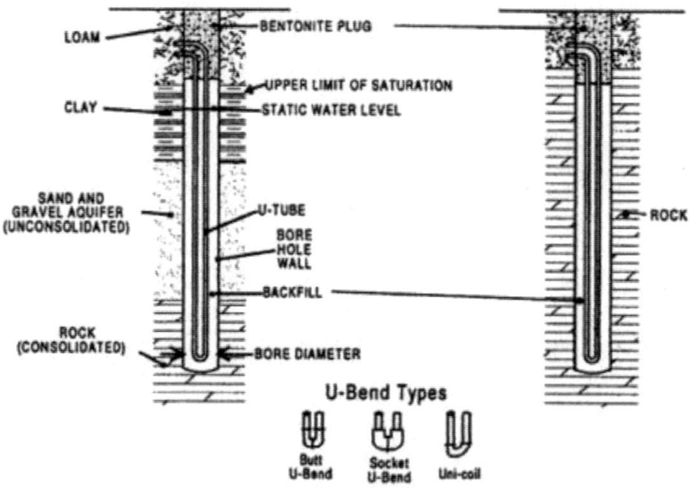

Figure 3 Ground loop typical installation [1]

The purpose of this study is to find an optimal ground loop pipe diameter and depth that maximizes heat dissipation within the soil taking into consideration the limited projected space.

Loop Heat Transfer Calculation

The heat transfer calculated is based on the peak load that needs to be dissipated to the ground; in this case 1400 kW (4781251 BTU/h).

A computer software is used for simplification. By adopting the steady state model proposed by ASHRAE [4], the general equation for cooling will take the following form:

$$L_c = \frac{q_a R_{ga} + q_{lc}(R_b + PLF_m R_{gm} + R_{gd} F_{sc})}{t_g - \frac{t_{wi} + t_{wo}}{2} - t_p} \quad (1)$$

Water temperature at the geothermal coil inlet is selected to be 6 °C (42.8 °F) above the undisturbed ground temperature t_g=30 °C (86 °F). Geothermal coil outlet water temperature rise is adopted to be 5 °C (41 °F) in order to maximize heat transfer across the coil. Therefore; our system is as follows:

1. Coil entering water temperature: t_{wi}=36 °C (96.8 °F). This is based on a 30 °C (86 °F) ground temperature (to have an efficient heat transfer with ground, a 6 °C (42.8 °F) difference between exchanging water and soil is maintained).
2. Coil leaving temperature: t_{wo}= 41 °C (105.8 °F) (to minimize the geothermal coil size a 3 °C (37.4 °F) difference between leaving geothermal cooling water temperature and entering air temperature is maintained
3. Water flow rate per station: 66.7 L/s (1060 gpm) or 0.047 L/s/KW (2.659 gpm/ton).

A 1 in (0.18 ft) diameter SDR 9 U tube size has been considered, R_b=0.11h.ft.°F/Btu **as shown in Table 2.**

Table 2 Equivalent diameters and Thermal Resistances for polyethylene U-Tubes [1]

U-Tube Dia.	SDR or Schedule	Pipe (Bore) Thermal Resistance (h•ft•F°/Btu)			
		For Water Flows Above 2.0 US gpm	20% Prop. Glycol Flow 3.0 US gpm	20% Prop. Glycol Flow 5.0 US gpm	20% Prop. Glycol Flow 10.0 US gpm
¾ in. (0.15 ft)	SDR 11	0.09	0.12	NR	NR
	SDR 9	0.11	0.15	NR	NR
	Sch 40	0.10	0.14	NR	NR
1.0 in. (0.18 ft)	SDR 11	0.09	0.14	0.10	NR
	SDR 9	0.11	0.16	0.12	NR
	Sch 40	0.10	0.15	0.11	NR
1 1/4 in. (0.22 ft)	SDR 11	0.09	0.15	0.12	0.09
	SDR 9	0.11	0.17	0.15	0.11
	Sch 40	0.09	0.15	0.12	0.09
1 1/2 in. (0.25 ft)	SDR 11	0.09[1]	0.16	0.15	0.09
	SDR 9	0.11[1]	0.18	0.17	0.11
	Sch 40	0.08[1]	0.14	0.14	0.08

Thermal Resistance Calculation

Three values for equivalent thermal resistance (R_{ga}, R_{gm}, R_{gd}) must be determined from the expected operating patterns function of "G-factor"; each thermal resistance is correlated with one heat pulse perio: R_{ga} for the annual pulse, R_{gm} for the monthly pulse and R_{gd} for the daily pulse[5].

Most systems will be stable after 10 years of operation (3650 days), the loop length will be determined during peak summer month (30 days), during a peak four hour period (0.167 days) in the afternoon therefore:

τ_1=3650 τ_2=3650+30=3680 τ_f=3650+30+0.167=3680.167

The diffusivity is considered to be equal to 0.6 ft^2/day (100% Sand and 5% Moist) **as shown in Table 3**

Table 3 Thermal Conductivity & diffusivity of Sand and Clay Soils [1]

Soil Type	Dry Density	5% Moist		10% Moist		15% Moist		20% Moist	
		k	α	k	α	k	α	k	α
Coarse 100% Sand	120 lb/ft³	1.2-1.9	0.96-1.5	1.4-2.0	0.93-1.3	1.6-2.2	0.91-1.2	-	-
	100 lb/ft³	0.8-1.4	0.77-1.3	1.2-1.5	0.96-1.2	1.3-1.6	0.89-1.1	1.4-1.7	0.84-1.0
	80 lb/ft³	0.5-1.1	0.60-1.3	0.6-1.1	0.60-1.1	0.6-1.2	0.51-1.0	0.7-1.2	0.52-0.90
Fine Grain 100% Clay	120 lb/ft³	0.6-0.8	0.48-0.64	0.6-0.8	0.4-0.53	0.8-1.1	0.46-0.63	-	-
	100 lb/ft³	0.5-0.6	0.48-0.58	0.5-0.6	0.4-0.48	0.6-0.7	0.37-0.48	0.6-0.8	0.41-0.55
	80 lb/ft³	0.3-0.5	0.36-0.6	0.35-0.5	0.35-0.5	0.4-0.55	0.34-0.47	0.4-0.6	0.30-0.45

The thermal resistances for varying heat pulses are calculated using Fourier number (FO):

$Fo=\frac{4\alpha_g\tau}{d^2}$ (2)

Where d is the tube diameter=0.18ft

$Fo_f=\frac{4\alpha\tau_f}{d^2}$ (3)

$Fo_1=\frac{4\alpha(\tau_f-\tau_1)}{d^2}$ (4)

$Fo_2=\frac{4\alpha(\tau_f-\tau_2)}{d^2}$ (5)

Fo_f= 272604.9

Fo_1= 2234.59

Fo_2 = 12.37

The three equivalent thermal resistances during each heat 'pulse' are found from [6]:

$R_{ga}=\frac{G_f-G_1}{k_g}$ (6)

$R_{gm}=\frac{G_1-G_2}{k_g}$ (7)

$R_{gd}=\frac{G_2}{k_g}$ (8)

The 'G-factor' for each of the Fourier values is then determined **as shown in Figure 4**:

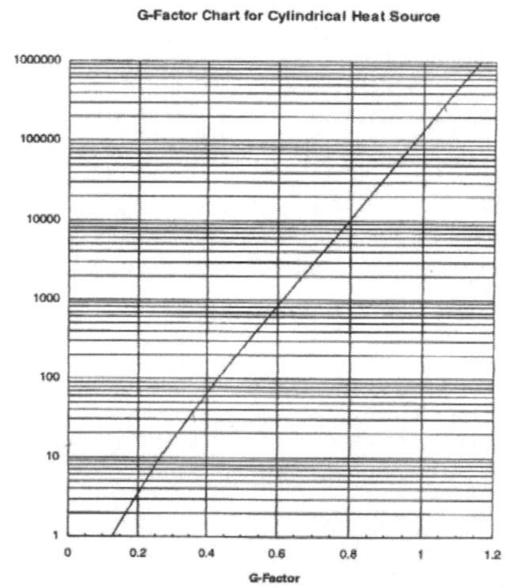

Figure 4 G-Factor Chart for cylindrical Heat source[1]

Where k_g is the soil conductivity in Qatar equal to 2.09 W/m.°K (1.21 Btu/h ft °F)

Therefore R_{ga}= 0.32 (h.ft.°F/Btu) (0.184 m. °K/W)

R_{gm}= 0.33 (h.ft.°F/Btu) (0.19 m. °K/W)

R_{gd} = 0.21 (h.ft.°F/Btu) (0.121 m. °K/W)

Annual Net Heat Rate

Taking into consideration the following assumptions: 8h/day cooling and 10 event/month for 4 month/year, then the Effective full load hours (EFL) is equal to 320hrs.

$Q_a=\frac{q_{lc}*EFL}{8760}$ (9)

Q_a= -174,657.57 Btu/h (-51.14 KW).

Monthly Part load factor:

$$PLF_m = \frac{\sum load * Hours}{Peak\ load * 24h} * \frac{Days\ occupied\ per\ Month}{Days\ per\ Month} \quad (10)$$

For $\sum load = Peak\ load$ and for 8/24h and 10 days/month

PLF_m=0.11

Required Length of Bore:

Assuming a temperature penalty of -3 °F (this temperature results from the thermal interference of adjacent bores) F_{sc} will be taken as 1.02 assuming 2 bores per parallel loop [6].

It is concluded that the total length is: L_c= 144,677.35 ft (44,100 m). This length can be accommodated with 482 bores considering 300 ft (91m) depth bore hole. Therefore, the average load per bore hole sums up to be around 0.82 ton/bore hole.

Cost Benefit Analysis

To identify the benefit from geothermal system a cost saving estimate has been developed considering one event day per month. **As shown in Table 4,** peak cooling geothermal exchange occurs in July with a capacity of 1400 kW, peaking for eight hours. Saved energy estimated to be 3186 kWh electric at peak day taking into consideration that the source of chilled water in this project is water-cooled chiller plant operating at 1 kW of electrical power per 1 tons of refrigeration. In order to identify the net energy saving, the effect of additional static pressure induced by the geothermal cooling coil on the fresh air fans is considered. The additional fan consumption is estimated to be 450 kWh at peak day. Therefore the Net Saved Energy at peak day from the Ground Vertical Loop system is 2736 kWh or 82 USD based on 0.03 USD/kWh (According to Qatar Electricity Tariff). For the months with negative cooling load The Air will be bypassed and will avoid going to the Geothermal unit. The large daily energy saving is minimized by Qatar low electric tariff that is much lower than the actual electric kWh production cost.

Table 4 Daily Savings in USD

Month	Precooled outdoor Air Dry Bulb Temperature (°C / °F)	Coil leaving Temperature (°C / °F)	Coil Entering Temperature (°C / °F)	Cooling Load (KW)	Total Energy Per Day (KWh)	Electric energy per day (KWh)	Net Energy Saving Per day (KWh)	Savings in USD per Day
December	35 / 95	32 / 89.6	36 / 96.8	-1115	-2549	450	0	0
January	34.4 / 93.92	31.3 / 88.34	36 / 96.8	-1310	-2995	450	0	0
February	35.5 / 95.9	32.5 / 90.5	36 / 96.8	-976	-2230	450	0	0
March	38.2 / 100.76	35.2 / 95.36	36 / 96.8	-223	-510	450	0	0
April	39.7 / 103.46	36.7 / 98.06	36 / 96.8	195	446	450	0	0
May	41.9 / 107.42	38.9 / 102.02	36 / 96.8	809	1848	450	1398	42
June	43.4 / 110.12	40.4 / 104.72	36 / 96.8	1227	2804	450	2354	71
July	44 / 111.2	41 / 105.8	36 / 96.8	1394	3186	450	2736	82
August	44 / 111.2	41 / 105.8	36 / 96.8	1394	3186	450	2736	82
September	42.9 / 109.22	39.9 / 103.82	36 / 96.8	1087	2485	450	2035	61
October	40.8 / 105.44	37.8 / 100.04	36 / 96.8	502	1147	450	697	21
November	37.6 / 99.68	34.6 / 94.28	36 / 96.8	-390	-892	450	0	0

GROUND WATER OPEN LOOP

The second alternative proposed is to use the groundwater open system. **As shown in Figure 5**, this system consists of a production well, an injection well and cooling coils. Therefore, the water coming from the well will be heated and injected into an injection well. This procedure will cool the air from 44 °C (111.2 °F) to 36 °C (96.8 °F).

The Geothermal coils were selected to meet the following criteria.

1. Entering dry bulb air temperature: 44 °C (111.2 °F) (with the assumption that the Earth-tube has been initially pre-cooling the outdoor air temperature from 46 (114.8 °F) to 44 °C (111.2 °F)).
2. Leaving dry bulb air temperature: 36 °C (96.8 °F).
3. Total load on geothermal coil: 1400 KW (4781251 BTU/h)
4. Sensible load on geothermal coil: 1400 KW (4781251 BTU/h)
5. Coil entering water temperature: 30 °C (86 °F) based on a 30 °C (86 °F) ground temperature
6. Coil leaving temperature: 38 °C (100.4 °F).

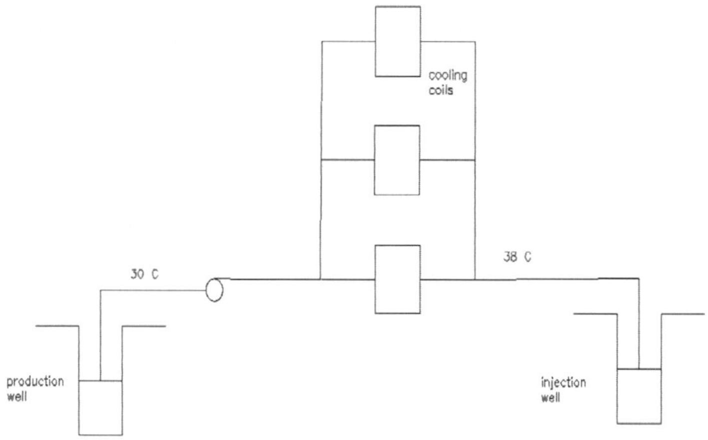

Figure 5 Ground water loop Sketch.

Ground water flow requirements

Ground water used for geothermal cooling need to be pumped from the aquifer, this issue is critical and needs authority approval for that exact pumping quantities needs to be calculated. To estimate the water quantities needed ASHRAE [7] provides the below calculation method.

$$G_w = \frac{qc * 12000 \frac{Btu}{ton\,hr}}{500 * (tgwo - tgwi)} \quad (11)$$

Where:

q_c: Cooling block load =400 tons (1400 KW)

$tgwo$ = 38 °C (100.4 °F)

$tgwi$ = 30 °C (86 °F)

Therefore the system peak flow is 667.6 gpm (42.12 L/s).

Required Spacing between the production and injection well

The system has a peak flow of 667.6 gpm (42.12 L/s) and the cooling is done over 4 months per year (124 days/year). However, the effective full load hours are 320hr/year. Therefore, we can calculate the average flow rate over the 4 months period. At 320 full load hours the 667.6 gpm (42.12 L/s) rate would be multiplied by 320/ (4*744) to result in an average flow of 71.67 gpm (4.52 L/s).

Assuming a 40 ft (12.19 m) aquifer thickness, the required spacing will be around 200 ft (60.96 m) **as shown in Table 5.**

Table 5 Required Spacing between twin wells [8]

Aquifer Thickness:	10 ft				20 ft				30 ft				40 ft			
Time/ Days:	100	140	210	280†	100	140	210	280†	100	140	210	280†	100	140	210	280†
q gpm																
10	114	127	145	163	98	109	127	145	86	100	117	134	79	94	108	122
20	158	177	200	223	135	153	177	201	122	140	164	188	109	126	152	176
40	227	251	289	327	195	220	253	286	174	195	231	267	154	178	213	248
60	280	310	358	406	237	270	311	352	208	241	280	319	189	218	262	306
80	325	360	418	476	276	315	364	413	242	278	325	372	220	254	301	348
100	365	403	472	541	309	353	408	463	268	310	362	414	246	285	338	391
200									385	445	515	585	353	411	485	559

Aquifer Thickness:	50 ft				80 ft				100 ft			
Time/Days:	100	140	210	280†	100	140	210	280†	100	140	210	280†
q gpm												
10	72	83	99	115	55	70	83	96	55	65	78	91
20	103	117	138	159	81	98	118	138	77	90	106	122
40	144	165	197	229	118	139	168	197	108	125	150	175
60	176	205	245	285	147	169	206	243	132	155	185	215
80	202	235	282	329	170	196	237	278	154	178	214	250
100	224	263	308	353	189	219	266	313	170	200	240	280
200	310	369	436	503	267	310	378	446	240	283	343	403
500	510	585	697	809	420	495	585	675	382	445	530	615
1000	712	824	986	1148	590	682	826	970	540	632	756	880

CONCLUSION

This study compared between two geothermal systems used to dissipate 1400 KW (400 ton) within the Doha soil. The vertical loops method requires a total of 482 bore holes each spaced by a distance of 20 ft (6 m). The ground area needed to implement the vertical closed loop system is about 16,351 m^2 (176,001 ft^2) Bore holes are distributed beneath the parking spaces.

Whereas, the Ground water open loop method consists of few simple equipment and has lower initial cost since it requires four production wells and four injection wells each with a capacity of 167gpm (10.53 L/s). Vertical closed loop geothermal system needs high cost involved with the implementation of the 482 bore holes, however; since the Doha deep aquifer poses a major constraint, it will be more beneficial to use the vertical loop method to preserve this important asset.

NOMENCLATURE

L_c : required bore length for cooling (ft)

F_{sc} : short-circuit loss factor.

PLF_m : Part load factor during design month

q_a : Net annual average heat transfer to the ground (Btu/h)

q_{lc}: Building design cooling block load (Btu/h)

R_{ga}: Effective thermal resistance of the ground, annual pulse (h.ft.°F/Btu)

R_{gd}: Effective thermal resistance of the ground, daily pulse (h.ft.°F/Btu)

R_{gm}: Effective thermal resistance of the ground, monthly pulse (h.ft.°F/Btu).

R_b: Thermal resistance of bore (h.ft.°F/Btu).

t_p: Temperature penalty for interference of adjacent bores (°F)

t_g: Undisturbed ground temperature (°F)

t_{wi}: Liquid temperature at coil inlet (°F)

t_{wo}: Liquid temperature at coil outlet (°F)

qc: cooling block load (Btu/h)

$tgwo$: building return temperature (°F)

$tgwi$: Water entering building temperature (°F)

REFERENCES

[1] Stephen P. Kavanaugh, and Kevin Rafferty. 1997. Ground-Source Heat pumps: Design of Geothermal Systems for Commercial and Institutional Buildings.

[2] Long Period Means& Extremes of climatologically elements Doha International Airport period (1962-1999), state of Qatar ministry of Comm& Transport Dept. of civil Aviation& Meteorology

[3] Geothermal Heat Pump Design Manual Application guide AG 31-008 McQuay air conditioning 2002.

[4] ASHRAE. 2009. ASHRAE Handbook-Fundamentals. Atlanta: American Society of Heating Refrigeration and Air Conditioning Engineers, Inc.

[5] ARI Standard 330-98 – Ground Source Closed Loop Heat pumps – Air-Conditioning and Refrigeration Institute. 1998. Arlington, Va.

[6] Geology and Drilling Methods for Ground-Source Heat pump Installations: An Introduction For Engineers. Sachs, H. 2002. ASHRAE. Atlanta Ga.

[7] ASHRAE. 2012. ASHRAE Handbook- HVAC Systems and Equipment Handbook. Atlanta: American Society of Heating Refrigeration and Air Conditioning Engineers, Inc.

[8] R.G. Kazemann and W.R. Whithead, " the spacing of Heat Pum supply and discharge Wells" Groundwater Heat pump Journal Summer 1980, Vol,1No2. Columbus Ohio: Water Well Journal publishing, 1980.

Ground Source Heat Pump Systems for Residential Buildings in Warm Climates: Energy, Environmental and Economic Considerations

Apostolos Michopoulos, PhD.
Member ASHRAE

Katerina Tsikaloudaki, PhD.

Vasiliki Voulgari

Theodoros Zachariadis, PhD.

ABSTRACT HEADING

In this paper the heating and cooling energy consumption of a typical single-family house is calculated for five selected locations in Cyprus. Hourly calculations are performed using the EnergyPlus software. The results are translated in oil or LPG and electricity consumption, assuming that the typical building in Cyprus is equipped with an oil-fired or LPG-fired boiler for heating and local air-to-air split-type heat pumps for cooling. The same energy needs are assumed to be covered by an alternative system, i.e. a vertical closed loop ground heat exchanger combined with a water-to-water heat pump system for heating and cooling. The ground source heat pump system is dimensioned with the aid of the EED 3.0 software, analyzed using an in-house developed and validated simulation code, and as a result the electricity consumption of the system is calculated. Based on the resulting fuel consumption of the alternative system, the primary energy consumption and the corresponding emissions are determined, while a financial analysis is also performed. The results prove that significant energy, environmental and economic benefits can be achieved.

INTRODUCTION

In Cyprus, the residential building sector consumes 322 ktoe, which accounts for the 16.8% of the total final energy consumption (2010). These energy needs are mainly covered by electricity (46.6%), oil (24.8%) and LPG (11.8%), while the remaining 17% is served by renewable energy sources mainly for the production of warm water with solar thermal water heaters. It is interesting that the 27.5% of the final energy is consumed for space heating, where oil and LPG are mainly used. Although for the residential sector the energy consumption has been increased by 2.9% since 1995, the CO_2 emissions remain constant and equal to 0.504 Mt (Kitsios and Zachariadis, 2012). Within this framework, Cyprus, as a member country of the E.U., is obliged to adopt measures in order to improve the energy efficiency, decrease the CO_2 emissions and increase the share of RES by 2020.

Ground source heat pump (GSHP) systems are recognized as one of the most efficient technologies for space heating and cooling. In recent years the worldwide installed capacity has been growth rapidly (Lund et al.,

Apostolos Michopoulos is a senior researcher in the Department of Environmental Science and Technology, Cyprus University of Technology, Limassol, Cyprus. Aikaterini Tsikaloudaki is an Assistant Professor in the School of Civil Engineering, Aristotle University of Thessaloniki, Thessaloniki, Greece. Vasiliki Voulgari is a researcher in the Department of Environmental Science and Technology, Cyprus University of Technology, Limassol, Cyprus. Theodoros Zachariadis is an Assistant Professor in the Department of Environmental Science and Technology, Cyprus University of Technology, Limassol, Cyprus.

2011) due to their advantages. In contrast, in Cyprus the installation of GSHP systems still remains limited. It is estimated that today there are only 160 installations on the island and the majority of them are employed in private single-family buildings. However, it is expected that a wide application of GSHP systems in Cyprus will be beneficial, as it will not only improve the energy efficiency of the building sector but also increase the energy independency of the country. This is attributed to the small and isolated energy system of Cyprus, which is not interconnected to other energy networks, as well as to the fact that there is no production of fossil fuels on the island and the energy needs are primarily covered by the imported oil products.

The purpose of this study is to evaluate the energy, environmental and economic benefits derived by the application of GSHP systems on single-family residential buildings in Cyprus. The main objective is to estimate the actual impacts of the GSHP systems under the warm climate conditions of the Mediterranean area following a holistic simulation approach. A typical single-family building of the country was defined and its energy performance was modeled when a GSHP system and two common conventional heating and cooling systems were alternatively employed. The primary energy consumption, the CO_2 emissions and the net present value (NPV) index derived for each different system were comparatively assessed in order to highlight the most efficient system. It would be ideal if this study attracted the interest of government authorizes, researchers, as well as private constructors and land developers not only from Cyprus but also from the other Mediterranean countries or countries from the wider area with similar climate conditions, in order to promote the wider application of the GSHP systems through the presentation of their benefits.

DESCRIPTION OF THE REFERENCE BUILDING

The first step of the current study was the establishment of the typical single family residential building for Cyprus. For this purpose, information regarding the building stock in Cyprus was retrieved from the Cypriot statistical service. The parameters that were used for designing the typical single family building were the construction materials, the total floor area of the building, as well as the number of floors and rooms.

With regard to these characteristics and taken into account the local architectural features, a typical single family building was defined, the architectural plans of which are presented in Figure 1. It is developed in two levels, which are connected with an internal staircase. In the ground floor a spacious living room, a kitchen and a study are located, while the upper floor accommodates three bedrooms and a sitting area. In total the building covers an area of 202.2 m^2. With a floor to floor height equal to 3.05 m, the total heating volume of the building is equal to 616.71 m^3. The building is freestanding on the plot, therefore windows are located on every façade; the majority is positioned on the main façade, which is orientated due south. More specifically, southern windows account for the 33% of the southern façade, whereas the transparent elements on the remaining facades are substantially limited, i.e. the northern and the western windows cover 11% and 15% of the respective façade, while the windows on the eastern facade occupy only 1% of its area.

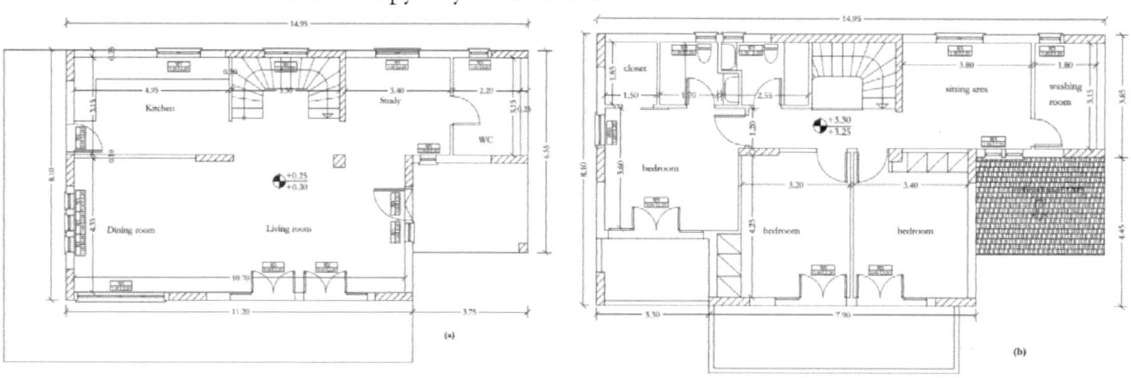

Figure 1 The architectural plans of: (a) ground floor and (b) upper floor of typical single family building.

In Cyprus, the majority of new constructions have a load bearing structure made of reinforced concrete and masonries built with hollow clay bricks. The external opaque vertical elements are usually insulated with expanded polystyrene, which is positioned on their external surface. The building roofs are flat, thermally protected with expanded polystyrene, which lies above the water protection layer (bitumen layer) and is covered with gravels. The floor is in contact with the ground and is thermally insulated with expanded polystyrene below the concrete slab. The thickness of the thermal insulation and the resulting thermal transmittance of each building element were calculated in compliance with the recent building energy performance regulation of Cyprus (Table 1). Accordingly, all windows comply with the current regulation for building energy performance, which sets a maximum value for thermal transmittance equal to 3.8 W/(m²·K).

Table 1. The thermal transmittance of the building elements

Building Element	Thermal transmittance	Maximum thermal transmittance value
Masonry	0.691 W/(m²·K) (0.12 Btu/(hr.·ft.²·°F))	0.72 W/(m²·K) (0.126 Btu/(hr.·ft.²·°F))
Reinforced concrete	0.586 W/(m²·K) (0.103 Btu/(hr.·ft.²·°F))	0.72 W/(m²·K) (0.126 Btu/(hr.·ft.²·°F))
Flat roof	0.61 W/(m²·K) (0.107 Btu/(hr.·ft.²·°F))	0.63 W/(m²·K) (0.11 Btu/(hr.·ft.²·°F))
Floor above ground	0.766 W/(m²·K) (0.135 Btu/(hr.·ft.²·°F))	2.0 W/(m²·K) (0.352 Btu/(hr.·ft.²·°F))

DESCRIPTION OF THE LOCAL CONDITIONS

Cyprus is the eastern country of the Europe and lies between 34.5° N to 35.8° N latitude and 32.2° E to 34.6° E longitude. The climate of the island is Mediterranean and is characterized by hot dry summers from mid-May to mid-September and rainy winters form November to mid-March. There are also short autumn and spring seasons of rapid change weather conditions. The geomorphology of the island is distinguished by two main mountain ranges, Troodos -1,952 m (6,273 ft.)- and Pentadaktylos -1,024 m (3,360 ft.)- which cover the central-west and northern-east regions of the island. The lowland occupies the 63% of the island and it is mainly located along the southern coastline areas as well as at the central area of the island, where the capital city, Nicosia, is positioned.

In order to estimate the benefits of the GSHP installations in Cyprus, 5 characteristic and representative of the local climate areas have been selected: Nicosia, Larnaca, Limassol, Paphos and Saittas. Larnaca, Limassol and Paphos are located at the southern coast, Nicosia in the central mainland and Saittas, which is a characteristic village of Troodos, represents the mountainous climate of the island.

CALCULATION OF THE BUILDING'S DESIGN LOADS

The heating and cooling design loads of the typical single family building in the cases that is located at the 5 selected locations of the island have been calculated using the Elite Chvac® software. The information regarding the geometrical characteristics of the building is retrieved from the architectural plans while the thermal characteristics of the building elements are displayed in Table 1. The building has been divided into 11 independent thermal zones according to the zone usage, i.e. bedroom, kitchen, living room, WC, etc. The desirable indoor temperature was set equal to 22°C (71.6°F) and 25°C (77°F) -50% RH- for the heating and cooling period respectively, following the ASHRAE's recommendations (ASHRAE, 2013). In order to calculate the cooling load, the internal heat gains have been considered in each zone with regard to ASHRAE's recommendations (ASHRAE, 2013). The design external conditions in the cases of Larnaca and Paphos were retrieved from the ASHRAE's climatic database, while for the remaining areas they have been estimated from the available statistical records provided by the Cypriot Meteorological Service, due to the fact that these are not included in the national or international databases. In Table 2 the design heating and cooling load in each characteristic location is presented.

Table 2. The design heating and cooling loads

Location	Heating design load	Cooling design load

Nicosia	9.3 kW (31.7 kBtu/hr.)	11.9 kW (40.6 kBtu/hr.)
Larnaca	8.5 kW (29.0 kBtu/hr.)	8.9 kW (30.4 kBtu/hr.)
Limassol	7.6 kW (25.9 kBtu/hr.)	8.8 kW (30.0 kBtu/hr.)
Paphos	8.1 kW (27.6 kBtu/hr.)	9.7 kW (30.1 kBtu/hr.)
Saittas	10.1 kW (34.5 kBtu/hr.)	7.8 kW (26.6 kBtu/hr.)

ENERGY ANALYSIS OF THE BUILDING

The energy requirements of the single family building located on the 5 representative areas of Cyprus have been calculated with the aid of EnergyPlus software using the typical meteorological years (TMY-2) provided by the Meteonorm database. The building has been divided into eleven independent thermal zones, the indoor conditions of which were considered in hourly basis according to ASHRAE's recommendations (ASHRAE, 2013). In Table 3 the parameters used in the energy simulations are presented.

Table 3. Energy analysis parameters

Parameter	Value
Operation period	24-h per day
Heating period	16/11-15/05
Winter indoor temperature	22°C (71.6°F) [07:00-21:00]/20°C (68°F) [22:00-06:00]
Cooling period	16/05-15/11
Summer indoor temperature	25°C (77°F) [07:00-21:00]/27°C (80.6°F) [22:00-06:00]
Air changes	0.8 ach [1:00-24:00]
Lighting power	6 W/m^2 (1.9 Btu/(hr.·ft.^{2})), 3.5 W/m^2 (1.1 Btu/(hr.·ft.^{2})) in WC
Number of occupants	0 to 2 depending on the zone's use
Equipment capacity	0 to 700 W (0 to 2.39 kBtu/hr.) depending on the zone's use

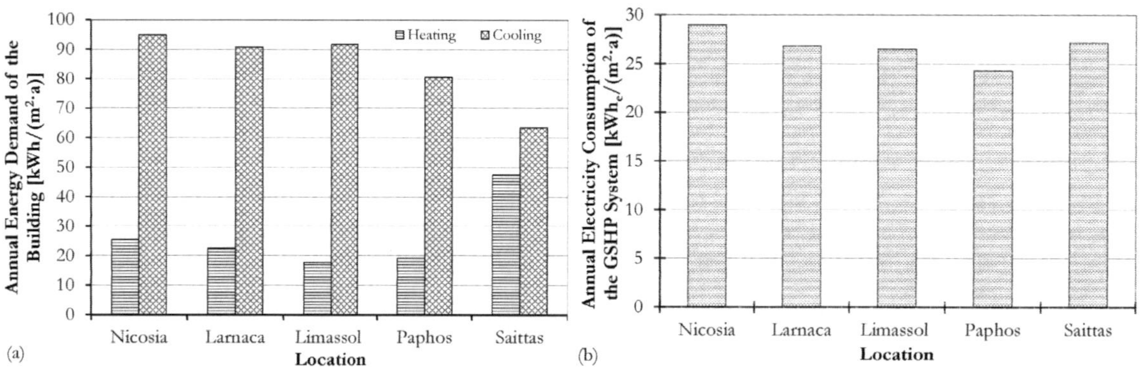

Figure 2 Annual: (a) energy demand of the building, (b) electricity consumption of the GSHP system.

The annual heating and cooling energy demands of the typical single-family building estimated for the 5 representative locations in Cyprus are presented in Figure 2a. Obviously, the cooling energy demands are in every case higher than the heating ones, as this study is focused on the warm European climates. The highest annual heating demand is observed for the mountainous area of Saittas, 47.5 kWh/(m^2·a) (15.06 kBtu/(ft.^{2}·a)), while the lowest for the southern coastal area of Limassol, 17.8 kWh/(m^2·a) (5.65 kBtu/(ft.^{2}·a)). Conversely, the highest and lowest annual cooling demand occurs for the central lowland area of the island, Nicosia 94.9 kWh/(m^2·a) (30.1 kBtu/(ft.^{2}·a)) and for the mountainous area, 63.4 kWh/(m^2·a) (20.1 kBtu/(ft.^{2}·a)), respectively. It is worth mentioning that the ratio between the heating and cooling energy demand varies form 1:1.4 to 1:5.2 depending on the climate conditions of the examined area.

DESIGN AND SIMULATION OF THE GSHP SYSTEM

The design of a GSHP system involves the dimensioning and the selection of the GSHP, as well as the dimensioning of the vertical GHEx. The dimensioning of the GSHP is based on the design heating and cooling loads of the building described earlier. For this study, a high performance GSHP was selected for each location, manufactured by WaterFurnace, type Envision NSKW12. The dimensioning of the vertical GHEx was performed for the 5 representative locations separately with the aid of the EED software. The required GHEx length was calculated using the monthly heating and cooling energy demands calculated from the EnergyPlus software. Moreover, it was assumed that the GHEx consists of double U-tube high density polyethylene pipes (Φ 32x2.9 mm) and the boreholes are filled with thermal enhanced grout (Fischer GeoSolid 240), with a thermal conductivity equal to 2.4 W/(m·K) (1.39 Btu/(hr.·ft.^{2}·°F)), in order to improve the heat flux between the ground heat exchanger and the surrounding soil. The thermo-physical ground properties have been selected from the existing literature (Morgan, 1973) and remained constant for all locations: ground thermal conductivity 1.75 W/(m·K) (1.01 Btu/(hr.·ft.^{2}·°F)), ground heat capacity 2.1 MJ/(m^{3}·K) (31.3 kBtu/(ft.^{3}·°F)). The number and the length of the boreholes as well as the GHEx configuration have been calculated and optimized with the EED software after a 20-years analysis of the system's operation, assuming that the minimum and maximum water temperature in the GHEx should be between 5°C (41°F) and 38°C (100.4°F), respectively. The results are presented in Table 4.

Table 4. Ground heat exchanger construction's characteristics

Location	Number of boreholes x Length / Configuration
Nicosia	4 x 65 m (213.2 ft.) / L: 2 x 2
Larnaca	4 x 67 m (219.8 ft.) / L: 2 x 2
Limassol	4 x 64 m (209.9 ft.) / L: 2 x 2
Paphos	4 x 52 m (170.6 ft.) / L: 2 x 2
Saittas	3 x 55 m (180.4 ft.) / line: 1 x 3

The electricity consumption of the geothermal heat pumps was calculated using an in-house developed and validated algorithm (Michopoulos and Kyriakis, 2009). Moreover, the amount of electricity consumed by the circulation pump on the GHEx loop was calculated according to the methodology for decentralized pumping systems proposed by Sfeir et al. (Sfeir et al., 2005). Finally the overall electricity consumption of the GSHP system was calculated (Figure 2b). Based on these findings, the highest electricity consumption of the GSHP system is observed for the locations of Nicosia and Saittas, and is equal to 29 kWh$_{e}$/(m^{2}·a) (9.2 kBtu/(ft.^{2}·a)) and 27.2 kWh$_{e}$/(m^{2}·a) (8.6 kBtu/(ft.^{2}·a)) respectively. The lowest electricity consumption is expected when the building is located at Paphos, 24.3 kWh$_{e}$/(m^{2}·a) (7.7 kBtu/(ft.^{2}·a)), an amount which is decreased by 19.2% compared to the one found for Nicosia. In addition, for the remaining locations, Larnaca and Limassol, the electricity consumption ranges in similar levels, i.e. 26.9 kWh$_{e}$/(m^{2}·a) (8.5 kBtu/(ft.^{2}·a)) and 26.6 kWh$_{e}$/(m^{2}·a) (8.4 kBtu/(ft.^{2}·a)) respectively.

ENERGY EVALUATION

The energy evaluation of the GSHP system has been performed by comparing its primary energy consumption with the corresponding consumption derived when the conventional system, i.e. the one that is primarily used on private households of Cyprus, is employed. The conventional system consists of an oil- or LPG-fired boiler and air-to-air split type heat pumps. In order to calculate the energy consumption of the conventional boiler system, the heating energy requirements of the building are converted into final heating energy consumption assuming the minimum efficiency of the oil- or LPG-fired boiler (0.92), according to the European directive 92/42/EC (EU, 1992). In addition, the final heating energy consumption has been turned into primary energy consumption using the national reduction factor, which is equal to 1.1 for both fuels.

The selection of the air-to-air heat pumps was performed using the design cooling load of each thermal zone. Four high efficiency air source heat pumps were selected, manufactured by Toshiba, type Suzumi Plus RAS-10SAV2-E. The electricity consumption of the conventional cooling system was calculated on an hourly basis, using the energy requirements of the building, taking into account the EnergyPlus simulations (Figure 2a) and the energy efficiency ratio (EER) of the heat pump. The energy efficiency ratio was calculated according to the performance data provided by the manufacturer and the corresponding hourly ambient air temperature, which was retrieved from the TMY-2 weather files. Furthermore the electricity consumption of the heat pumps was converted into primary energy consumption using the national reduction factor for electricity (2.7). Finally, the total primary energy consumption of the conventional system was calculated by adding together the primary energy consumption of heating and cooling.

The annual primary energy consumption per heating/cooling area for both systems, the GSHP and the conventional, is presented in Figure 3. From Figure 3 it is indicated that the conventional system consumes more energy than the GSHP in the locations of Nicosia, Larnaca, Paphos and Saittas. Moreover, it can be noticed that the discrepancy on the primary energy consumption between the GSHP and the conventional system increases when moving from hot climates towards the colder ones. More specifically, in Saittas, the coldest area of Cyprus, the conventional system consumes 23.4% more energy than the GSHP, while for the hottest areas of the island, Paphos and Limassol, the discrepancy is limited to 1.6% and 0% respectively. In addition, areas with mild climate conditions, i.e. Nicosia and Larnaca, demonstrate an intermediate behavior, resulting to a relevant discrepancy equal to 7.5% and 7.7% respectively.

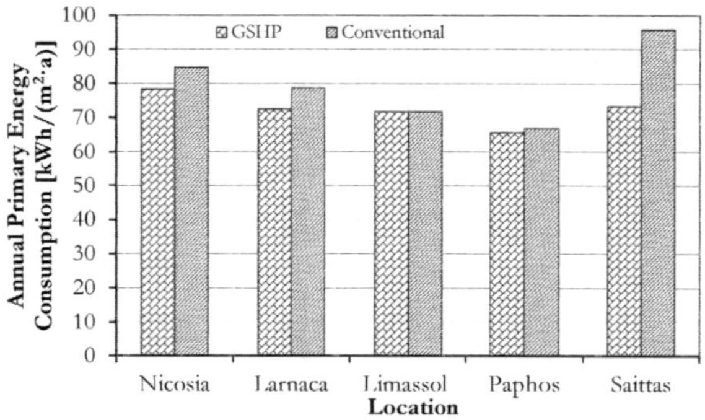

Figure 3 Annual primary energy consumption of the building.

ENVIRONMENTAL EVALUATION

The environmental evaluation of the GSHP and the conventional systems were performed by the calculation and the comparison of the amount of carbon dioxide emitted by each system. The calculation of the carbon dioxide emissions was based on the fuels' national emission factors per unit of primary energy, which are characteristic of the Cypriot energy system. More specifically the annual primary energy consumption of the building has been multiplied by 0.794, or 0.266, or 0.249 in the case of the electricity source -for the GSHP system and the air-source heat pumps (A/A)-, heating oil (HO), or LPG sources -for the conventional boilers- respectively (Figure 4a). From Figure 4a it is depicted that the GSHP system produces higher CO_2 emissions than the conventional ones. In addition, the emissions of the conventional oil-fired system are slightly increased than that of the alternative LPG-fired one. This is primarily based on the higher emission factor that the heating oil has instead of LPG. More specifically, the CO_2 emissions of the GSHP system range from 19.15% to 28.95% higher levels than the conventional systems. The highest discrepancy is observed for the colder area of Saittas, while the lowest one for the mild climate areas of Nicosia and Larnaca. It must be noted that this trend could not be considered as common for all areas with similar climate conditions, as it is strongly affected by the energy mixture of the power production. Furthermore, it is expected that the increase of the renewables' share as well as the introduction of natural gas in the Cypriot power production system will reduce the environmental advantage of the conventional systems and improve the environmental behaviour of the GSHP systems.

ECONOMIC EVALUATION

The economic benefits gained from installing vertical GSHP systems in the private household sector of Cyprus rather than the conventional oil- or LPG-fired boiler and air source heat pump systems was performed with the aid of the NPV index. In accordance with the European Union's Regulation No 244/2012/EU (EU,

2012), the initial construction cost and the annual operation cost have been calculated assuming that the economic lifetime of the investment is equal to 15 years (EN 15459, 2007). The construction cost of the conventional systems has been calculated based on the real marker prices of January-February 2014 (Table 5). It is assumed that the GSHP system and the oil- or LPG-fired boiler employ the same distribution and emission heating systems, therefore their construction costs could be excluded from the analysis for both systems.

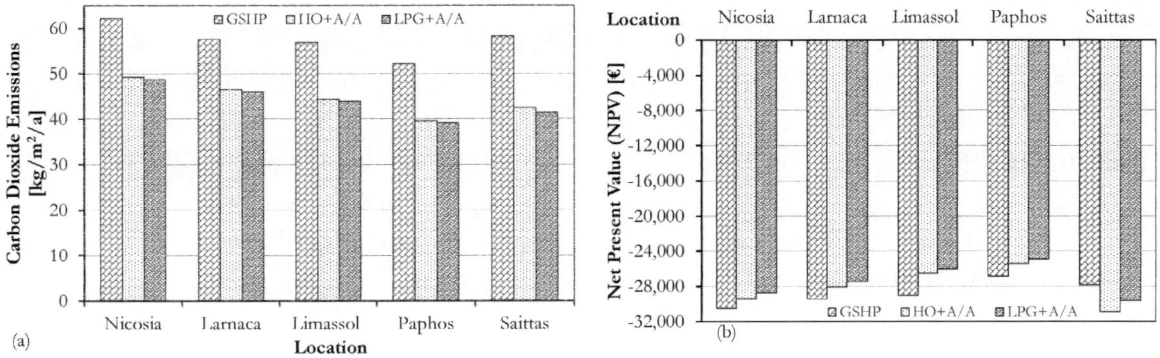

Figure 4 (a) annual carbon dioxide emissions of the alternative systems, (b) the net present value of the alternative systems.

The calculation of the annual fuel cost for each system and characteristic location was based on the annual electricity, the heating oil and LPG consumption, calculated on the energy analysis. The annual fuel consumption was multiplied with the forecast of the fuel price for the Cypriot energy system, which has been recently published (Zachariadis and Michael, 2014) and the annual fuel cost has been calculated. Moreover the annual preventive maintenance costs were set equal to 2% and 4% of the initial investment for the boiler and heat pumps respectively. Finally, the NPV for each alternative system has been calculated for a 6% real discount rate.

Table 5. The construction cost of the two alternative systems

Location	GSHP system [€]	Conventional system [€]
Nicosia	12,700	8,500
Larnaca	13,000	8,500
Limassol	12,700	8,500
Paphos	11,500	8,500
Saittas	10,500	7,700

Figure 4b illustrates the results of the NPV calculations for the examined systems for the 5 representative locations in Cyprus. It is shown that the GSHP system presents the lower NPV when the building is located at the southern and mainland areas of the island, where the conventional LPG+A/A system is the most economically attractive. More especially, the NPV of the GSHP system is reduced by 8.8% (2,550 €) to 3.6% (1,090 €) or 10.3% (3,000 €) to 5.7% (1,750 €) in comparison to HO+A/A and LPG+A/A systems respectively. Conversely, the GSHP system becomes economically beneficial on the mountainous and colder areas of the island, as the NPV is higher by 10.9% (3,050 €) or 6.4% (1,785 €) than the alternative systems.

CONCLUSION

In this study a holistic analysis on the installation of a GSHP system on single-family households in Cyprus in place of the conventional ones has been performed. It was found that a typical single-family buildings requires 17.8 kWh/(m²·a) (5.65 kBtu/(ft.²·a)) to 47.5 kWh/(m²·a) (15.06 kBtu/(ft.²·a)) for space heating and 63.4 kWh/(m²·a) (20.1 kBtu/(ft.²·a)) to 94.9 kWh/(m²·a) (30.1 kBtu/(ft.²·a)) for space cooling with regard to the climate conditions of the site. Based on these demands, the GSHP system constitutes a beneficial solution as regards the primary energy consumption, since it consumes up to 23.4% less energy compared to the conventional oil- or LPG-fired boiler and air-to-air heat pump systems. On the other hand, and taking into account the specific characteristics of the energy mix in the studied area, the use of conventional systems for space heating and cooling leads to 28.95% less CO_2 emissions compared to the vertical GSHP systems. The results of the economic analysis indicate that the GSHP system is financially attractive for the colder area of the study, while for mild- and hot-climate areas the conventional systems are more favorable as their NPV index is 3.6% to 10.3% higher.

ACKNOWLEDGMENTS

Work reported here was partly funded by the Research Promotion Foundation of Cyprus under the programme 'DESMI 2009-2010' (project no. DIDAKTOR/0311/37), which is co-funded by the Republic of Cyprus and the European Regional Development Fund.

NOMENCLATURE

A/A = Air-to-Air heat pump
ASHRAE = American Society of Heating Refrigerating and Air-conditioning Engineers
GHEx = Ground Heat Exchanger
GSHP = Ground Source Heat Pump
HO = Heating oil
LPG = Liquefied Petroleum Gas
NPV = Net Present Value
RES = Renewable Energy Sources
RH = Relative Humidity
TMY = Typical Meteorological Year

Subscripts

e = electricity

REFERENCES

ASHRAE. 2013. ASHRAE Handbook-Fundamentals. Atlanta: American Society of Heating Refrigeration and Air Conditioning Engineers, Inc.

EN 15429. 2007. Energy performance of buildings - Economic evaluation procedure for energy systems in buildings. European Committee for Standarization.

EU. 1992. Council Directive 92/42/EEC of the 21 May 1992 on efficiency requirements for new hot-water boilers fired with liquid or gaseous fuels.

EU. 2012. Regulation No. 244/2012 supplementing Directive 2010/31/EU of the European Parliament and of the Council on the energy performance of buildings by establishing a comparative methodology framework for calculating cost-optimal levels of minimum energy performance requirements for buildings and building elements.

Kitsios, K., Zachariadis, T. 2012. Energy Efficiency Policies and Measures in Cyprus. Report to the Odyssee-Mure 2010.

Lund, J.W., Freeston, D.H., Boyd, T.L. 2011. Direct utilization of geothermal energy 2010 worldwide review, Geothermics 40:159-180.

Michopoulos, A., Kyriakis, N. 2009. A new energy analysis tool for ground source heat pump systems, Energy and Buildings 41:937-941.

Morgan, P. 1973. Terrestrial heat flow studies in Cyprus and Kenya. PhD thesis published by Imperial College, University of London.

Sfeir, A., Bernier, M., Million, T., Joly, A. 2005. A methodology to evaluate pumping energy consumption in GCHP systems, ASHRAE Transaction 111(1):714-729.

Zachariadis, T., Michael, M. (2014). Update of national energy forecasts for the Republic of Cyprus. Report to the Ministry of Energy, Commerce, Industry and Tourism.

Passive Solar Heating

Subhas Prasad **Somjit Sahani**

ABSTRACT

In any solar energy system for space heating there are three functions performed: **collection** *of solar energy,* **storage,** *and* **distribution** *of that energy (heat) from storage to living space. The two basic categories of solar systems for space heating - active and passive - perform these three functions, but in different ways. Active solar heating systems incorporate mechanical devices to circulate air or fluids through collectors and into thermal storage units such as water tanks or rock beds. Additional fans or pumps are required to bring stored heat to areas where it is needed. Active systems can be quite complex and rely on external sources of energy to operate. The passive heating in cold climates through the thermal storage like Trombe wall can be available option.*

Solar heating and cooling (SHC) can play a significant role in providing an economically viable and environmentally sustainable long-term solution to these essential heating and cooling needs. *SHC draws from an inexhaustible energy source while displacing fossil fuels and electricity otherwise needed for heating and cooling. The basic design of building involving right orientation, external obstructions to solar gain, right internal layout, optimum level of insulation, right window sizing, window type, right amount, placement & color of thermal mass & Trombe wall can lead us to a better sustainable building design.*

INTRODUCTION

Since ancient times, people have used thick walls of adobe or stone to trap the sun's heat during the day and release it slowly and evenly at night to heat their buildings. Today's low-energy buildings often improve on this ancient technique by incorporating a thermal storage and delivery system called a Trombe wall. Named after French inventor Felix Trombe in the late 1950s, the Trombe wall continues to serve as an effective feature of passive solar design.

The concept of passive solar systems is well-known method for use of solar energy as a source of heating in buildings. There is a vast literature on this technology, but real objects (houses) with passive solar systems are still rare. The major impediments to increase market penetration of passive solar systems is the lack of available information and experience data for the efficiency and constructive parameters of passive solar elements.

Passive systems perform the three functions of collection, storage, and distribution of solar energy in the following way. Sunlight enters the clear or translucent section of a wall. This section (known as glazing) should be on the south side of a building (in the Northern Hemisphere) to collect the maximum amount of solar radiation available. Solar radiation is then absorbed by the storage medium behind the glazing. This stored heat is distributed into the living space by means of the heat transfer mechanisms.

Subhas Prasad & Somjit Sahani are the lead designer in Mechanical & Electrical Domain.

THERMAL STORAGE WALLS

Thermal storage walls fall into three general categories: those utilizing a massive wall to store heat -these are known as Trombe walls; those utilizing a water wall to store heat; and the more experimental type in which heat is stored in eutectic salts or salt hydrates. Because Trombe walls are the most used type of thermal storage wall, much of our discussion will focus on them. Five elements of a thermal storage wall can be identified: glazing, air space between glazing and wall, the mass or storage wall, vents (in some thermal storage walls), and roof overhang (especially in warm climates).

A typical unvented Trombe wall consists of a 4- to 16-in (10- to 41-cm)-thick, south facing masonry wall with a dark, heat-absorbing material on the exterior surface and faced with a single or double layer of glass. The glass is placed from ¾ to 2 in. (2 to 5 cm) from the masonry wall to create a small airspace. Heat from sunlight passing through the glass is absorbed by the dark surface, stored in the wall, and conducted slowly inward through the masonry. High transmission glass maximizes solar gains to the masonry wall. As an architectural detail, patterned glass can limit the exterior visibility of the dark concrete wall without sacrificing transmissivity. With vented thermal storage walls, the vents can provide an important control mechanism both in heating and cooling the building. They can facilitate heat transfer into the building during a winter day. The use of vents through the **glazing** while upper vents through the mass wall are closed reduces heat gain by the mass wall and thus keeps the building cool in the summer. A roof overhang can also reduce heat gain during the warm months when the sun is high by shading the thermal storage wall. And, as will be shown, there are many other ways to adapt thermal storage walls to increase or decrease solar gain and reduce heat loss.

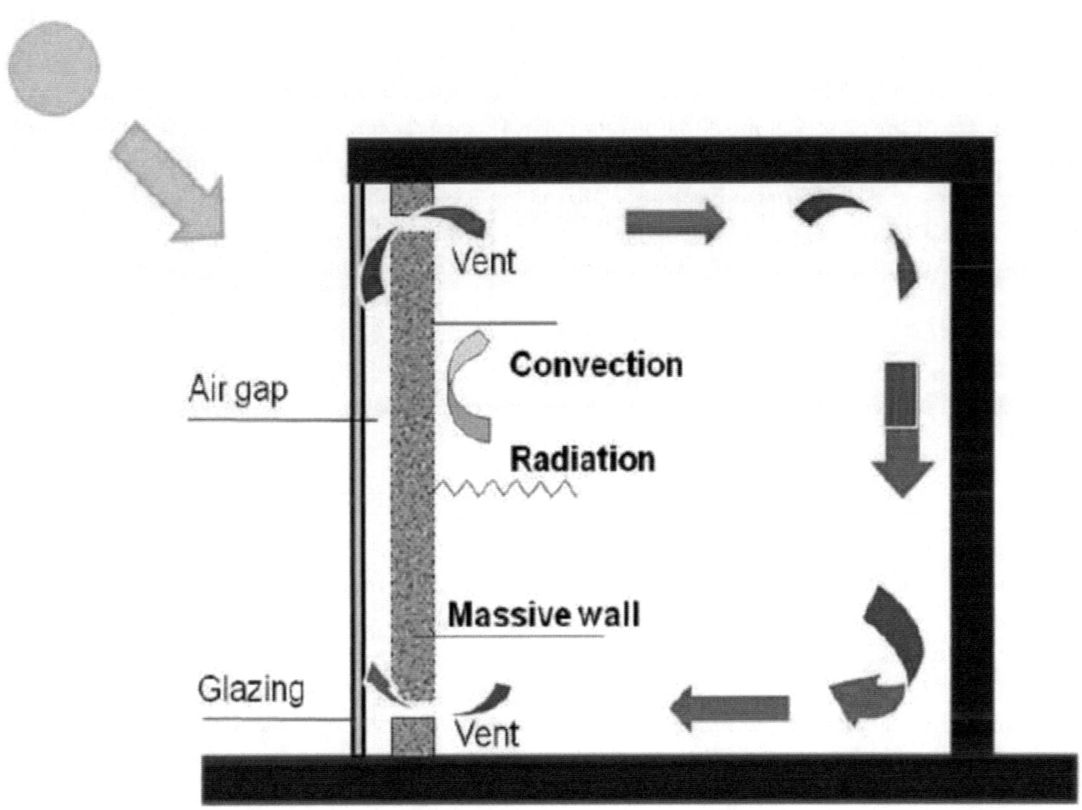

Figure 1. Working principle of a traditional Trombe wall and energy gain.

GLAZING

-Glazings are critical components of most solar collection systems. The purpose of the clear translucent coverings IS to trap heat from the incoming solar radiation. The heat-trapping ability of glazings arises largely from their wavelength dependent transmission. That is, they allow radiation of certain wavelengths to pass through while blocking the passage of others.

A good glazing material should allow **maximum** transmission of solar (short wave) radiation (expressed as the percentage of incident light that passes through). And it should keep heat loss to a **minimum** by preventing long-wave transmission and by serving as a barrier to heat loss. Long wave radiation or heat is radiated out from surfaces that absorb light in any collector system. By preventing the escape of this long wave radiation, the collector heats up. This process is the familiar "greenhouse effect". Additionally, an ideal solar glazing should possess resistance to ultraviolet ray deterioration, good thermal stability, a high resistance to abrasion and weather, low maintenance and purchase costs, high fracture and Impact resistance, and ease of handling.

THERMAL MASS WALL

The mass wall is the most crucial component of a Trombe wall type thermal storage wall. In it the solar heat will be stored and transmitted to the inside of the building. The material used for a mass wall is, therefore, very important and is discussed in some detail below. Also important with a mass wall is the surface exposed to the sun. It is necessary that the surface of the mass wall absorb nearly all the light energy passing through the glazing. To do this, the surface of the mass wall should be a **dark** color. If using paint on the mass wall, it should be black or a very dark color and should be able to withstand the high temperatures reached in a Trombe wall collector. Darkening agents other than paints may be used, depending on the wall material. Wood stains have been used to darken adobe and concrete block. Cement stucco can easily be darkened with added pigments. Counter too much previously published information, there is apparently very little difference in absorption between flat and glossy paints, glossy paints being, in fact better as they tend to pick up less dirt and dust.

With thermal characteristics, we are interested in 1) how much heat a material can store, and 2) how rapidly that heat can be transmitted (by conduction) through the material and released to the inside air. These characteristics are determined by four physical properties of a material: density, conductivity, specific heat, and heat capacity.

Density, p, is a measure of how heavy a given volume of a material is?. In general, heavier (more dense) materials tend to absorb and store more heat than lighter ones.

Thermal conductivity is a measure of how rapidly and easily heat can move through a material. The movement of heat is always due to a difference in temperature - heat moves from warmer to cooler parts of any material.

Specific heat Cp is a measure of the amount of heat needed to raise the temperature of a given mass of material.

Volumetric heat capacity is a measure of how much heat can be stored in a cubic foot (cubic meter) of material when being raised in temperature 1°F(1°C). It can be found by multiplying the density (p) of a material by the specific heat (Cp).

DESIGNING & SIZING A THERMAL STORAGE WALL

Site Selection:

For any solar collector to operate effectively, it must be placed in a good location. The most important criterion for selecting a site for a solar collector system (such as a Trombe wall) is access to the sun. To insure that the site will have enough solar gain, the following must be determined: 1) where south is, 2) the sun's changing position in the sky during the day and year, and 3) potential shading of the site by existing structures, trees, etc.

The best site is one that faces due south. Some variation is possible, however, in most climates. The sun changes position in the sky both during the day and throughout the year with the winter sun being lower in the sky than the summer sun. In selecting a site for any type of solar collector, one must determine whether nearby buildings or trees will provide unwanted shade and when that shading will occur. A large cottonwood tree that shades part of one's house in the summer may not be a problem in winter when the sun is lower and the leaves are off the tree. On the other hand, a neighbour's house may not shade your house at all in summer, but in winter may block your south wall completely. ***The building should be roughly rectangular, with the longest axis running roughly east to west, such that the largest face points roughly towards "true south". The south-facing side can be up to plus or minus 15 degrees true south without incurring a significant penalty to performance. Large, exposed west-facing walls should especially be avoided.***

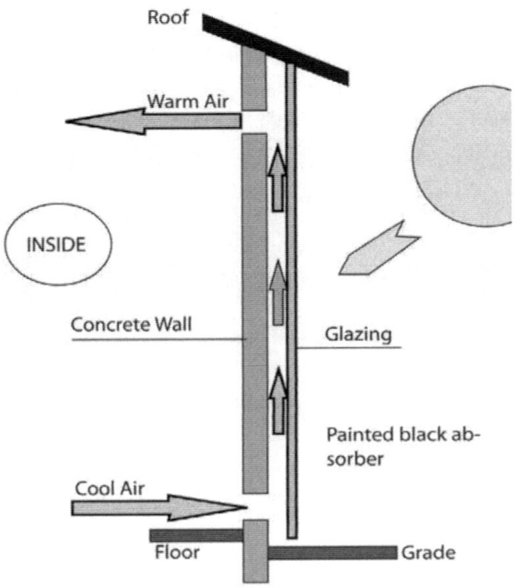

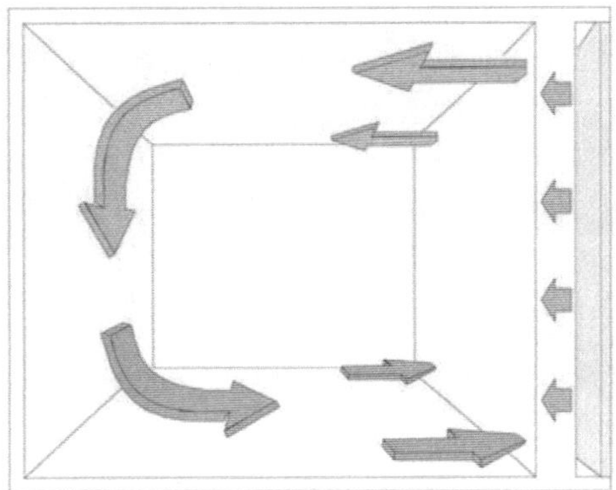

The basic goal of winter solar access is to allow unobstructed sunlight on the south side from 9 am to 3 pm in the winter months.

As a general rule of thumb, this means that obstructions, such as trees, other houses, etc, should be absent from within 60 degrees horizontally from due south from both south corners of the house if possible, and minimally from within 45 degrees.

Trees: Contrary to a persistent but misguided belief, deciduous trees on the south side should be avoided if at all possible. Deciduous trees on the east and west, however, can be very advantageous. They will tend to allow some additional heating from the east and west into late spring (because they haven't yet fully leafed out), when heat is still needed in substantial amounts but southern solar gain is diminished. Likewise, deciduous trees will keep their leaves well into fall, and so help shade the east and western windows at this time when you don't want heating.

SIZING A THERMAL WALL

Heat Balance:

One of the most important parts of planning any heating system for a building is the determination of the size of the system. To do this, we must take into account the following:

1) Weather conditions (temperature and wind)
2) Desired inside temperature;
3) What the building is made of (wood frame, adobe, brick, etc.) and how well it is made;
4) Areas of wall, ceiling, floors, windows, and doors;
5) Volume of each room;
6) The placement of exterior windows and doors, and
7) Heating system specifications.

This information forms the basis of what is called a heat balance which will help us size the heating system of virtually any building.

A heat balance consists of first determining the amount of heat needed to keep the building at a specified temperature (e.g. 65°F for a residential building). This is done by calculating the heat lost from a building directly through its "skin" (walls, ceiling, etc.) and through cracks and vents (infiltration). This heat loss is then balanced by designing a heating system to replace the losses. In the case of a solar heating system this takes into account the available energy from the sun, the orientation of the collector surface to south, the tilt of the collector surface and the efficiency of the system.

Once the total heating load of a building has been calculated, one can determine either: case a) for a given sized Trombe wall, what percentage of the total heating load can be obtained, or case b) how large a Trombe wall is necessary to provide the desired heating demand. There are a number of methods for determining these values and there is working being done to simplify and standardize the calculations.

DESIGNING A THERMAL WALL

Thermal Storage Wall Glazing:

The outer glazing should be durable. It will be directly exposed to weather and harsh ultra-violet rays. But temperatures reached by the outer glazing will not be as high as those that the inner glazing is exposed to. Although glass is the **best** Trombe wall glazing material, various considerations such as cost, ease of handling (especially for a non-professional builder), and possibility of breakage, often make it preferable to use plastic and/or fiberglass glazings.

Polycarbonates, fluorocarbons, and polyvinyl fluorides are all good inner glazing materials in terms of heat resistance. Polyethylene and fiberglass, on the other hand, will degrade much more quickly than their projected lifetimes if used as an inner glazing. Because of fairly good durability properties and low cost "greenhouse quality" fiberglass makes a good outer glazing.

Air Space between Glazing & Mass Wall:

Several factors should be taken into account when deciding on the distance between the inner glazing and the mass wall of a Trombe or thermal storage wall. If the air space is too thin (less than one-half inch/ 32mm), heat loss will be large, and, with vented Trombe walls, air circulation will be restricted - preventing good heat transfer into the building. With a very large air space (greater than about six inches(150 mm), heat losses could be large at the top and sides and the large framing members could cause a shading problem.

Using standard two-by-fours for framing the glazing of a thermal storage wall has proven quite successful (allowing a 3 ½ inch (90mm) air space between the glazing and mass wall) and minimizes lumber expense. If heavy glass is being used for the glazing, sturdier lumber than two-by-fours may be necessary. If unsure about the structural framing requirements for the glazing area of a Trombe wall, it may be worthwhile to contact a local builder, contractor, architect, or engineer.

Venting:

Trombe walls may be of two types: vented or unvented. While an unvented Trombe delivers heat to the inside of a building by conduction and radiation, a vented Trombe additionally heats a building during the day by means of hot air **convection**. The circulation will continue as long as the air in the collector is heated above the room temperature. While the daytime heat is being delivered to the building by means of air convection, the mass wall is steadily being charged with heat which will be radiated into the building at night as with all Trombe walls. In vented Trombes, convection of air provides about 30% of the heat supplied to the living area by the Trombe wall, while conduction and radiation supply the other 70%.

A vented Trombe will result in more temperature variation in building than an unvented. Inside temperatures will get higher during the day and temperatures may drop lower at night with a vented Trombe wall. Overly high inside temperatures can be reduced by providing enough thermal storage materials **inside** the building such as brick floors, concrete or brick partition walls, etc. These materials will absorb extra daytime heat. Thus evening out the daily temperature swing. Vents are generally desirable when heat is most needed during the day - (e.g. in a school or office building).

On the subject of vent sizing and placement, it is necessary to clearly differentiate between two types of vents: inner vents through the mass wall, and outer vents through the glazing. If overheating is expected to be a problem (e.g. in much of the Southwest), outer vents should be included in any Trombe wall design (unless removable insulators are used). If inner vents are **not** being built, both **upper and lower** outer vents would be necessary to adequately vent the collector area during the summer to keep the mass wall from heating up too much. If inner vents **are** being used, only high outer vents would be necessary because the lower inside vents could serve to bring cool air into the collector area.

If too much wall area is taken up by vents, too little heat will go into the wall. If vents are too small, there will be less convection of daytime heat into the room. Several upper and lower vents spaced evenly across the wall will operate more effectively than one large localized vent because the convective airflow will be more even. This is true for both inner and outer vents. For the inner vents; vertically offsetting the vents (so that upper vents are not directly above lower vents) allows for better transfer of heat into the building than if upper vents are directly above lower vents. This is because the flow of air into the building (through upper vents) and out of the building (through lower vents) cannot be as direct, resulting in heat moving further into the room.

Roof Overhang:

The optimal overhang distance can best be calculated using

1) **Sun Angle Chart** which is specific for given latitude,
2) A scale drawing of your Trombe wall house,
3) An architect's scale and
4) A protractor.

From a Sun Angle Chart one can easily find the altitude of the sun (height above the horizon in degrees) at different times of the day and year. With the maximum summer sun altitude and minimum winter sun altitude, one can draw lines through the tip of the overhang at those prescribed degrees using a protractor and ruler and see where the sun will strike the Trombe wall. If a great deal of the wall is exposed to the sun in summer, lengthen or change the angle of the overhang. If too little sun falls on the wall in winter, shorten or tilt up the overhang in the drawing.

Other systems such as movable or adjustable overhangs and "organic shades" (climbing plants, sunflowers, grapes, etc.) can allow more control. Fall is the most difficult season to design for in terms of overheating because of warm temperatures and a fairly low-altitude sun. Spring is a difficult season for heating. Some form of adjustable shading can be very important for these times.

CONCLUSION

The Trombe wall system presents many advantages in comparison with the other passive solar systems, because of the simplicity of its construction, the relative low cost and the significant contribution on heating. Offers heat gains to the inside, from the noon hours (if vents are present) but especially on late evening, night, and the first morning hours (depends from its thickness) when heating is most needed. The differences on the inside temperature are smaller than in the direct gain systems or in a conservatory. Moreover, this system could be applied easily to existing buildings, to the southern concrete walls with small modifications.

The disadvantages of this system are:

The outside surface of the Thermal storage wall is relatively warm. This leads to heat losses, by convection and radiation to the cold night sky and consequently to the reduction of the effectiveness of the system. This could be avoided by providing night insulation or a second glazing.

A Trombe wall practically consists of two separate walls, one from glass and another one from the mass. In that case, the obvious higher cost of that wall in comparison with a typical one is the main disadvantage.

A secondary disadvantage is the space loss on the inside part of the wall that has to be free from furniture or other objects in order to operate effectively.

However, in a building with Thermal walls, it has to be considered from the designer, not only the balance between the heating requirements and the thermal mass but also the demand for outside view and daylight, because the Trombe wall take up a large amount of the external wall (usually the entire wall) and it is by rule "blind" (without windows).

The Trombe wall in particular, has to be designed with the possibility of cleaning the glazing of the wall. In addition problems may occur from the vapour concentration on the glazing and dust accumulation between the thermal mass and the glazing (because of the openings). Last but not least, for the correct operation of the Trombe wall, the contribution of the users/residents is essential. This, often result as the main disadvantage of the system, when the contribution from the owners is incorrect or absent.

REFERENCES

ASHRAE. 2005. ASHRAE Handbook-Fundamentals. American Society of Heating Refrigeration and Air Conditioning Engineers, Inc.

ASTM E891-87. 1987. Standard Tables for Terrestrial Direct Normal Solar Spectral Irradiance for Air Mass 1.5. Philadelphia: American Society for Testing and Materials.

ASTM E903-96. 1996. Standard Test Method for Solar Absorptance, Reflectance, and Transmittance of Materials Using Integrating Spheres. Philadelphia: American Society for Testing and Materials.

Thermal Storage Wall Design, 1979, by Alex Wilson

Green Architecture, Advance Technologies & Materials, by Osman Attmann.2010

HVAC System Optimization and energy savings at Hot Dry Climate Zones in Middle East

Mohie Eldin M. Elsayed Badawi [1]

[1] *Free-lance Mechanical Engineer, Mechanical Consultant*
Chairman of CEC-Cairo Engineering Consultant office
Tel, +2233443353, +201001102238 Email : mohy@kma-dc.com, mohybadawy2002@yahoo.com

ABSTRACT

In approach of optimization the HVAC system to make the system much more efficient and minimizing the initial and running cost, a lot of variables has to be considered during the system design, the building application, the project location and the end users specific requirements are the major concerns to be respected. This article focused in the large chilled water system application that might applicable in the mega projects such as the airports, the industrial hangers, the Holy Haram extensions, assembly areas and all the applications need the large amount of fresh air.

The amount of fresh air should meet the minimum requirements of the IAQ specified in ASHRAE standard 62.1-2010 or the latest published version. The larger the amount of the fresh air the great cooling capacity and electrical power consumption shall be expected.

In deed the fresh air load has the great impact in the thermal load and the cooling capacity of the HVAC system equipments.

In such buildings applications, the power consumption of the HVAC system represents at least 60% of the total connecting power input, the percentage shall increased in the hot climate zones. The water cooled central chilled water system is normally the suitable system for such applications when considering the higher COP of the centrifugal chillers. In hot dry climate zone in US western cities, such as lass Vegas, los angelus, Albuquerque, Tocson, the indirect / direct water evaporation system integrated with the traditional cooling coil is the most energy efficient air cooling system. The engineering technical analysis performed for the cities located in the Middle East such as Riyadh, Madinah in KSA and Aswan in Upper Egypt gives significant energy savings in both of the capital cost and interested benefits in the running savings.

At least 20% saving of the initial cooling capacity and 60% annual savings in the electrical power consumption as resulting from the technical engineering analysis, when considering to provide indirect – direct evaporation sections to the traditional full fresh air handling units, moreover the saving of the makeup water needed for the cooling towers due to the 20% saving in the chillers capacity shall used to the indirect – direct evaporation sections in the air handling units.

The technical analysis performed based on the ASHRAE handbook, fundamentals, system and equipments and code of practice referenced to the brand name manufacturer's in this field.

INTRODUCTION

In the large projects in middle east, where hot climate zones in most of the cities, the selection of the HVAC system have a significant impact on the electrical power and water consumption specifically in the cities have shortage of the natural water sources.

Mohie Eldin M. Elsaed Badawi , Mechanical Consultant , Chairman Of Cairo Engineering Consultant Office, Cairo ,

In some projects, the designers intended to use air cooled water chillers to avoid the large amount of makeup water needed in case of using water cooled water chillers, however dramatically high rate of the power consumption shall expected due to the higher rate of the electrical power consumption per refrigerant ton (Kw/R.T) of the air cooled chillers comparing with the water cooled chillers in additional of the less life cycle and durability.
Thus, in mega projects, the water cooled water chiller system is still the optimized system, the centrifugal chillers that have a best COP in comparing with the other screw, scroll and reciprocating chillers, but for the sustainable buildings it is not only required to concentrate in the energy saving, but also to limit the water consumption as well. In respecting to achieve the two terms, this article investigate the HVAC system optimization in hot dry climate like in Madinah City in KSA in application that consumed large amount of fresh air like in Holy Prophet mosque, The optimized system is to integrate the indirect / direct evaporating systems with the regular central water cooled chilled water system.

SYSTEM DESCRIPTION

The Regular water cooled chilled water system consist of water cooled chiller, cooling tower, chilled water pumps and central full fresh air handling units. The adapted proposed system is the indirect – direct evaporative cooling sections integrated with regular chilled water cooling coil in the full fresh air handling units. IDEC: Indirect Evaporative Cooling, **Figure-1** (a), secondary (scavenger) air stream is cooled by water. The cooled secondary air stream goes through a heat exchanger, where it cools the primary air stream. The cooling primary air stream is circulated by a blower. Indirect evaporative cooling does not add moisture to the primary air stream as there is no direct contact between the air and the water. Both the dry bulb and wet bulb temperatures are reduced, the processes in psychometric is sensible cooling at constant humidity ratio, The heat exchanger is normally stainless steel or polymer tubes, the lower of the air flow velocity through the heat exchanger the higher of the efficiency of the exchanger shall be expected, DEC: Direct Evaporative cooling, **Figure-1** (b), outside air is blown through a water-saturated medium (usually cellulose) and cooled by evaporation. The cooled air is circulated by a blower, the processes in psychometric is sensible cooling and adiabatically heat gain.

(a) (b)

Figure 1 (a) Indirect Evaporative cooling (b) Direct Evaporative cooling

Figure-2 is showing the basics of the process air of the proposed system in psychometric chart, F**igure -2** (a) illustrate the process air when the humidity ratio of the outside air entering condition is lower than the targeted coil leaving condition, the IDEC shall sensible cool the outside air and the DEC section shall adiabatically cool and humidify the IDEC leaving air, the cooling magnitude shall based on the sections effectiveness, in general, 60 % sensible effectiveness for the polymer heat exchanger for the IDEC section , while 80 % saturation effectiveness for the DEC section as an average values specified by ASHRAE, Systems & Equipment, thus the chilled water cooling coil "C. Coil" shall cater the rest of cooling and de-humiliation to achieve the targeted air leaving condition of the air handling unit. When comparing this process with the regular cooling coil without the IDEC-DEC sections, the process shall sensible cooling to the desired leaving dry bulb temperature, and still the wet bulb temperature not yet achieved as the process need kind of humidification, moreover the surplus enthalpy difference due to the pure sensible process is expected.

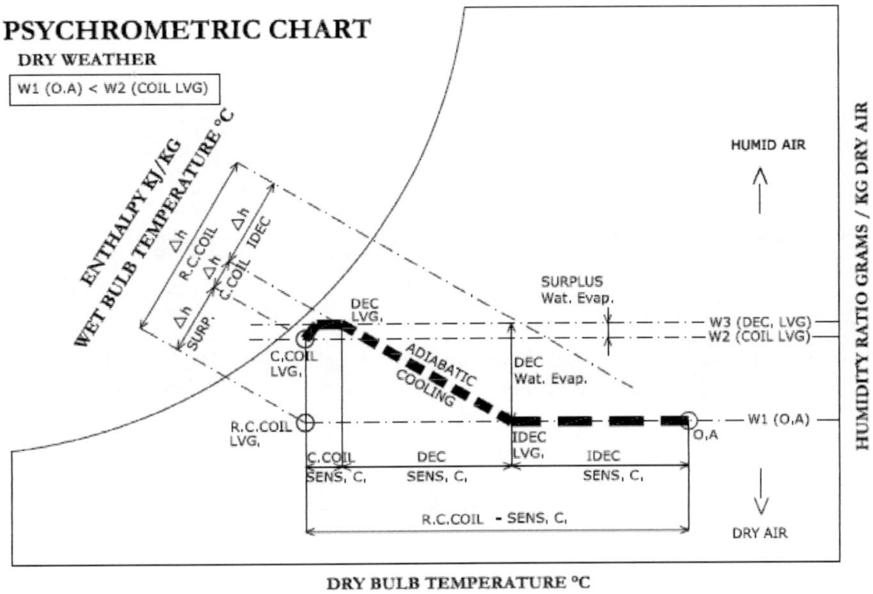

Figure 2 (a) Process air in dry weather

Figure -2 (b) is showing the same but when the humidity ratio of the outside air entering condition is higher than the targeted coil leaving condition, it worth that the DEC section shall not activated as no chance of more humidification founded, only the IDEC section should active. That meaning the portion of the overall process cooling through the IDEC and DEC sections shall depend up on the outdoor design conditions in relation to the requested coil leaving condition.

SYSTEM METHODOLOGY

The evaluation of the energy savings might gained when using both of IDEC and DEC in full fresh air handling unit against the regular full fresh air units, in addition of the consideration of the water evaporation consumption, it shall applied on MADINAH city in KSA, where considering 472 LPS (1000 CFM) sample full fresh air handling unit

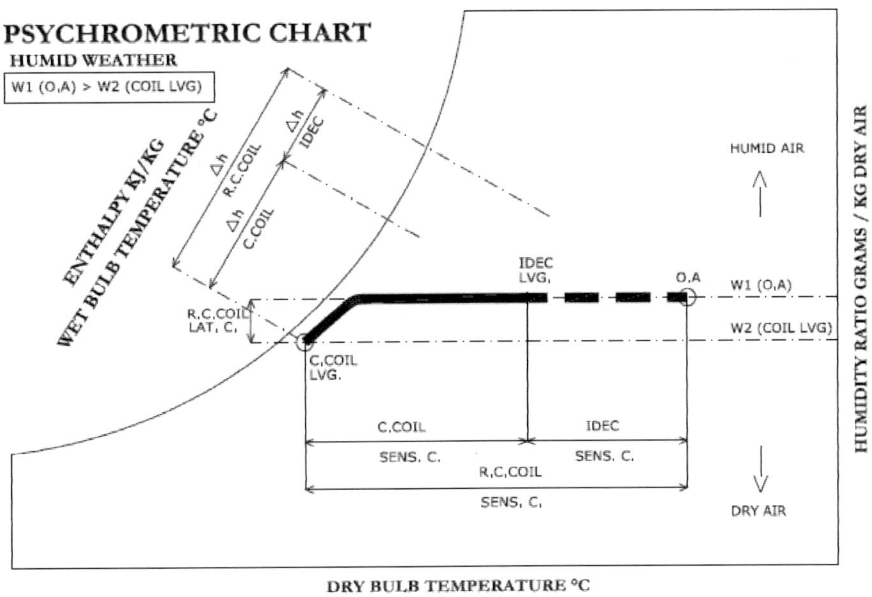

Figure 2 (b) Process air in humid weather

capacity to brought the outside air cooled to the design supply air temperature or leaving coil conditions at 13.00 / 12.50 (DB / WB).

The study shall investigate the cooling and evaporation rate in annual basics where considering the fluctuation of the outdoor conditions in the different months, in reference to ASHRAE Climate Conditions –Madinah City – KSA, the study will considering two scenarios, first scenario is based on the worst monthly Dry bulb with Mean Coincident Wet Bulb temperature, (DB-MCWB), while the second scenario is based on the worst monthly Wet Bulb with Mean Coincident Dry Bulb temperature, (WB-MCDB). The average results of the two scenarios represent the average cooling load required and reflecting what could be gained when applying the proposed IDEC-DEC - HVAC system.

Twelve numbers of graphs plotted in the ASHRAE Psychometric chart for scenario-1 and similar graphs for scenario-2, two samples are illustrated in **Figure-3** (a) Illustrate the cooling process on the Psychometric chart in July at DB-MCWB. While as a sample plot graph for scenario-1, while **Figure-3** (b), illustrates the cooling process in August at WB-MCDB as a sample plot graph but for scenario-2.

In **Figure-3** (a) , IDEC-DEC-CC – AHU arrangement process described as in the IDEC: The intake outside air at (46.1/18.9) sensibly cooled through the IDEC heat exchanger to (29.8/13.0) based on 60% efficiency. (9.4 Kw sensible cooling effect), the water evaporation through the scavenger side is 17.6 l/hr. DEC: The IDEC leaving air cooled through the DEC in adiabatically cooled process at constant wet bulb temperature to (14.3/13.0) based on 80% saturation effectiveness, 8.9 Kw sensible cooling effect. The water evaporation in the wet deck is 13 l/hr. C.C: The cooling coil shall cool the DEC leaving air to the design leaving temperature (13.0/12.5). (0.8 Kw sensible cooling effect). R.C.C.: The intake air at (46.1/18.9) shall directly cool to the desired leaving condition (13.0/5.4). (19.1 Kw sensible cooling effect)

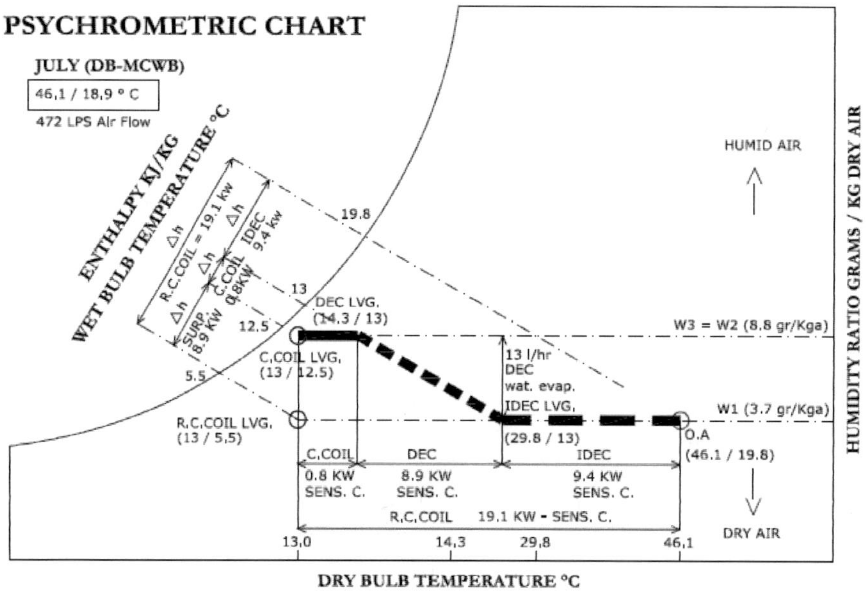

Figure 3 (a) Process air in July (DB-MCWB), Senario-1

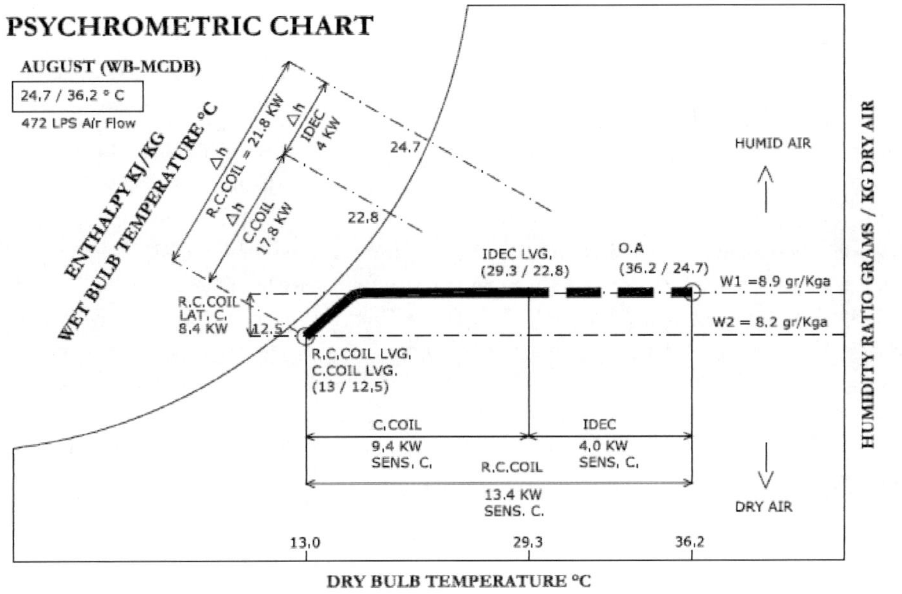

Figure 3 (b) Process air in July (DB-MCWB) , Senario-2

In-**Figure3** (b), IDEC-DEC-CC – AHU arrangement process described as the followings: IDEC: The intake air at (36.2/24.7) sensibly cooled through the IDEC heat exchanger to (29.3/22.8) based on 60% efficiency. (3.9 Kw sensible cooling effect), the water evaporation through the scavenger side is 8 l/hr. DEC: The air passed through the DEC section without activation of the water nozzles as the level of the desired humidity ratio is lower than the inlet air level. (No cooling effect). C.C: The cooling coil shall cool and dehumidify the DEC leaving air to the design leaving temperature (13.0/12.5). (9.4 Kw sensible cooling effect and 18.0 Kw total cooling).R.C.C.: The intake air at (36.2/2.7) shall directly cool and dehumidify the air to the desired leaving condition (13.0/12.5). (13.4 Kw sensible cooling effect and 21.8 Kw total cooling).

Table. 1 Monthly Comparison between IDEC-DEC C.Coil and Regular C.Coil Arrangement for Scenario -1

	IDEC-DEC Integrated with C.C					Regular C.C		IDEC-DEC	
	IDEC		DEC		C. Coil	Regular Cooling		Total	Water
Month wb-mcdb	Sensible cooling (kw)	Evapo-ration (l/hr)	Sensible Cooling (kw)	Evapo-ration (l/hr)	Total (kw)	Sensible (kw)	Total (kw)	Cooling (kw)	Evapo-ration (l/hr)
January	2.1	4.1			6.9	6.5	9	2.1	4.1
February	3.1	6.2			6.5	8.5	9.6	3.1	6.2
March	3.4	6.9			8	9.6	11.4	3.4	6.9
April	3.6	7.2			9.7	10.5	13.3	3.6	7.2
May	4.8	9.6			9.6	12.8	14.4	4.8	9.6
June	6.6	13.2	2.5	3.7	6.2	15.3	15.4	9.1	16.9
July	5.2	10.4			10.6	13.9	15.8	5.2	10.4
August	3.9	8			18	13.4	21.8	3.9	8
September	5.3	10.7			9.4	13.8	14.8	5.3	10.7
October	3.2	6.4			10.3	9.8	13.5	3.2	6.4
November	2.7	5.4			8.9	8.4	11.7	2.7	5.4
December	2.5	5			7.3	7.5	9.8	2.5	5
Monthly Average	**3.9**	**7.8**	**2.5**	**3.7**	**9.3**	**10.8**	**13.4**	**4.1**	**8.1**

SYSTEM ANALYSES

The system analysis performed based up on the information of the **Table-1**, which represent the scenario -1, where comparing between the proposed IDEC-DEC integrated with C.C against the R.C.C arrangement, the sensible, total cooling and the water evaporation per each section where applicable, monthly average values per year is including in the last raw of the table. **Table-2**, represent the same information but for scenario -2. **Table-3**, represent the average values between scenario-1 and scenario-2. The summary of the extracted information is summarized as the maximum cooling capacity that could be installed is 18 Kw at August month in the first scenario, while the Regular cooling coil is 21.8 Kw at the same month, that meaning about 18 % less cooling capacity of the capital equipment shall be expected when using IDEC-DEC – CC arrangement. The average monthly cooling capacity consumed in the C.C is 5.1 Kw for the two scenarios, against 14.1 Kw in the R.C.C arrangement, which is represent 64 % average saving in the running cooling capacity. The average monthly of the water evaporation consumed in the IDE-DEC sections shall be 16.3 l/hr to introduce 9.1 Kw sensible cooling capacity, which represent 6.2 l/hr per refrigerant ton or per 3.5 KW cooling energy. The rule of thumb for the cooling tower water evaporation rate is represent about 0.01 GPM / Condenser Water Flow, if we are considering 3 GPM / R.T standard condensing water flow per R.T, accordingly 0.03 GPM / R.T water evaporation rate or 6.8 l/hr per R.T cooling or 1.94 l/hr per Kw cooling shall be respected.

Considering our project have 3500 kw cooling load and it shall utilized water cooled water chiller system, and full fresh air handling units, the average of the monthly cooling capacity is 14.1/21.8 x 3500 = 2264 kw, the cooling tower make up water evaporation shall be 2264x1.94 = **4390 l/hr,** When applying the IDEC-DEC integrated with cooling coil, the chilled water system that shall be installed will be 0.18 x 3500 kw = 2870 KW, the IDEC-DEC section shall cater the rest of cooling is 630 kw, the monthly cooling capacity of the cooling coil is 5.1/18 x 2870 = 813 kw, the makeup water evaporation shall be 813x1.94 = 1577 l/hr, the IDEC-DEC monthly cooling capacity shall be 2264 – 813 = 1451 kw, the water evaporation shall be 1451 x 1.78 = 2582 l/hr , the total water evaporation shall be 1577 + 2582 = **4160 l/hr** almost the same consumed water evaporation.

Table. 2 Monthly Comparisons between IDEC-DEC C. Coil and Regular C. Coil Arrangement for Scenario -2

	IDEC-DEC Integrated with C.C					Regular C.C		IDEC-DEC	
	IDEC		DEC		C. Coil	Regular Cooling		Total	Water
Month	Sensible cooling	Evapo-ration	Sensible Cooling	Evapo-ration	Total	Sensible	Total	Cooling	Evapo-ration
wb-mcdb	(kw)	(l/hr)	(kw)	(l/hr)	(kw)	(kw)	(kw)	(kw)	(l/hr)
January	4.6	9	4.7	6.9		9.3	9.2	9.3	15.9
February	5.3	10.5	4.9	7.2	0.7	10.9	11	10.2	17.7
March	6.2	12.1	5.7	8.3	0.8	12.7	12.7	11.9	20.4
April	7.8	15.3	7	10.3	0.6	15.5	15.5	14.8	25.6
May	8.7	17	8.7	12.8		17.4	17.4	17.4	29.8
June	9.2	18.1	9.5	13.9		18.7	18.5	18.7	32
July	9.4	18.3	8.9	13	0.8	19.1	19.1	18.3	31.3
August	9	17.8	7.2	10.6	2.7	18.9	19.1	16.2	28.4
September	8.8	17.3	7.9	11.6	1.3	18.1	18.1	16.7	28.9
October	7.9	15.5	7.2	10.5	0.4	15.5	15.6	15.1	26
November	6.2	12.2	6.3	9.3		12.5	12.4	12.5	21.5
December	5.2	10.2	4.4	6.4	0.2	9.8	9.9	9.6	16.6
Monthly Average	7.4	14.4	6.9	10.1	1	14.9	14.9	14.2	24.5

Power consumption

The water cooled water chiller system consume 1 kw electrical power / 3.5 Kw cooling, The power plant shall be 1.0 MW , annual power consumption shall be 2264 x 12 x 10 x 30 / 3.5 = 2.3 MW, when applying the IDEC-DEC integration system with the cooling coil, the power plant shall be 0.8 MW, the annual power consumption shall be 813 x 12 x 10 x 30 /3.5 = 0.84 MW, in

Table.3 Average Rates Per Month for the Two scenarios

	IDEC-DEC Integrated with C.C					Regular C.C		IDEC-DEC	
	IDEC		DEC		C. Coil	Regular Cooling		Total	Water
Month wb-mcdb	Sensible cooling (kw)	Evapo-ration (l/hr)	Sensible Cooling (kw)	Evapo-ration (l/hr)	Total (kw)	Sensible (kw)	Total (kw)	Cooling (kw)	Evapo-ration (l/hr)
Monthly Average Two Scenarios	5.6	11.1	4.7	6.9	5.1	12.8	14.1	9.1	16.3

addition to the fan power consumption of the IDEC-DEC section, about 0.3 kw fan power consumption per ach 472 LPS full fresh air, 3500 / 21.8 x 472 = 75780 LPS the power consumption shall be 0.3 x 75780 / 472 = 48 kw. The total power consumption shall be 0.884 Mw, thus the annual power saving percentage is about **61%.**

CONCLUSION

The indirect direct evaporative section integrated with the cooling coil system for full fresh air applications has major cooling saving reach to 60 to 70 % comparing with the traditional full fresh air units in hot dry climate zones. For the project utilized water cooled water chillers and cooling towers, the optimization of the fresh air handling units by integrating the indirect direct evaporative cooling section shall lead to decrease the installed capacity of the chilled water system equipments by not less than 18 %, the saving in the makeup water evaporation through the cooling tower shall provided to the indirect – direct evaporative cooling section. The saving in the electrical power consumption shall expect to be 61% comparing with the regular system. The optimized system is not only applied in the hot – dry climate, it is valuable, the designers to investigate and make similar analysis in semi-hot dry climate, it might gain some benefits but with less saving impact. The indirect direct evaporative section integrated with the cooling coil air handling unit is bigger size than the regular AHU units and required almost double foot print space. The selection should be sized at the worst outdoor wet bulb temperature as specified in the ANSI/ASHRAE/IES Standard 90.1-2010. The Evaporative cooling wet deck does not provide suitable growth conditions for the bacteria like occurred in cooling towers and generally do not release an aerosol. A good maintenance program eliminates potential microbial problems and reduces the concern for disease transmittal (ASHRAE 1998, 2000; Puckorius et al. 1995). The designers has to investigate the project local market to find out the suppliers of such proposed system as it has no wide range of manufacturer's in middle east employed with, the import products available in US manufacturer's such as Munter's, Energy LAB and Speck air.

NOMENCLATURE

CC : Cooling Coil integrated with IDEC-DEC section.
CFM: Cubic Feet Per Minute
CHW: Chilled Water
COP : Coefficient Of Performance
DBT : Dry bulb Temperature
DEC : Direct Evaporative Cooling
Evap : Water Evaporation
FFA: Full Fresh Air

IAQ : Indoor Air Quality
IDEC : Indirect Evaporative Cooling
KSA : King Dom Of Saudi Arabia
Kw : Kilo Watt
LPS : Liter Per Second
LVG: Leaving Air
MCDB: Mean Coincident Dry Bulb Temperature
MCWB: Mean Coincident Wet Bulb Temperature
O.A: Outside Air
RCC : Regular Cooling Coil Arrangement.
R.T : Refrigerant Ton
WBT: Wet Bulb Temperature
W: Humidity Ratio
Δ h : Enthalpy Difference

REFERENCES

- 2005 ASHRAE – Handbook Fundamentals. -ANSI/ASHRAE/IES Standard 90.1.
- 2010 . Energy Standard for Buildings except Low-Rise Residential Buildings
- ANSI/ASHRAE/IES Standard 62.1-2007 Ventilation for Acceptable Indoor Air Quality.
- 2008 ASHRAE Handbook—HVAC Systems and Equipment Evaporative Air Cooling Equipment.
- ASHRAE. 1998. Legionellosis: Position statement.
- ASHRAE. 2000. Minimizing the risk of Legionnaires' disease.-ASHRAE. 2000. Minimizing the risk of legionellosis associated with building water systems, Guideline 12-2000.
- US Munters, Corporations Products. http://www.Munters.com
- US Energy Labs Inc., products. http://www.EnergyLabs.com

Impact and Feasibility Analysis of an Alternate Illumination Strategy to Cut-Down on Energy Loads

Gaurav Chaudhary
Student Member ASHRAE

Tanmay Jain
Student Member ASHRAE

ABSTRACT

Recent studies have found that prolonged exposure to artificial lighting can be damaging to one's physical and psychological health. Whereas providing natural daylight can increase productivity up to 10%.These benefits are more noticeable in official, retail and educational buildings. This paper deals with improving illumination in our work spaces, specially targeting office and educational spaces where illumination lux is as high as 450 and reducing electricity consumption of the building through the use of central illumination in the roof cavity. The developed retrofit would be applicable only in 12-hr working building such as retail, offices, factories and educational spaces. In all these buildings recommended illumination lux is of range 300450.

Retrofit would be core right through the roof of the building and it would be like a hollow cuboid with its two sides made of a special translucent material and the other two sides concave mirrors facing inside the core. The translucent material opted is polycarbonate after considering all its properties like durability over a long period of time and that it is a natural ultraviolet filter.

Experimental outcomes were taken from a miniature form of a scaled down actual building. The readings were taken using a well calibrated lux meter. The movement of sun was replicated using a Ring Heliodon. By doing the feasibility analysis of the device it was found that this device can work on par with other day lighting techniques. The payback period of this device was around 5 yrs.

INTRODUCTION

Daylight systems collect natural light and deliver deep into the heart of buildings. They use collectors in the roof to harvest light, and then transport it to diffusers into interior spaces. The quality of this natural light tends to be alive and vibrant, and diffusers can be placed strategically to allow the best distribution of light within the interior.

There are three main types of day lighting system:

- Tubular daylight devices.
- Vertical systems.
- Horizontal systems.

TUBULAR DAYLIGHT DEVICES

Also called TDDs, tubular daylight devices consist of a fixed ocular that collects light from the roof of a building and

Gaurav Chaudhary is a student of B.Arch in the Department of Architecture and Planning , Indian Institute of Technology, Roorkee, India
Tanmay Jain is a student of B.Tech in the Department of Mechanical and Industrial Engineering , Indian Institute of Technology, Roorkee, India

directs it into a tubular 'pipe' lined with a highly-reflective surface. The light is reflected down the tube by this surface to diffusers in the interior of the building that direct it into occupied spaces.

VERTICAL SYSTEMS

Vertical systems are very similar to TDDs, but they use a powered tracking system to point a light 'collector' towards the sun, and a series of mirrors and lenses that concentrate the light before directing it into distributing tubes.

HORIZONTAL SYSTEMS

Horizontal systems use lenses to collect natural light from external walls and transport it through flat light ducts above the ceiling to diffusers placed deep inside the building. They make use of a polymer material that combines a high reflectivity with the practicality of a hollow duct similar to a ventilation duct. Horizontal systems are a fairly recent development and as such are relatively untested and expensive.

In all of the above design light is captured from the roof and transferred to target location via internal reflections and then through translucent surface through diffusion. In this paper a vertical straight light shaft is developed and tested in which the light is diffused and internally reflected all along its way.

ABOUT THE IDEA

WORKING

The shaft would be right through the roof of the building and it would be a hollow cuboid with its two sides made of a special translucent material and the other two sides concave mirrors facing inside the core. Sunlight would enter the core from top, i.e. from the roof, get reflected from the concave mirrors and hence after multiple reflections, diffuse evenly from the translucent sides inside the work spaces of the building.

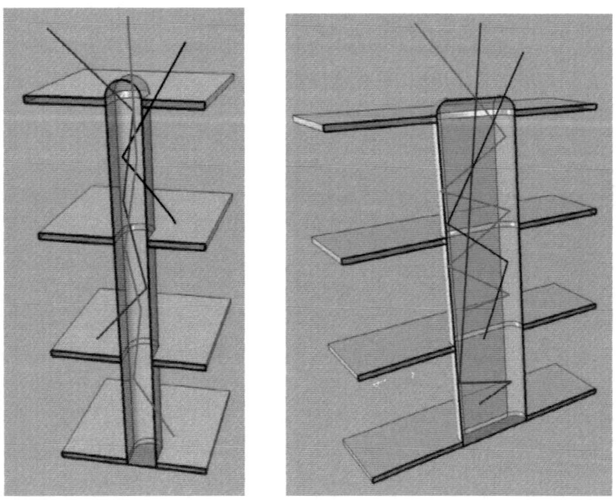

Figure 1.1: Sectional views showing the internal reflection

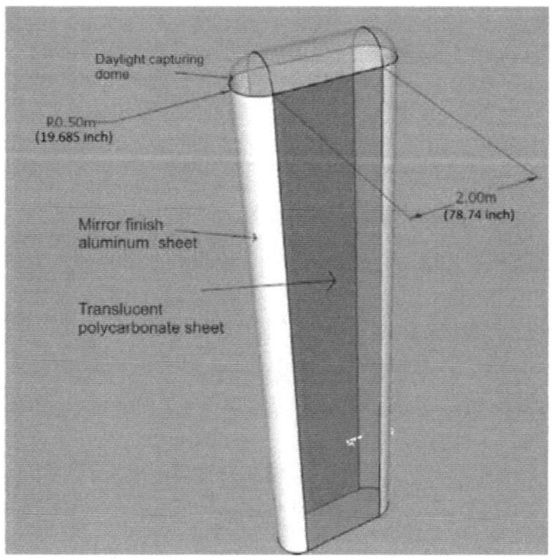

Figure 1.2: Materials and dimensions

MATERIALS

Advanced materials are used to give highest efficiency.

1. A daylight-capturing dome lens that: redirects low-angle sunlight for maximum light capture, rejects overpowering summer midday sunlight, and provides consistent day lighting throughout the day.
2. Tubing made of the mirror finish aluminum sheet.

3. Translucent Polycarbonate Sheets: Also known as: structural polycarbonate, polycarbonate channel, hollow polycarbonate. The air contained in the voids between the layers of boards – "honeycombs", provides its high insulation properties, and the ribs – great structural strength of polycarbonate. This sheet is resistant to UV radiation, has good insulating properties, high light transmission and impact resistance, and exceptional insulating properties.

PHYSICAL EXPERIMENT

EXPERIMENT SET-UP

The experimental study was carried out both in a laboratory and in bright afternoon sun to examine the behavior of light through the light shaft. The experimental model was a 3 storey building of room size 5m*10m (196.85 inch*393.7 inch) scaled down to 1:25.

In place of concave mirrors aluminum foil of reflectance 88% was used while in place of polycarbonate sheets translucent sheets were used. Building model was made from mount board of thickness 2mm.

EXPERIMENT PROCEDURE

The whole setup was first put under a ring heliodon and then under sun. With the help of a lux meter 3 lux value were noted down i.e. lux outside the model, lux on the 2nd floor and lux on the 1st floor.

ECOTECT STIMULATION

To get the average and accurate lux value model was also made on **Autodesk®** Ecotect® **Analysis software** and stimulation was done for 2nd and 1st floor.

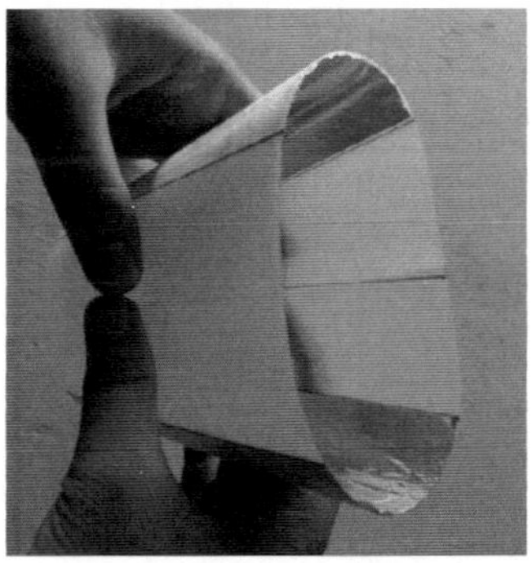

Figures 2: Model of shaft

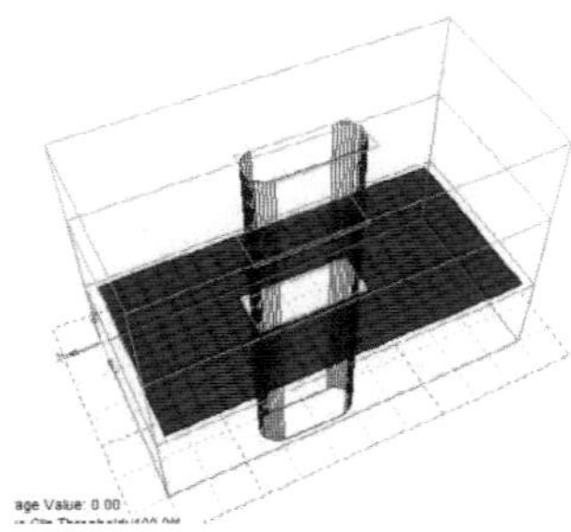

Figures 3: Lighting Analysis window in Ecotect

ECOTECT STIMULATION

1. UNDER RING HELIODON

Figures 4: Light analysis under heliodon

Figures 5: Model on open roof under bright afternoon sun

This experiment was conducted under ring heliodon. Lux meter's sensor was put inside the model and readings were noted down.

Table 1. Reading under ring heliodon

Serial No.	Floor	Lux Value
1	1st floor	315.2 lux
2	2nd floor	395.8 lux

2. UNDER SUN

At around 3 PM the model was taken in the sun and readings were again taken.

Table 2. Reading under sun

Serial No.	Floor	Lux Value
1	1st floor	320.7 lux
2	2nd floor	400.5 lux

3. IN ECOTECT

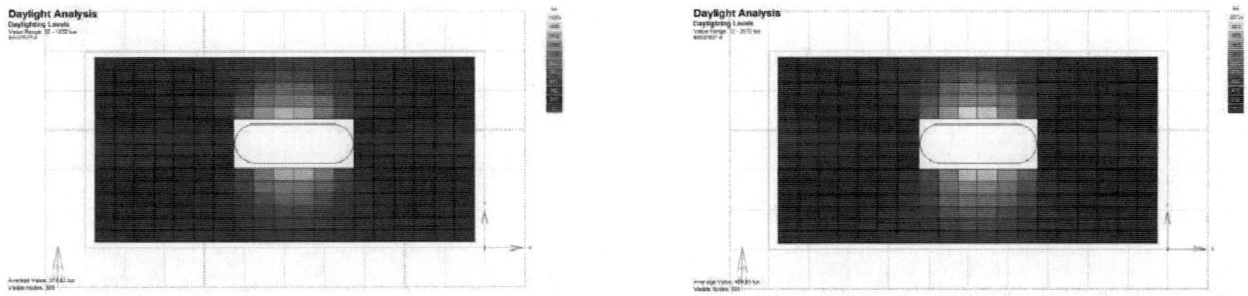

(a) (b)

Figures 6: Light analysis on Ecotect (a) First floor (b) Second floor

Table 3. Reading in Ecotect

Serial No.	Floor	Lux Value
1	1st floor	374.63 lux
2	2nd floor	484.93 lux

As output values from Ecotect were average while those from physical model were at single point so we will consider these values for further calculations.
So this shaft would give an approx. of 320-480 lux on normal day.

FINANCIAL FEASIBILITY

Under this we will calculate how much electricity the shaft would be able to save and that in turn how much money would be saved. After that taking an approx. cost of the setup and comparing it with the money saved payback period of the setup would also be calculated.

Step 1: Converting lux into lumens [3]

The illuminance E_v in lux (lx) is equal to the luminous flux Φ_V in lumens (lm) divided by the surface area A in square meters (m^2):

$$E_{v(lx)} = \Phi_{V(lm)} / A_{(m^2)}$$

lux = lumens / (square meters) lumens = lux *(square meters) lm = lx * m^2

Luminous surface area per floor(3.5 mtrs) = (3.5m * 2)*2 = 14 m^2 or 150.69ft^2

So lumens per floor = 420 lux * 14 m^2 =5880 lm

Step 2: Converting lumens into watt

The power P in watts (W) is equal to the luminous flux Φ_V in lumens (lm), divided by the luminous efficacy η in lumens per watt (lm/W):

$$P_{(W)} = \Phi_{V(lm)} / \eta_{(lm/W)}$$

watts = lumens / (lumens per watt) W = lm / (lm/W)

Energy saving lamps have high luminous efficacy (more lumens per watt). In offices and schools where reading and writing is the major activity, fluorescent light bulbs are used so we will use value of fluorescent lamp from the above table to move forward in our calculations.

Compact Fluorescent lamp = 60 lm/W Total lumens = 5880 lm

So total power per floor = 5880 W/ 60 lm/w = 98 W

Step 3: Converting watt into no. of electricity units

Unit of electricity is kWh i.e. kilowatt hour. To calculate the amount of electricity saved we will take power in watts (W) and multiply it with number of hours (hr) for which shaft would be lit and divide by 1000 to.

$$E_{(kWh)} = P_{(W)} \times t_{(hr)} \;/\; 1000$$

So total electricity per floor = 98 W/ 1000 = 0.098 kWh

Step 4: Cost of electricity units

INDIA- \$0.088/KWH(Approximate cost)

So in a day total money the shaft would be saving

Table 5: Total money recovered per day

Country	No. of hours (a)	Units of electricity (kWh) (b)	Cost per kWh (c)	Total money recovered = a*b*c
India	12hr	0.098 kWh	\$0.088	\$0.103

COST OF SETUP

We will calculate the **approximate** cost of shaft if it is installed in one floor of height 3.5 mtrs.

Total cost of the shaft = Total Cost of polycarbonate sheet + Total Cost of reflective surface + LightTracker™ Reflector + Misc. costs [4]

1. Approximate Cost of polycarbonate sheet = US \$5 / Square Meter

   Total area of sheet per floor = 3.5 * 2 = 7 m^2 or 75.347ft^2

   Total Cost of polycarbonate sheet = \$5/m^2 * 7m^2 = **\$ 35**
2. Approximate Cost of high grade mirror finish aluminum sheet = US \$6.5 / Square Meter

   Total area of sheet per floor = (3.5 * π * 1) * 2 = ~22 m^2 or 236.81ft^2

   Total Cost of polycarbonate sheet = \$6/m^2 * 22m^2 = **\$ 132**
3. Cost of LightTracker™ Reflector= **\$15**
4. Misc. cost would include fixation charges which would include cost of nuts and bolts and screws= **\$ 5**

   Total cost= \$35 + \$132 + \$15 + \$5 = \$ 187

PAYBACK PERIOD

Total cost of setup per floor = \$187(Approximate cost)

Table 7: Payback period

Country	Cost of setup per floor (a)	Total money recovered per day (b)	Payback period (days) = a/b	Number of years
India	\$187	\$0.103	1815	5yrs

EXPERIMENTAL UNCERTAINITY

Because the experiment was not done using the exact same materials the device would be finally made up of, there would be some uncertainty in the above outcomes and that from the real scale device. Following are the reasons uncertainty errors would be seen.

1. Instead of mirror finish aluminum, 13 μm (0.5 mils) aluminum foil was used. Uncertainty would come because reflectivity of mirror finish aluminum used would be 95-99% and that of aluminum foil used while doing experiment was 88%.
2. While doing experiment light capturing dome was not used. But in the real device it would be used and hence more sunlight would enter the device thereby increasing the efficiency.

CONCLUSION

At the end of all tests and stimulation it was found out that along with a little amount of electrical illumination this device can illuminate an office or a classroom effectively for reading and writing. It can easily provide more than 400 lux on any given normal day.

By doing the feasibility analysis of the device it was found that this device can work on par with other daylighting techniques. The payback period of this device was around 5 yrs.

Further as this device is an all new idea more research is possible and it can be developed more to make it more effective and feasible.

ACKNOWLEDGMENTS

We would like to express our very great appreciation to Dr. Mahua Mukherjee, Associate Professor, Department of Architecture and Planning, IIT Roorkee for her valuable and constructive suggestions during the planning and development of this research work. Her willingness to give her time and lab equipments so generously has been very much appreciated.

REFERENCES

Energy and Work Conversion – Unit Converter: 10 April 2014<http://www.unitconversion.org/unit_converter>

M.M. Aman, G.B. Jasmon, H. Mokhlis, A.H.A. Bakar, 2013. Analysis of the performance of domestic lighting lamps. Energy Policy, 52: 482-500

Pricing and materials: 1 July 2014 <http://www.tradeindia.com>

SP 41: Handbook on Functional Requirements of Buildings (Other than Industrial Buildings) (1987), Bureau of Indian Standards: Recommended values of Illumination, pp.112

Improving Performance of a Variable Speed Refrigeration System Using Smart Control Strategy

M. H. Salama, PhD
Member ASHRAE

N. A. Mohamed, PhD

N. A. Mahmoud, PhD

ABSTRACT:

A refrigeration system was used in cooling and freezing applications controlled by a smart control strategy. This was done using an inverter driven scroll compressor, an inverter driven evaporator fan motors, an inverter driven condenser water pump and an electronic expansion valve. The smart control strategy depends on different controllers; it controls the compressor by artificial neural networks (ANN), the evaporator fan motors by proportional controller, the condenser water pump and the electronic expansion valve by proportional integral differential (PID) controller. Controlling the evaporator fan motors speed was done to maintain a constant supply air temperature, while controlling the condenser water pump speed was done to maintain a constant condensing pressure. The compressor speed was controlled using artificial neural networks (ANN) based on the analyzed data from a preceding experimental results. It was found that operating the refrigeration system using smart control strategy to control the motors speeds improved the performance of the refrigeration system compared to traditional on/off control strategy. Using smart control strategy showed a reduction in the starting operation time compared to the traditional on/off control strategy by 55% in order to achieve the desired room temperature set value. The reduction in operation time to achieve the desired room temperature set value results in total energy saving 60% by using smart control strategy compared to traditional on/off control strategy. The energy saving during system operation in the starting operation period is found to be 11.3%, while the energy saving during the steady state operation time is 11%.

1. INTRODUCTION

Traditional on/off control strategy is one of the widely used and mostly spread refrigeration system control. Many new control strategies were invented and showed better performance than the traditional on/off control strategy. Thambanchacheep and Kuntanapreeda, implemented an artificial neural network (ANN) controller on a split type air conditioning. They found that ANN control system consumes less energy than that of the on/off control's by 17%. Ekren et al, studied the effect of three different control methods on a variable speed compressor and electronic expansion valve in an air cooled chiller system. They used proportional integral differential, fuzzy logic and ANN controllers. They showed that ANN controller achieved minimum power consumption 8.1% and 6.6% lower than both PID and Fuzzy controllers, respectively. Lazzarin and Noro, used a PID controller to control an electronic expansion valve in an air conditioner. They obtained an increase in the

M. H. Salama is a head of refrigeration department, Carrier, 6th October, Egypt. N. A. Mohamed is a lecturer in mechanical power department, faculty of engineering, Ain Shams University, Cairo, Egypt. N. A. Mahmoud, is a professor in mechanical power department, faculty of engineering, Ain Shams University, Cairo, Egypt.

EER range from 8.7% to 22% in high and medium ambient temperature, respectively, when using the electronic expansion valve with PID control instead of thermostatic expansion valve. Marcinichen et al, used a dual single input single output (SISO) controller using two individual PI controllers to control an expansion valve and a reciprocating compressor speed in a water cooled chiller. They showed that the EER reached a maximum value of 7 (COP: 2.08) when the superheat reference values were within the range of 42.8 °F (6 °C) to 46.4 °F (8 °C). Hua et al [e], used a PI controller to control an electronic expansion valve and a variable speed reciprocating compressor, they found that the independent control of electronic expansion valve from compressor control can guarantee not only precise control performance but also high COP for the variable speed operation. Benamer and Clodic, compared the performance of a fixed speed scroll compressor and a variable speed one fitted in a water chiller. They indicated that at 20% part load, the power consumption is reduced by 40% when using a variable speed compressor, while at full load, the power consumption is the same for both fixed and variable speed compressors. In this research, the objective is to maximize the energy saving using a combination of smart control strategies to control variable speed motors and an electronic expansion valve in a direct expansion refrigeration system using refrigerant R-507.

2. EXPERIMENTAL TEST RIG

The experimental test rig is mainly composed of refrigerant side, an air side and a condensing water side. A schematic diagram of the experimental test rig is shown in figure 1.

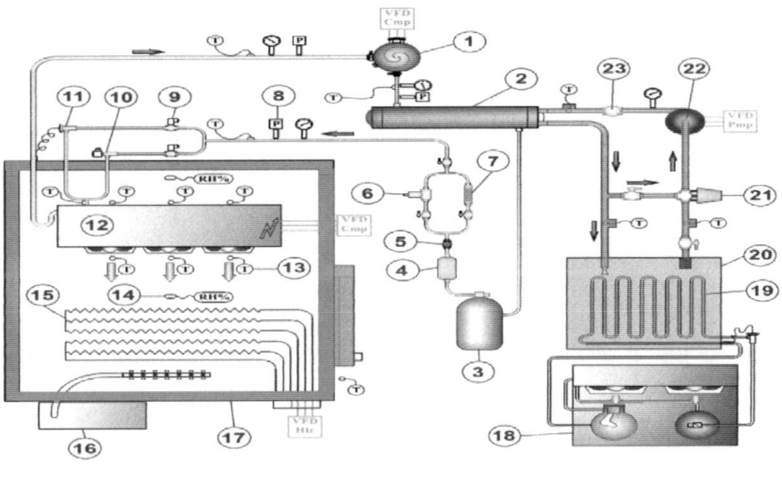

Figure (1) Schematic diagram of the experimental test rig, Salama.

1- Scroll compressor.	2- Water cooled condenser.	3- Liquid receiver.
4- Filter Drier.	5- Sight Glass.	6- Turbine flow-meter.
7- Rotameter.	8- Pressure transducer.	9- Solenoid valve.
10- Electronic expansion valve.	11- Thermostatic expansion valve.	12- Evaporator.
13- Temperature sensor.	14- Humidity sensor.	15- Electric heaters.
16- Humidifier.	17- Insulated psychrometric room.	18- Air cooled condensing unit.
19- Evaporator coil.	20- Condenser water tank.	21- Three way valve.
22- Condenser water pump.	23- Water turbine flow-meter.	

The figure illustrates major components of the refrigerant side, which are: a scroll compressor (1) Model ZF11K4E-TFD made by Copeland, an electronic expansion valve (10) model E2V-18 made by Carel, a fin and tube evaporator (12) model CAB3367 made by Profroid, and a water cooled shell & tube condenser (2) model CR-14 made by Onda. The air side consists of a prefabricated cold storage room (17) which is thermally insulated with 10 cm polyurethane foam, a fin and tube evaporator (12), electric heaters (15) acts as an artificial sensible heat load, a steam humidifier (16) made by Nordmann acts as an artificial latent heat load, humidity sensors (14) made by Sauter for relative humidity measurements and IC temperature sensors (13) made by TSic. Pressure measurements were done by means of pressure transducers (8) made by Omega.

The condensing water side consists of the shell and tube condenser (2), a water tank (20) with capacity 1 m^3 fitted with an immersed cooling coil (19), a recirculation water pump (22) model CDA-1.50-M, made by Ebara, a turbine flow meter model FTB-4806, made by Omega. An air cooled condensing unit (18) made by Bock; used for cooling the water in the water tank (20). A three way modulating valve (21), model VB-3000, made by Nanjing Acmelec Electricis, used to mix the condenser hot water outlet with the cold water from the tank to precisely adjust the condenser water inlet temperature. Four inverters were used to drive the compressor, the evaporator fan motors, the condenser water pump and the electric heaters. These inverters were also used as watt meters.

3. SMART CONTROL STRATEGY

Similar to the biological neural cell, the artificial neural networks (ANN) unit of structure is the neuron which can be classified into some as inputs, which receive signals from the outside world, some as outputs, which delivers signals to the outside world and others are considered computational neurons in a hidden layer. The most common type of ANN in control applications is the feed forward artificial neural network (FFANN) which consists of a group of interconnected neurons organized in layers: input layer, hidden layer and output layer, where each layer consists of a group of neurons. The feed forward propagates signals only in a forward direction from the input nodes to the output nodes.

A commonly used error function tries to minimize the average error between the network's output and the target value is called the back propagation. Using feed forward artificial neural network with back propagation is called the recurrent neural network (RNN) which differs from the feed forward structure by having feedback connections which propagate the outputs of some neurons back to the inputs of other neurons in order to perform repeated computations on the signal. Thus, the chosen control algorithm in controlling the compressor speed in this study is the recurrent neural networks (RNN).

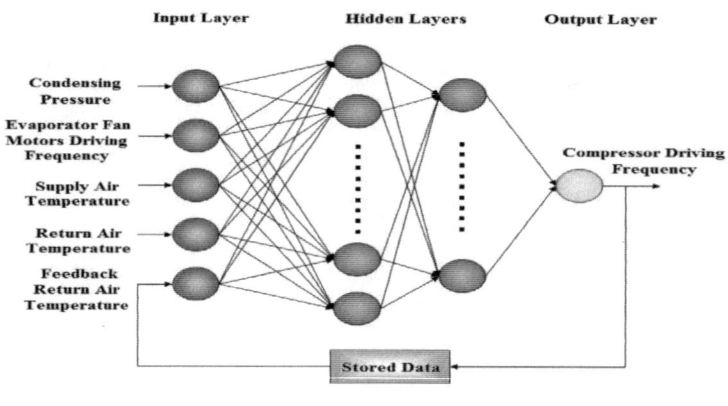

Figure (2) Architecture diagram of recurrent artificial neural network

Figure (2) shows the structure of the used control algorithm which includes the input layer which consists of 5 inputs; the condensing pressure, the supply air temperature, the return air temperature, the evaporating fan motors driving frequency and the feedback return air temperature from stored data. The hidden layers include the neurons which perform the calculations based on the stored data and the output layer releases the compressor driving frequency control signal.

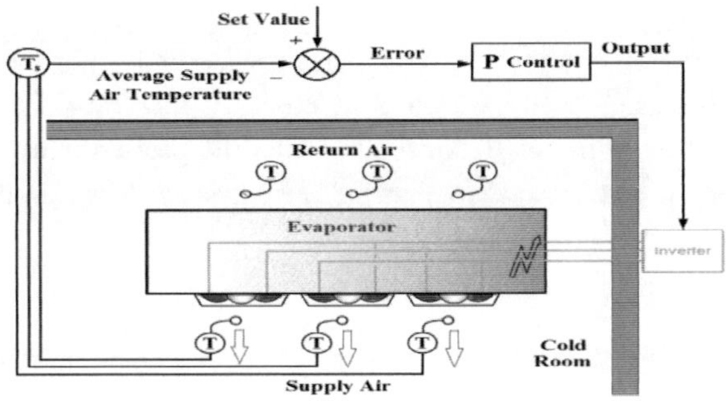

Figure (3) Schematic diagram of evaporator fans proportional control

Figure (3) shows a schematic diagram of the evaporator fan motors controller. The desired supply air temperature set value is applied via the data acquisition software and the actual supply air temperature values are measured using temperature sensors. The error released from the difference between the set value and the measured value is then multiplied by the proportional gain, to release the analogue output signal to the evaporator fan motors inverter.

The change in the condenser water pump driving frequency has a direct effect on the water mass flow rate of the condenser water circuit. It affects the condensing pressure of the refrigeration system. By measuring the condensing pressure and comparing it with a preset value by proportional integral differential (PID) controller, the required set pressure is achieved.

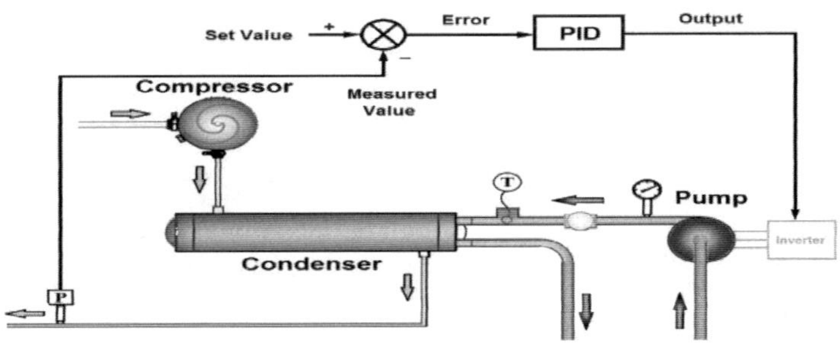

Figure (4) Schematic diagram of the condenser water circuit with proportional integral differential control

Figure (4) shows a schematic diagram of the condensing water circuit with the proportional integral differential (PID) controller. The measured condensing pressure is measured by a pressure transmitter which gives an analogue feedback signal to the data acquisition system. The desired condensing pressure is set via the data acquisition software and the PID algorithm releases an analogue output signal, which is sent to the condenser water pump inverter for pump speed variation. This action will keep the condensing pressure at a fixed value equals to the set value.

The electronic expansion valve (EEV) is controlled via a smart driver which acts as an PID controller, it uses the measured data collected from the suction pressure and suction temperature and calculates the current state of refrigerant at evaporator outlet, then compares the result with stored data of saturation vapor of the used refrigerant, and thus the actual superheat is obtained while the desired superheat is programmed into the driver. The difference between the measured and the desired superheat, is then multiplied by 3 separated parameters; P, I & D to release the present error from proportional parameter, the past error from the integral parameter and the future predicted error from the differential parameter, respectively. The resultant of this calculation process releases the output signal to the EEV.

4. EXPERIMENTAL RESULTS

4.1. Traditional Control Strategy results

In traditional control strategy, the compressor operates continuously at its nominal speed till the return air temperature reaches the set point temperature. Figure (5) shows the decrease in supply and return air temperatures when starting to operate the system with traditional control strategy. The return air temperature is set to 39.2 °F (4 °C) while the initial room air temperature was 82.4 °F (28 °C). Electric heaters were adjusted to 5 kW and started operation from the beginning of the refrigeration system operation to act as a continuous heating load. It was found that the compressor operated continuously for 33 minutes to decrease the return room air temperature from its initial value to the set point temperature. The figure indicates that the supply air temperature decreases below the set point.

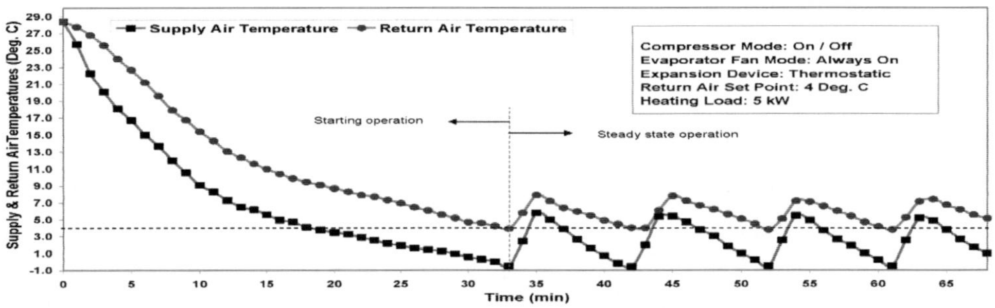

Figure (5) Effect of using traditional control strategy on the supply and return air temperature

After achieving the set point temperature at the end of the starting operation period, the system operated with on/off control, this results in temperature fluctuations due to the dependence of the control system on a temperature differential value of 3.6 °F (2 °C). It is found that the return air temperature fluctuates between 39.2 °F (4 °C) and 46.4 °F (8 °C), while the supply air temperature fluctuates between 31 °F (-0.6 °C) to 42.4 °F (5.8 °C).

The total input power of the refrigeration system and the total consumed energy during the starting period and the steady state operation are shown in figure (6). It is shown that the total consumed energy is gradually increased to reach 2.3 kWh after 33 minutes of operation, with an average total input power 4.24 kW. Then the consumed energy continue increasing in the steady state operation from the 33rd minutes to reach 4.4 kWh after 68 minutes of operation with average total input power 3.34 kW.

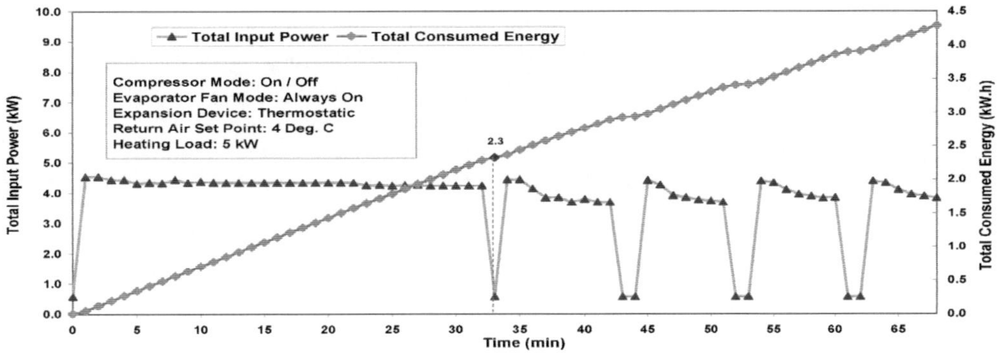

Figure (6) Effect of using traditional control strategy on the system input power and consumed energy during starting

4.2. Smart Control Strategy Results

Figure (7) shows a comparison between the supply air temperatures when using the smart control strategy and the traditional on/off control strategy during the starting period. It is shown that when using the traditional on/off control strategy the supply air temperature was achieved after 27 minutes, while when using the smart control strategy, the supply air temperature achieved its set value before 13 minutes.

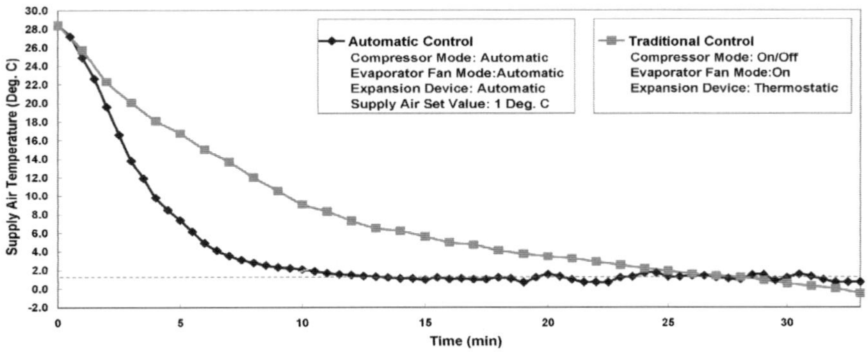

Figure (7) Comparison between traditional and smart control strategies to achieve supply air set value

Figure (8) shows a comparison between the return air temperatures when using the smart control strategy and the traditional on/off control strategy during the starting period. It is found that by using traditional control strategy, the return air temperature is achieved after 33 minutes, while when using the smart control strategy, the return air temperature is achieved after 15 minutes, this gives 55% saving in starting time of operation.

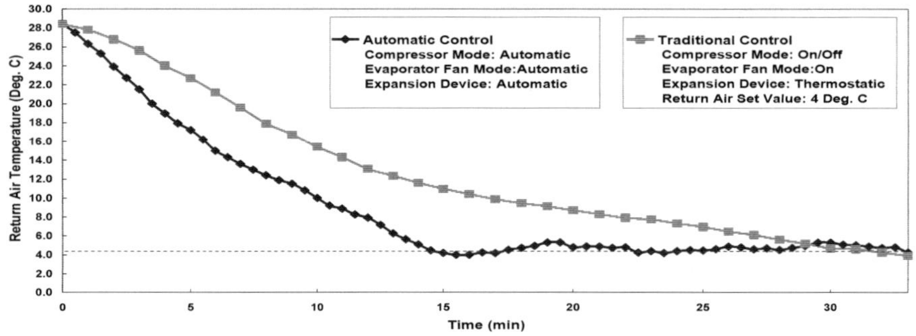

Figure (8) Comparison between traditional and smart control strategies to achieve return air set value

Figure (9) shows the effect of using smart control strategy on the total consumed energy. The total consumed energy is found to be 0.93 kWh after 15 minutes from starting operation with an average total input power 3.76 kW. The total consumed energy then increased to 1.82 kWh after 33 minutes. This means that operating the refrigeration system in steady state condition consumes 0.89 kWh in 18 minutes with an average input power 2.97 kW.

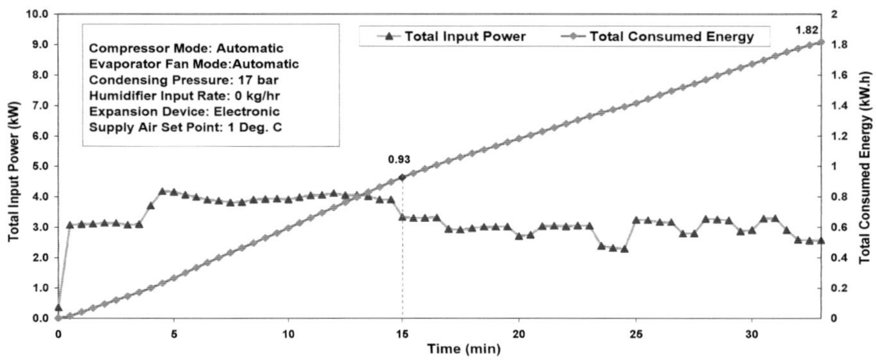

Figure (9) Effect of using smart control strategy on the system input power and consumed energy

Figure (10) shows a comparison between the smart control strategy's total consumed energy and the traditional control strategy's total consumed energy during the starting period. It is shown that the traditional control strategy consumed 2.3 kWh to achieve the set temperature value, figure (6), while it takes 0.93 kWh when using the smart control strategy, figure (9). This means that the reduction in the total input power during starting operation is 11.3%. The obtained energy saving when using the smart control strategy in the starting operation period is 60% due to operating the refrigeration system 15 minutes only instead of 33 minutes.

During the steady state operation, the average total input power is found to be 3.34 kW, figure (6), when using the traditional control strategy, while it is found to be 2.97 kW, figure (9), when using the smart control strategy. This means that by operating the refrigeration system using the smart control strategy, there is a reduction in the total input power by 11% during steady state operation.

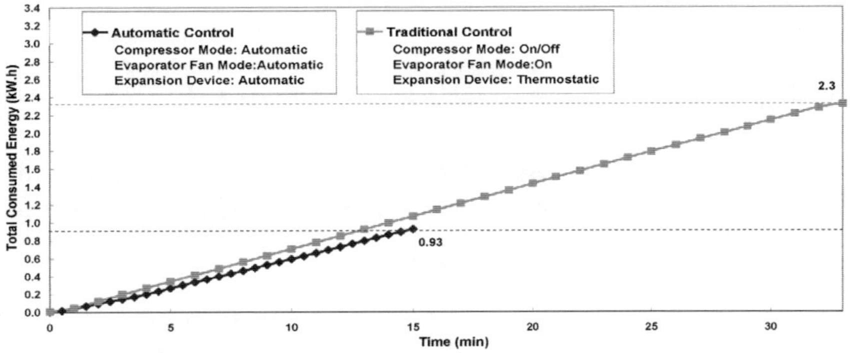

Figure (10) Comparison between consumed energy in traditional and smart control strategies

5. CONCLUSIONS

It was found that operating the refrigeration system using smart control strategy to control the motors speeds improved the performance of the refrigeration system compared to traditional on/off fixed speed control strategy. Controlling the evaporator fan motors speed was done using proportional controller to maintain a constant supply air temperature. Controlling the condenser water pump speed was done using proportional integral differential controller to maintain a constant condensing pressure, while the compressor speed was controlled using artificial neural network based on the analyzed data from the experimental results.

Using smart control strategy showed a reduction in the starting operation time compared to the traditional on/off control strategy by 55% in order to achieve the desired room temperature set value. The reduction in operation time to achieve the desired room temperature set value results in total energy saving 60% by using smart control strategy compared to traditional on/off fixed speed control strategy. The energy saving during system operation in the starting operation period is found to be 11.3%, while the energy saving during the steady state operation time is 11%.

6. REFERENCES

Benamer, A. and D. Clodic. 1999. Comparison of energy efficiency between variable and fixed speed scroll compressors in refrigerating systems. 8th European Conference on Technological Innovations in Refrigeration, Air conditioning and Food Industry, Milano, Italy.

Ekren, O., Sahin, S. and Y. Isler. 2010. Comparison of different controllers for variable speed compressor and electronic expansion valve. International Journal of Refrigeration 33:1161-1168.

Hua, L., Jeong S.K. and S.S. You. 2009. Feed forward control of capacity and superheat for a variable speed refrigeration system. Applied Thermal Engineering 29:1067-1074.

Lazzarin, R. and M. Noro. 2008. Experimental comparison of electronic and thermostatic expansion valves performances in an air conditioning plant. International Journal of Refrigeration 31:113-118

Marcinichen, J., Holanda T.N. and C. Melo. 2008. A dual siso controller for a vapor compression refrigeration system. International Refrigeration and Air Conditioning Conference, paper 922.

Salama, M.H. 2012. Performance improvement of a cooling/freezing system using variable speed motors. Ph.D. thesis, Mechanical Power Engineering Department, Faculty of Engineering, Ain Shams University, Cairo, Egypt.

Thambanchacheep, W. and S. Kuntanapreeda. 2002. Model reference neural network control for a variable-speed air conditioner. Third WSEAS International Conference, Switzerland.

Optimizing Energy Efficiency in PNU Campus

Toufic Itani
Member ASHRAE

Suhail Shatila

Sharbel El Haber
Member ASHRAE

ABSTRACT

Princess Nora Bint Abdul-Rahman University (PNU) is currently the largest construction project in the Kingdom. Initiated in 2009 and opened in May 2011, the built-up area totalled 30 million ft² (2.8 million²) with an ultimate population of 40,000 students and 12,000 staff. PNU represents almost one-quarter of current LEED registered projects in the Kingdom. With 38 separate buildings having a total BUA in excess of 11 million ft² (1 million m²), it is the largest higher education facility in the world to target LEED on such a scale. All its campus buildings aspire to be LEED NC v.2.2 Certified with the exception of: the Female Administration building and the Central Library, which aspire to be LEED Gold certified, the Academic Campus which has been awarded LEED Gold Certification, and the PNU Medical Center and Hospital that will be the world's largest LEED certified healthcare facilities.

In optimizing sustainability and, specifically, energy efficiency, the design process integrated numerous energy conservation measures among all engineering disciplines. In addition to an air cooled district cooling plant with a 100,000 ton-hr thermal storage tank, a 391,171 ft² (36,341 m²) solar panel array comprising over 3,600 separate flat plate collectors provides about 10% of the campus's requirements for hot water. Following optimization of building orientation, the envelope is designed to demonstrate an energy efficient response to a harsh and challenging arid climate. The curtain wall system includes high performance vision glazing and insulated shadow-box construction at column and spandrel zones. The incorporation of shading strategies into the design of PNU's building facades consist of musharabiah-inspired glass fibre reinforced concrete panels and screens that shade the building from direct solar gain. Within the academic colleges, 861,000 ft² (80,000 m²) of this shading has been deployed, contributing a 3.5% reduction in overall energy consumption and a 13% reduction in heat gain on the exterior envelope.

Additional energy conservation measures (ECMs) include: decreased lighting power densities, occupancy and day-lighting controls, blow through fans for higher supply air temperature differential, demand control ventilation (DCV), variable frequency drives (VFDs) on all equipment, high equipment and motor efficiencies, dedicated outdoor air system (DOAS), underfloor air distribution in auditoriums, evaporative cooling, heat recovery, pressure reset and dual-maximum-control sequence for VAV boxes, and last but not least supply air temperature reset during winter and unoccupied periods. A whole building simulation for the campus with over 30 energy models using Integrated Environmental Solutions IES VE indicated annual savings of 17.6 to 25.6% in energy and 16.3 to 22.8% in cost.

INTRODUCTION

The 30 million ft² (2.8 million²) built up area campus is the largest woman-only university in the world, **shown in Figure 1**. Inaugurated by the Custodian of the Two Holy Mosques exactly two years and two months from the date construction contracts were signed in February 2009. The campus's population is estimated to be around 60,000 and it consists of: Academic Campus with 15 buildings five of which are medical, Health Sciences and Research Labs, a 700-bed Medical Center, Hospital, Residential Quarters including 1,436 villas and apartments for faculty and staff and capacity to house 12,200 female students, Female and Male Administration Buildings, Main Library holding 5-million book collection with fully automated retrieval and storage system and a 4,000 occupant capacity, Main Mosque with 3,000 worshiper capacity, Convention Center with a 4,000 occupant capacity, Sports and Recreational Facilities including a separate male and

Toufic A. Itani, M.E., LEED AP BD+C and O+M, Member ASHRAE, is a project/lead mechanical and energy modeling engineer at Dar Al Handasah (Shair). Suheil Shatila, M.E., LEED AP BD+C is a mechanical and energy modeling engineer at Dar Al-Handasah (Shair). Sharbel El Haber, M.E.M., M.A, LEED AP BD+C, Member ASHRAE, is a senior mechanical engineer and previous associate at Dar Al-Handasah (Shair).

female community centers including a 7,000 multi-functional sports stadium, a 2.8 mile (4.5 km) underground Utility and Infrastructure tunnel housing electro-mechanical, water, telecom and IP utility services, an on-campus waste water and water re-use treatment plant, a utility zone comprising of a solar farm, print shop, maintenance and workshop facilities and a K-12 school with 800 student capacity. The entire campus is LEED NC v.2.2 certified [A], except for the Female Administration building and the Central Library that aspire to be LEED Gold Certified. Additionally, the Academic Campus was awarded LEED Gold Certification after the team working on the project managed to elevate the project's status from LEED Certified to LEED Gold by applying minor modifications to the design and construction. The fast track nature of the design phase was maintained for the construction schedule to meet the project completion target date. During the months of construction, 75,000 workers, over 225 cranes, and more than 5,000 construction equipment worked around the clock to beat the set 25-months construction.

The Kingdom's ministry of Finance and PNU's governing body have made sure the most important feature of this campus is sustainability. The design methodology ensured the commitment to sustainability goals by aligning the design strategy with the site's arid climate, energy and water savings, enhanced environment for the occupants, and future ecological requirements. PNU's sustainability plan is performance driven, consisting of strategies that are adaptable enough to forecast conservation gains and harbor future requirements. Taking into consideration the social fabric and the environmental needs, the multi-disciplinary team worked synergistically to develop the ecological footprint necessary for a long-term environmental policy that started at the project's onset [B].

The energy and cost saving strategies are grouped under five categories: sustainable sites, water efficiency, energy and atmosphere, materials and resources and indoor environmental quality. Sustainability was also ensured during the construction of the project by capping the site's carbon emission and reducing the building works' overall energy demands [C]. The challenging arid desert climate left its imprint on the degree of sustainability of the campus. Energy conservation measures, shading and cooling programs had to be craftily designed to meet the sustainability goals. Buildings' envelopes, high performance low-e glazing profiles, GRFC musharabiah-inspired cladding over approximately 861,000 ft^2 (80,000 m^2) of shading facades accounted a large reduction in building energy consumption of around 3.5%.

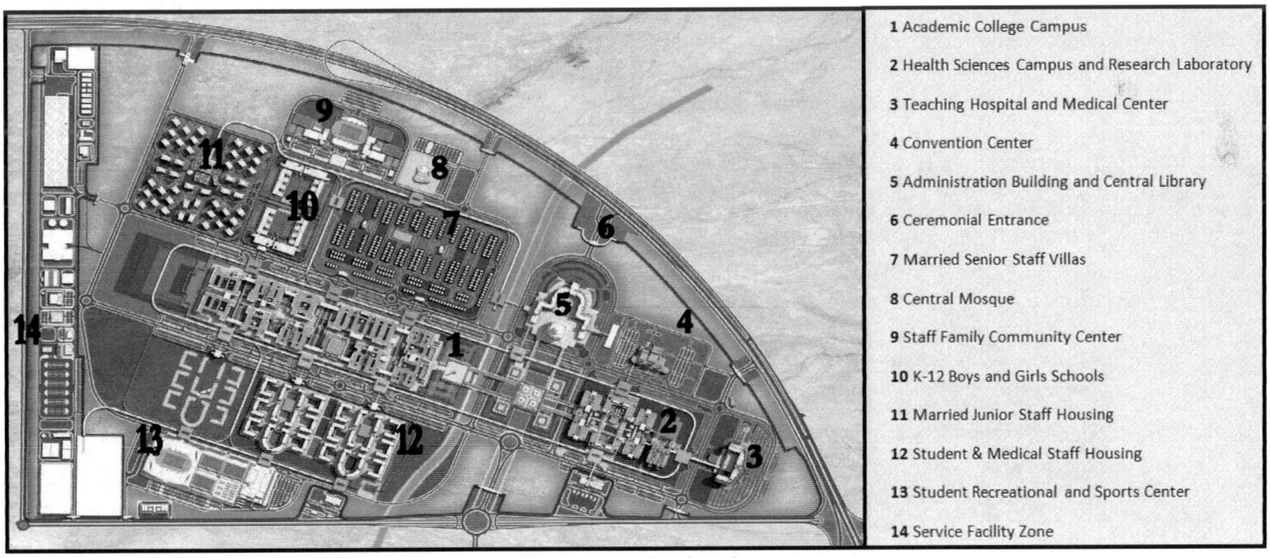

Figure 1 PNU Site plan

METHODOLOGY TO ACHIEVE SUSTAINABILITY

Achieving sustainability in a project of PNU's scale is always a challenging task. The design was done in a very short period of time while adhering to LEED NC v.2.2 requirements. The workload had to be distributed prudently among several teams while ensuring efficient collaboration within a synergistic approach. Under contractual constraints, time pressure, meetings with the main contractors to introduce the ECMs and any changes in design to address all parties' concerns, sustainability was successfully attained. The energy modeling team persisted on submitting a detailed energy modeling analysis by testing several design options to reach an optimized design while researching innovative green technologies.

Strategies that supported sustainable design include: drafting a basis of design document that included the sustainability goals and expectations for the project, collaboration between the diverse project teams, training and assisting engineers to integrate LEED requirements at the early stages of the design and seek synergistic solutions, and encouraging a culture of innovation, environmental awareness and a passion for redefining what is possible.

The vision was not just to meet the LEED certification standards for PNU, but rather to build a base that will promote sustainable design, encourage and facilitate upgrading the conventional building practices. Most important of all was to integrate the LEED process into concept designs while focusing on synergistic interactions among various teams, a task that requires a lot of dedication from all parties.

ENERGY MODELING

Taking into consideration the Arabian Desert arid climate, the design teams had to achieve the desired levels of energy efficiency by providing the appropriate architectural and mechanical energy conservation methods. These ECMs include: enhancement of the building envelope, external shading, decreased lighting power densities with occupancy and daylight controls, usage of variable frequency drive motors, evaporative cooling, VAV pressure reset with dual maximum control sequence, underfloor air distribution (UFAD) [D], blow thru-fans, dedicated outdoor air systems (DOAS) [E], [F], exhaust air recovery, cooling coil supply temperature reset during the winter and demand control ventilation (DCV).

A detailed building energy performance analysis using IES VE 6.0.3 [G] has been conducted to evaluate the effectiveness of the proposed building design in comparison with the Baseline model, and in accordance with ASHRAE 90.1-2004, Appendix G performance rating method including Addenda (a) and (am) [H]. Since the registration consists of multiple buildings, a weighted average based on the buildings' conditioned areas was calculated as per the document, "LEED NC Application Guide for Multiple Buildings and On-Campus Building Projects", October 2005. Additionally, as the campus buildings are connected to a dedicated district cooling and heating plant, a building stand-alone scenario is modeled following the guidelines of the document "Treatment of District or Campus Thermal Energy in LEED V2 and LEED 2009 - Design & Construction", August 13, 2010. A comparison between the conceptual IES model, **shown in Figure 2**, versus the actual building of the Central Library, **shown in Figure 3**, is shown below.

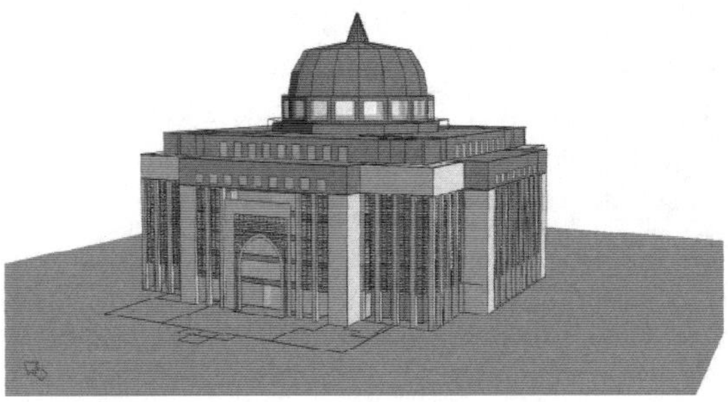

Figure 2 IES Model of Central Library

Figure 3 Actual Model of Central Library

Overview of ECMs in Campus Buildings

Buildings are well insulated with wall, roof and glazing U-values that are lower than those of ASHRAE 90.1-2004 Baseline for Climate Zone 1B. Please see **Table 1** for the building envelope performance for the Central Library.

Table 1. PNU Library Envelope Performance Summary

Envelope	Value Btu/ft².h.°F (W/m².K)
Wall U-value	0.046 to 0.147 (0.261 to 0.832)
Roof U-value	0.006 & 0.046 (0.037 & 0.261)
Window-to-wall ratio (%)	25
U-value	0.356 (2.02)
Shading Coefficient	0.28 & 0.3

The external shading devices, **shown in Figures 4 and 5**, in the proposed model consist mainly of canopies, overhangs and vertical screens with void area of about 35%. These shade the building fenestration and reduce the amount of solar incident radiation, thus, contributing to a reduction in cooling energy loads.

Figure 4 External Shading in Academic Colleges

Figure 5 External Shading in Health Sciences Colleges

Most of the spaces have occupant sensor controls to switch off the lighting whenever the spaces are not occupied, such as wall mounted PIR occupancy sensors, wall mounted or ceiling mounted ultrasonic sensors and photocell sensors.

Variable frequency drives are used for all pumps and fans (all motor sizes) and save energy by allowing the volume of fluid moved to match the system demand.

Evaporative cooling is used in the Baseline and proposed models to provide humidification which also reduces the mechanical cooling load [I].

VAV air handling unit efficiency is increased by reducing duct pressure at part-load conditions. VAV with pressure reset method reduces energy consumption by adjusting fan speeds to lower the supply static pressure, which reduces supply fan-power requirements. Lower fan power also reduces waste heat and the energy required for cooling. The expected reduction in VAV fan energy is around 20%.

By using a single-maximum-control sequence, the minimum-airflow set point will be high enough to provide design heating airflow at a reasonable supply-air temperature e.g. 90°F (32.2°C), **shown in Figure 6**. Thus, the minimum typically is set at 30 to 50 percent of the maximum, which means that at low load, the space is overcooled in dead band and driven into reheat. However, in the dual-maximum-control sequence, as the load goes from full cooling to full heating, the cfm set point is reset from maximum to minimum before the supply air- temperature set point is reset from minimum e.g. 55°F (12.8°C) to maximum e.g. 90°F (32.2°C), **shown in Figure 7.** If more heat is needed, the cfm set point is reset from minimum to heating maximum. Because it is decoupled from the heating needs, the minimum-airflow set point can be as low as the controllable minimum of the VAV box (as long as ventilation requirements are met). When a dual-maximum-control scheme is used, the minimum-airflow set point typically is 10 to 20 percent of design airflow ensuring considerable levels of energy savings [J].

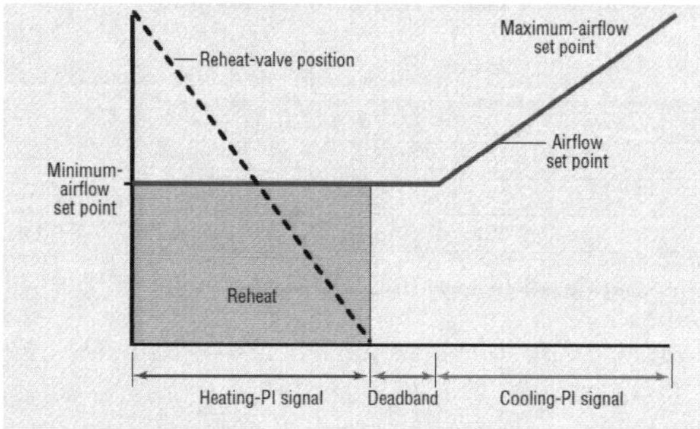

Figure 6 Single-maximum-control sequence

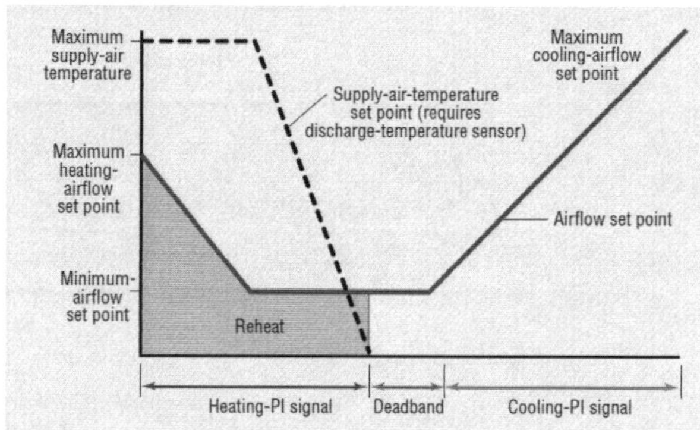

Figure 7 Dual-maximum-control sequence

Additionally, heat stratification is achieved in the libraries as well as the main entrances. For this purpose, a computational fluid dynamics (CFD) analysis was performed to improve the air distribution scheme in the Grand Lecture Hall, Library and Convention Center [K], [L]. In the assembly hall of the convention center, an under-floor distribution system (UAFD) was coupled with a conventional ceiling air supply [D]. The basic design (**as shown in Figure 8a**) consisted of 902 diffusers beneath the audience seats and 62 flow bar diffusers distributed throughout the levels. The design was optimized (**as shown in Figure 8b**) by decreasing the supply air flow and directing it at predefined angles while ensuring proper air delivery in the 79 ft (24 m) high assembly hall and avoiding disturbance of the thermal stratification of the air. CFD analysis was crucial in optimizing the HVAC design to ensure an ideal level of occupant comfort [M], [N].

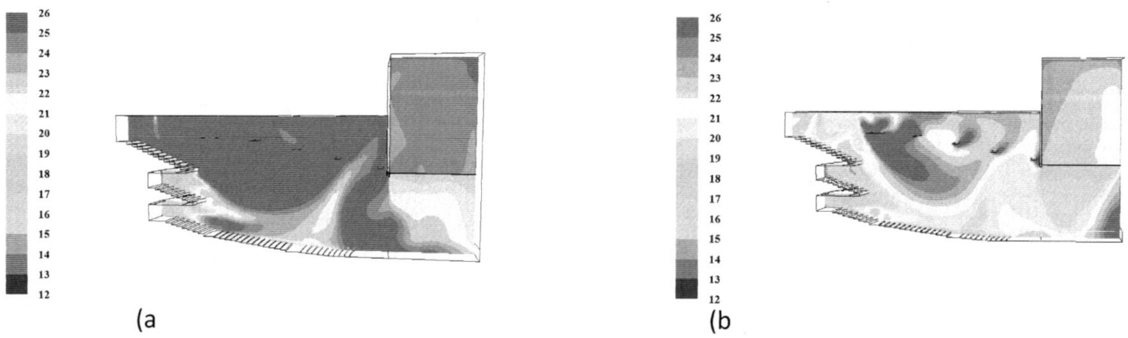

Figure 8 Convention Center temperature distribution in original design (a) versus optimized design (b)

Specifying blow-thru fans in lieu of draw-thru fans is an additional energy conservation measure. Our proposed strategy was to have the cooling coil leaving air temperature in a blow-thru unit less than 56°F (13°C) to gain at least a 33.8°F (1°C) reduction in the supply air temperature to the space. The lower supply air volume resulted in annual fan energy savings.

DOAS is used for almost 50% of the building zones including all classrooms, multi-purpose rooms, conferences & labs. This greatly reduces the fresh air quantities as it resolves the problem of having to serve multiple spaces with different ventilation needs and avoids having to unnecessarily increase the fresh air flow rates due to the mandatory correction measures of ASHARE 62.1-2004 [E], [O]. In addition, a DOAS offers better indoor air quality and eliminates the requirement of reheat coils, as treated outdoor air at 72°F (22°C) is supplied with the minimum setting of VAV boxes set at zero flow.

Heat recovery wheels are used in all treated outdoor air handling units. For VAV air handling units, cooling coil supply air temperature is reset in winter (December, January & February) to reduce chiller load.

DCV is used for all spaces with variable and high occupant density such as conference rooms, classrooms and prayer rooms. This scheme requires monitoring of CO_2 concentration to dynamically reset the ventilation rate, reducing outdoor air intake rates when zones are not occupied at their design occupancy.

The proposed energy conservation measures (ECMs) were analyzed and the individual energy savings computed for each. These are summarized in **Figure 9**. The energy conservation measures that were separately analyzed include demand controlled ventilation (DCV), night purging (NP) and VAV pressure reset (VAV-PR).

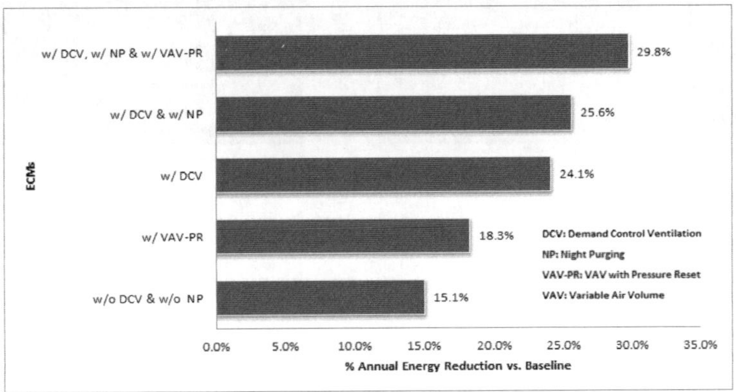

Figure 9 Percentage annual energy reduction for systems used in classrooms

A summary of the annual energy savings for the campus buildings is listed in **Table 2**.

Table 2. Annual Energy Savings in PNU Campus buildings

Building	Area ft^2 (m^2)	Annual Energy Savings (%)
Academic Campus	5,564,693 (516,960)	19.3%
Health Sciences	2,068,923 (192,203)	17.7%
Conference Center	350,301 (32,543)	21.6%
Research Center	389,311 (36,167)	23.0%
Hospital	1,051,442 (97,679)	19.4%
Sports & Recreation Center	406,264 (37,742)	27.0%
K-12 Schools	346,749 (32,213)	24.4%
Female Administration	676,200 (62,819)	25.9%
Library	438,040 (40,694)	24.5%

WIND TOWERS

Wind-towers have traditionally been used in buildings of the Arabian Gulf region as a natural ventilation tool to disperse the heat accumulating indoor amidst the prevailing hot, sunny days. They basically consist of a hollow square tower, constructed with the building fabric, with top side openings and internal partitioning to allow catching the wind from any direction, creating an air stream in the tower's shaft leading to the occupied building space to be ventilated. A comparison between the schematic of the wind tower and the actual wind tower is **shown in Figure 10 and 11**, respectively.

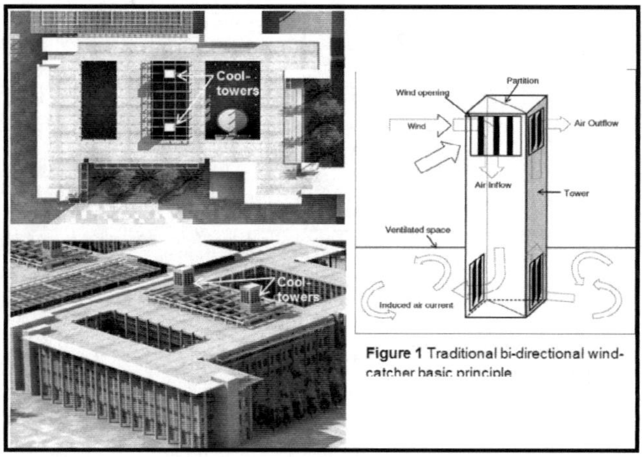

Figure 10 Concept of Wind Towers in Academic Colleges

Figure 11 Actual Wind Tower in Academic Colleges

The provision of cool-towers in the PNU campus was intended to add thermal comfort by supplying passively cooled outdoor air to the courtyard spaces normally occupied by students during open periods. A college building courtyard is typically a building-surrounded rectangular, open-to-sky space of dimensions 66-ft W x 197-ft L x 55-ft H (20-m W x 60-m L x 16.76-m H), with an array of wooden logs or ornaments crossing the top for shading. The objective was to remove stagnant air from the courtyard space by introducing outdoor air, and to passively cool the delivered air when its temperature is too high for thermal comfort. The cooled air is introduced to the courtyard at low level so as the courtyard space will become stratified, with cooler air layers at lower elevations.

Passive down draft evaporative cooling wind-towers, designed to provide cool air to the semi-open courtyards of the Princess Nora University (PNU) campus located in the hot, arid region of Riyadh, KSA, were simulated using ANSYS-FLUENT software [L]. The cool-towers were capable of generating a relatively thermally comfortable zone in the courtyard, by inducing adequate air movements with air speeds between 295-690 fpm (1.5-3.5 m/s) in the courtyard region, and passively cooling the 115°F DB/66.2°F WB (46°C DB/19°C WB) airstream captured from the prevailing wind down to 93°F DB/66.2°F WB (34 °C DB/19 °C WB), creating a temperature distribution in the courtyard central, longitudinal lower region mainly in the range of 93°F-100°F (34°C-38°C), while the normally low relative humidity is maintained below 30%. A minimum energy saving of 309 hp (230 kW) over standard refrigeration system was introduced per tower, which is equivalent to reducing carbon emission by around 81571 pound (37 metric tons) per hot season.

SOLAR FARM

The University Campus is inhabited by around 40,000 students, 13 faculties, lecturers, university personnel, dormitories, research facilities and a gymnasium. The area supplied with hot water and space heating also includes a hospital and hotel along with all other necessary infrastructure for living, working and studying. The implementation of the solar system was done in a way to lower the capital cost, operational and maintenance cost and to support the saving of conventional fuels [I].

The size of the project was challenging in terms of space allocation and uncertainty of solar input due to weather conditions. With a solid background and engineering experience, the customer needs for the project were identified and analyzed. The supply of domestic hot water and space heating during the 40 heating days in Riyadh was calculated to estimate the size of the solar supported district heating network to be installed. The demand of heat is covered by oil fired boilers with a peak load capacity of 239 MBTU/hr (70 MWth) together with a 85 MBTU/hr (25 MWth) generated from

391,171 ft^2 (36,341 m^2) of flat plate collectors, especially designed to withstand harsh desert conditions with low maintenance needs. The collectors are placed on the rooftop of a 646,000 ft^2 (60,000 m^2) warehouse. During the summer period the thermal load of the district heating network is around 103 MBTU/hr (30MWth).

CHILLED WATER CENTRAL PLANT

Located in the arid climate of Riyadh, Kingdom of Saudi Arabia, where water is not only scarce and therefore very expensive, the university selected a cooling system design that will minimize water consumption. Compared to a plant that uses evaporative cooling towers, air-cooled radiators, that removes heat from the closed-loop, condenser-water circuit, saved an estimated 2 million gallons (7 million liters) of water per day of chiller operation.

Princess Noura bint Abdulrahman University has the world's largest chiller plant using air-cooled radiators with 26 YORK model CYK high-lift chillers at a capacity of 2070 tons of refrigeration (7,280 kW). When they are operated together, the plant will boast a total cooling capacity of over 53,000 TR (189,000 kW).

The unique YORK model CYK chiller utilizes two centrifugal compressors piped in series to produce the high lift required for the PNU campus. With compounded compressors, the chiller is able to produce chilled water at 41°F (5°C), while the elevated condensing temperature due to the air-cooled radiators is 144.6°F (62.5°C). The high-lift ability of the chiller also allows it to be used for ice thermal-storage, heat-pump, and low-temperature-brine cooling applications. Princess Noura bint Abdulrahman University selected to use the option of cost saving high-voltage (13kV) TEWAC motors and reduced voltage starters. The elimination of step-down transformers, additional wiring, and associated installation assist with total cost and energy savings.

CHALLENGES and ACHIEVEMENTS

The fast track nature of the design phase was the major focus of the construction schedule to meet the project completion target date. It is because of this commitment that the role widened to include close coordination with the three main contractors, site supervision and construction management teams to bring this massive undertaking to fruition. Dar Al-Handasah and Perkins+Will design teams in Cairo, Beirut & Pune, San Francisco, Miami, Atlanta, Minneapolis, Chicago & Los Angeles offices in coordination with all Dar departments for architecture, civil, structural, transportation, electrical, mechanical, safety, water works, waste management, landscaping and sustainability have risen to the challenge. The coordination effort with guidance from Dar Al Handasah-Riyadh Office who, in addition to being in direct and constant contact with the Ministry of Finance and the University Board of Directors for the duration of design and construction phases, directed over 24 international sub-consultants working on the project. Hundreds of architects and engineers were on board working tight deadlines imposed and hectic schedules and communicating through a complex network of on-going coordination, conference calls, video conferencing, workshops and meetings.

There is a misconception that LEED probably is not appropriate for countries in the Gulf due to the arid climate. In contrast, it was quite appropriate and it led into different ways of saving energy. By integrating new mechanical systems such as evaporative cooling, energy consumption was reduced and optimized. Some of the energy conservation measures that are particularly effective include external shading and occupancy control. Solar impact was reduced by adopting the "Musharabiah" external facades while maintaining high daylighting levels. Under the guidance of LEED accredited professionals that carefully designed architectural and mechanical ECMs, PNU's sustainability strategy was translated by meeting LEED certification requirements for its campus buildings.

Energy management systems installed have radically reduced electricity consumption at the world's largest educational institution for women. Electricity used in the project was reduced by up to 40 per cent. On top of the electricity saving, the academic colleges and health sciences campuses have reduced their water consumption by 45 per cent, with the rest of the campus saving 35 per cent. The end use summary for the Health Sciences campus is shown in **Figure 13**. [P], [Q], [R], [S].

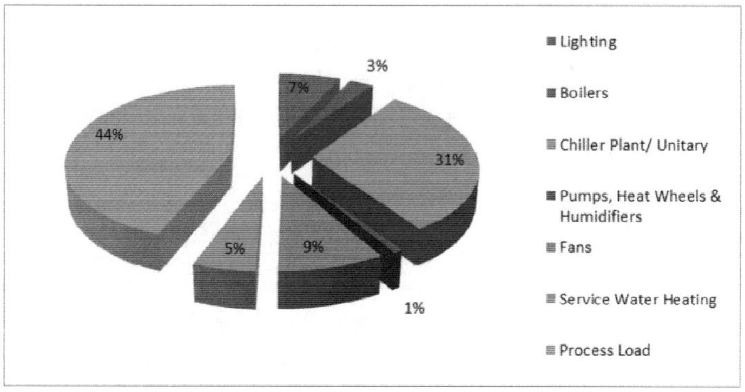

Figure 13 Health Science End use summary

Landscaping constituted a very large portion of how sustainability can be accomplished. Treated sewage blended with non-potable water is used for irrigation. Large portions of the site are also allocated over to local plant life.

An automated people mover was designed so that students and staff can travel anywhere in the university in privacy and they will never be more than 150 ft (500 m) away from where they want to get to. Almost always they are within 75 ft (250 m). It is an ideal way to use mass transit and free people from dependency on cars.

PNU campus has between 10-50 percent recycled content in steel rebar, structural steel, metal doors and other metals, curtain walls and false ceilings. The concept for construction waste management was to prevent waste becoming land fill. Contractors were inventive in developing ways of recycling waste.

CONCLUSION

PNU, operational in May 2011, is currently the largest higher education facility in the world to target LEED on such a scale. The campus includes LEED NC v.2.2 certified campus buildings, recreational buildings, healthcare facilities, a LEED Gold certification level for the Central Library, Female Administration and Academic Campus. Sustainability features include an air-cooled district cooling plant with thermal storage, a solar panel array, and wind towers. Mechanical and architectural ECMs were incorporated in order to achieve a high level of energy efficiency. This led to annual savings of 17.6 to 25.6% in energy and 16.3 to 22.8% in energy cost.

Optimizing energy efficiency in PNU campus was a very challenging task especially in the arid climate of Riyadh. By following a synergistic approach among the different teams, this task was achievable in a short period of time. Through incorporating inventive green technologies and energy efficient novelties, the PNU campus is now a landmark in energy efficiency by adhering to LEED standards. The campus now represents a cornerstone and a thrust for the green market to further expand in the MENA region.

REFERENCES

[A] Leadership in Energy & Environmental Design (LEED). Available from: http://www.usgbc.org

[B] ASHRAE. 2006. ASHRAE GreenGuide. Atlanta: American Society of Heating Refrigeration and Air Conditioning Engineers, Inc.

[C] Commercial Building Energy Consumption Survey (CBECS), 2003. Available from: http://www.eia.doc.gov

[D] Bauman, Fred S. 2003. Underfloor Air Distribution Design Guide. Atlanta: American Society of Heating Refrigeration and Air Conditioning Engineers, Inc.

[E] Mumma, Stanley A. 2010. DOAS & Building Pressurization. ASHRAE Journal, pp. 42.

[F] Larrañaga, Michael D., Beruvides, Mario G., Holder, H.W., Karunasena, Enusha, Straus, David C. 2008. DOAS & Humidity Control. ASHRAE Journal, pp. 34.

[G] Integrated Environmental Solutions. IES v. 6.0.3. [online]. Available from: http://www.iesve.com

[H] ASHRAE. 90.1-2004. Energy Standard for Buildings Except Low-Rise Residential Buildings. Atlanta: American Society of Heating Refrigeration and Air Conditioning Engineers, Inc.
[I] ASHRAE. 2007. ASHRAE Handbook-Applications. Atlanta: American Society of Heating Refrigeration and Air Conditioning Engineers, Inc.
[J] Stein, J. 2005. VAV Boxes, HPAC Engineering, pp. 40-44.
[K] Haber, S., Nehme, B., Akhdar, A. 2010. Meeting Green Building Design Goals With Engineering Simulation. ANSYS Journal.
[L] ANSYS Fluent Software v. 12.1.4 [online]. Available from: http://www.ansys.com
[M] ASHRAE. 55-2004. Thermal Environmental Conditions for Human Occupancy. Atlanta: American Society of Heating Refrigeration and Air Conditioning Engineers, Inc.
[N] ASHRAE. 2009. ASHRAE Handbook-Fundamentals. Atlanta: American Society of Heating Refrigeration and Air Conditioning Engineers, Inc.
[O] ASHRAE. 62.1-2004. Ventilation for Acceptable Indoor Air Quality. Atlanta: American Society of Heating Refrigeration and Air Conditioning Engineers, Inc.
[P] ASHRAE. 170-2008. Ventilation of Health Care Facilities. Atlanta: American Society of Heating Refrigeration and Air Conditioning Engineers, Inc.
[Q] ASHRAE. 2003. HVAC Design Manual for Hospitals And Clinics.
[R] VanGeet, O., Reilly, S. 2006. Ventilation Heat Recovery For Laboratories. ASHRAE Journal, pp. 44.
[S] Labs for the 21st Century (Labs21). Available from: http://www.labs21century.gov

Application of ASHRAE 90.2 for Middle East Regions

Mir Gayas Ali
Member ASHRAE

Walid.M. Chakroun PhD, PE
Fellow ASHRAE

Wayne Reedy
Fellow ASHRAE

ABSTRACT

ASHRAE 90.2 is a standard for Energy-Efficient Design of Low-Rise Residential Buildings. It provides minimum design and construction requirements for low-rise residential buildings and their systems. In the Prescriptive Requirements by Component, ASHRAE 90.2-2010 provides values for building envelope components such as wall, roof, and glazing and energy efficiency for air-conditioning systems for various climatic zones. It is often found that the values provided in the prescriptive requirements are not pertinent to the Middle Eastern regions because of its different geographical locations, climatic conditions, cultures, building practices and local regulations. Certain values are also often found to be less stringent than the local authority regulations and standard practices prevailing in the region. The objective of this Technical paper is to analyse and assess some of the prescriptive requirement values in ASHRAE 90.2 such as building envelope components wall, roof, glazing and energy efficiency of air-conditioning systems, using Building Energy Modelling Simulation techniques for different regions in the Middle East. It has been found that the building envelope forms a significant portion of the cooling or heating loads for low-rise residential buildings and therefore a major percentage of energy consumption for the air-conditioning systems. Optimising the building envelope values can provide substantial savings in energy. Similarly, the energy efficiency of the air-conditioning systems is a significant factor of energy consumption in the region. Optimising the energy efficiency of air-conditioning systems can also provide substantial savings on energy. A comparative analysis is made for thermal conductivity (U) values of walls, roofs, and glazing, along with solar heat gain coefficient (SHGC) for glazing and energy efficiency of air-conditioning systems from the building energy modelling simulation results, to analyse the energy consumption and potential energy savings and obtain optimised values for these parameters relevant to the Middle Eastern region. Comparative analysis is made using the ASHRAE 90.2 values against corresponding values in local regulations such as Ministry of Electricity & Water (MEW) Kuwait, Saudi Building Code (SBC), Dubai Municipality Regulations (DM) and Kharamaa, Qatar to form a wider spectrum of climatic regions. The optimised values of building envelope components and energy efficiency of systems can significantly improve and enhance the energy performance of the building and help in sustainable and efficient designs of buildings in the Middle Eastern region.

INTRODUCTION

Buildings represent the largest energy-consuming sector in the economy, with over one-third of all energy and half of global electricity consumed there. As a result, they are also responsible for approximately one-third of global carbon emissions. With improvements in economic development and living standards expected to increase as the planet's population grows by 2.5 billion by 2050, energy use in the buildings sector is also set to rise sharply, placing additional pressure on the energy system. The building envelope – the parts of a building that form the primary thermal barrier between interior and exterior – plays a key role in determining levels of comfort, natural lighting and ventilations, and how much energy is required to heat and cool a building. The building envelope's impact on energy consumption should not be underestimated especially in low-rise residential sector; where in the building envelope load forms a major portion of total heating and cooling requirements. Analysis of building envelopes is complicated by the extreme global diversity of building materials, climates, and standards and practices of building design and construction. There are vast differences in construction practices between traditional dwellings in developing countries and houses constructed in developed countries. As populations grow, housing demand also increases, and rapid increases in wealth usually drive greater increases in floor area per capita and in many cases higher land usage. It is vital to ensure new buildings use the most efficient technologies.

Energy efficiency revolves around reducing energy waste: this can be both on the demand side through power

Author A is a Senior Mechanical Engineer in SSH International Consultants, Kuwait. Author B is a Professor in Kuwait University, Kuwait. Author C is a Consultant in Monticello, IN.

demand management and on the supply side using a host of technologies to improve efficiency in power generation and distribution. A further analysis of these consumption patterns reveals that the GCC (Gulf Cooperative Council) countries consumption is driven largely by home use, with almost 47% of the energy consumption siphoned into residential use when compared to a global average of around 25%. What is worth noting here is that the per capita electricity demand in the GCC countries is expected to increase at a faster rate than their counterparts in the US. According to the '2010 International Energy Outlook' issued by the US Energy Information Administration (EIA), the per capita electricity consumption during the period 2007-2035 in the GCC is likely to increase at an annual rate of 2.5% , with a large part of this rate attributed to a growing population, as opposed to 0.8% in the US. So we can expect that in a matter of a few years, GCC residents may well be outright leaders in per capita residential electricity use. Given this scenario, one would expect effective demand management to be part of the energy solution for the GCC countries. This is also evident with some forms of energy efficiency policies and measures being implemented in some of the GCC countries – but the measures may need to be streamlined and perhaps reassessed in some cases.

The numbers suggest that effective demand management of residential electricity could help the GCC countries solve some of their power sector dilemmas in the coming years. Not only will this help in reducing some of the demand for installing more generation capacity, but it will also divert some of the saved energy towards the industrialization or commercialization agendas of these countries.

ASHRAE STANDARD 90.2 AND LOCAL REGULATIONS IN THE MIDDLE EAST

ASHRAE 90.2 is a prominent standard globally for energy efficiency of low-rise residential buildings. It provides minimum design and construction requirements for low-rise residential buildings and their systems. Table 4.1 Prescriptive Requirements by Component, provides values for building envelope components such as wall, roof, glazing, skylight and energy efficiency for air-conditioning systems for various climatic zones.

It is often found that the values provided in the prescriptive requirements are not pertinent to the Middle Eastern regions because of its different geographical locations, climatic conditions, cultures, building practices and local regulations. Certain values may also be found to be less stringent than the local authority regulations and standard practices prevailing in the region. Table 1, below shows a comparison of building envelope values and Energy Efficiency Ratios (EER's) for ASHRAE 90.2 and various local regulations in the Middle East.

Table 1. Building Envelope and EER Data from ASHRAE 90.2 and Local Regulations in the Middle East

Description	Units	ASHRAE 90.2-2010	MEW	DM	Kharamaa	SBC 601
		USA	Kuwait	Dubai	Qatar	KSA
Wall 'U' values	(Btu/hr/ft^2/ °F)	0.165	0.1	0.1	0.1	0.05
Roof 'U' values	(Btu/hr/ft^2/ °F)	0.035	0.06	0.05	0.08	0.02
Glazing'U' values	(Btu/hr/ft^2/ °F)	1.1	0.595	0.37	0.58	0.35
Glazing SHGC		0.3	0.245	0.35	0.35	0.52
EER	(Btu/Wh)	15*	7	9.5	8.5	9.2

- Converted from US SEER value, and at 95 deg. F (35 deg. C)
- EER for ASHRAE 90.2, DM, Kharamaa, and SBC 601 are at 95 deg. F (35 deg. C) ambient temperature and EER for MEW is at 118 deg. F (48 deg. C) ambient temperature.

An analysis of the above Table 2 indicates the following:

1. Wall 'U' values stipulated in most of the Local regulations are consistent except for SBC 601, KSA and are less than ASHRAE 90.2-2010 values.
2. Roof 'U' values for most of the local regulations except for SBC 601, KSA are higher than ASHRAE 90.2-2010.
3. Glazing 'U' values for most of the Local regulations are lower than ASHRAE 90.2-2010.

4. Glazing Solar Heat Gain Coefficient (SHGC) values for most of the Local regulations except for MEW, Kuwait are higher than ASHRAE 90.2-2010.
5. The Energy Efficiency Ratio for the HVAC systems indicated in ASHRAE 90.2-2010 is far more stringent than the Local regulations. However, The Energy Efficiency Ratio (EER) indicated in ASHRAE 90.2-2010 is far higher than normal possible value for most of the equipments in the region. Even the EER for same configuration of equipment in ASHRAE 90.1-2010 is less at 13 which seems to be reasonable.

CASE STUDIES

Considering the above analysis, four case studies were conducted to assess the energy efficiency potential for different cities in the region to form a wider spectrum of climatic regions. The four cities considered were Riyadh in Saudi Arabia, Kuwait City in Kuwait, Dubai in United Arab Emirates, and Doha in Qatar. The energy efficiency assessment was carried out using building energy modeling simulation techniques on a typical low rise residential building. Energy simulations were carried out using HAP 4.8 software. The assumed inputs for the building energy modeling are provided in below tables:

Table 2. Assumed Input data

Floor Description	Floor Area	Height	Lighting Power Density	Occu-pants	People Sensible Load	People Latent Load	Equipment Load	Fresh Air Rate
	$ft^2/(m^2)$	ft/(m)	watts/ft^2 / (watts/m^2)	(No.s)	(watts/ person)	(watts/ person)	watts/ft^2 / (watts/m^2)	cfm / (L/s)
Basement	2605/(242)	9/(2.74)	0.93/(10)	1	245	155	0.05/(0.54)	41/(19)
Ground	2605/(242)	11.5/(3.5)	0.93/(10)	2	245	155	0.09/(0.97)	160/(76)
First	2605/(242)	11.5/(3.5)	0.93/(10)	4	245	155	0.09/(0.97)	70/(33)
Second	1302.4/(121)	11.5/(3.5)	0.93/(10)	1	245	155	0.09/(0.97)	30/(14)

Table 3. Assumed Building Wall and Glazing dimensional data

Exposure Orientation	Ground Floor Wall Gross Area		Ground Floor Window Gross Area		First Floor Wall Gross Area		First Floor Window Gross Area		Second Floor Wall Gross Area		Second Floor Window Gross Area	
	(m^2)	(ft^2)	(m^2)	(ft^2)	(m^2)	(ft^2)	(m^2)	(ft^2)	(m^2)	(ft^2)	(m^2)	(ft^2)
West	71	760.7	18	190	71	760.7	18	190	39	414	10	104
North	66	708	16	177	66	708	16	177	39	414	10	104
East	71	760.7	18	190	71	760.7	18	190	39	414	10	104
South	66	708	16	177	66	708	16	177	39	414	10	104

Inside Design conditions assumed in the calculations were 75 ^{O}F (23.9 ^{O}C) dry bulb temperature and 50-55 % relative humidity. Mechanical ventilation rates were considered as per ASHRAE Standard 90.2-2010, Table A-1, Specifications for the Standard Design and the Proposed Design. People sensible and latent loads were considered as per Table 1, Chapter 18, ASHRAE Fundamentals 2009. Energy simulations were carried for each city using ASHRAE 90.2-2010 values, local regulations pertaining to the respective region, and optimized building envelope and EER values. Considering that the EER values in ASHRAE 90.2-2010 are higher than local requirements in the region, an additional simulation was performed using ASHRAE 90.2-2010 values for building envelope and the lower EER values prevalent in the region which is identified as ASHRAE 90.2 R(EER). The optimized ME building envelope values are the average values of different local regulations in the region. Optimised ME1 values are values with 10% more stringency and Optimised ME2 values are 20% more stringency compared to Optimum values on the building envelope.

The EER value for the HVAC systems is a very important parameter to be considered. The challenge in increasing the EER values in hot ambient temperature countries is more difficult when combined with meeting the HCFC Phase-out Management Plans (HPMPs). Industry representatives have voiced their concerns about meeting the freeze and reduction targets where alternatives to HCFC-22 in small/medium size air-conditioning applications are not yet introduced and verified by local markets. Additionally, governments have started to apply new energy saving requirements for placing air-conditioning units into markets, and increasing design temperature to 115°F (46°C) and in some countries to 118°F (48°C) calling to increase the ERR values at the same time will disqualify most of the commercially available alternatives. On the other hand, the continuation of HFCs, which are currently promoted as alternatives to HCFC over the last two decades, is currently doubtful due to its contribution to the global warming. HCFCs are used extensively in the refrigeration and air conditioning industry and in particular in the air-conditioning industry and some HFC's may not be the best efficient alternatives for many air-conditioning applications particularly in high-ambient operating conditions.

The EER value of 7.0 at 118°F (48°C) was kept constant for all three cases, since these values of EER are more consistent with the efficiencies currently available from most of the HVAC equipment manufactures in the region considering the harsh climatic conditions and the need to use zero ODP (Ozone depletion potential) and moderate GWP (Global Warming Potential) refrigerants. Building envelope and EER for optimum conditions are indicated in Table 4 below.

Table 4. Building Envelope and EER Data Optimization for the Middle East Region

Description	Units	Optimum ME/ (Average)	Optimum ME 1	Optimum ME 2
Wall 'U' values	(Btu/hr/ft^2/ °F)	0.1	0.09	0.08
Roof 'U' values	(Btu/hr/ft^2/ °F)	0.05	0.045	0.04
Glazing 'U' values	(Btu/hr/ft^2/ °F)	0.47	0.423	0.3768
Glazing SHGC		0.38	0.342	0.304
EER	(Btu/Wh)	7	7	7

CASE STUDY – 1: KUWAIT CITY, KUWAIT

Energy simulation and peak load results for Case Study 1 are summarized in Table 5, which indicates the annual energy consumption for HVAC, non-HVAC components and Total components and Peak cooling loads with various cases, and the associated percentage savings. Percentage savings are compared to ASHRAE 90.2 R(EER) for energy consumption and ASHRAE 90.2-2010 for peak cooling loads.

Table 5. Tabulated Summary of Energy Simulation and Peak Load Results for Case-Study 1 - (Kuwait City, Kuwait)

Component	ASHRAE 90.2	ASHRAE 90.2 R(EER)	MEW Kuwait	Savings (%)	Optimum ME	Savings (%)	Optimum ME 1	Savings (%)	Optimum ME 2	Savings (%)
HVAC	64,505	74,520	58,259	21.8%	64,866	13.0%	60,453	18.9%	56,513	24.2%
Non-HVAC	86,087	86,087	86,087	0.0%	86,087	0.0%	86,087	0.0%	86,087	0.0%
Total	150,592	160,606	144,346	10.1%	150,953	6.0%	146,540	8.8%	142,600	11.2%
Peak Cooling Load	221.10	-	165.00	25.4%	169.70	23.2%	157.90	28.6%	146.90	33.6%

Note: Energy Consumption is in KWH (Kilo-watt hour) and Peak Cooling Load in MBTUH (Million Btu/hr)

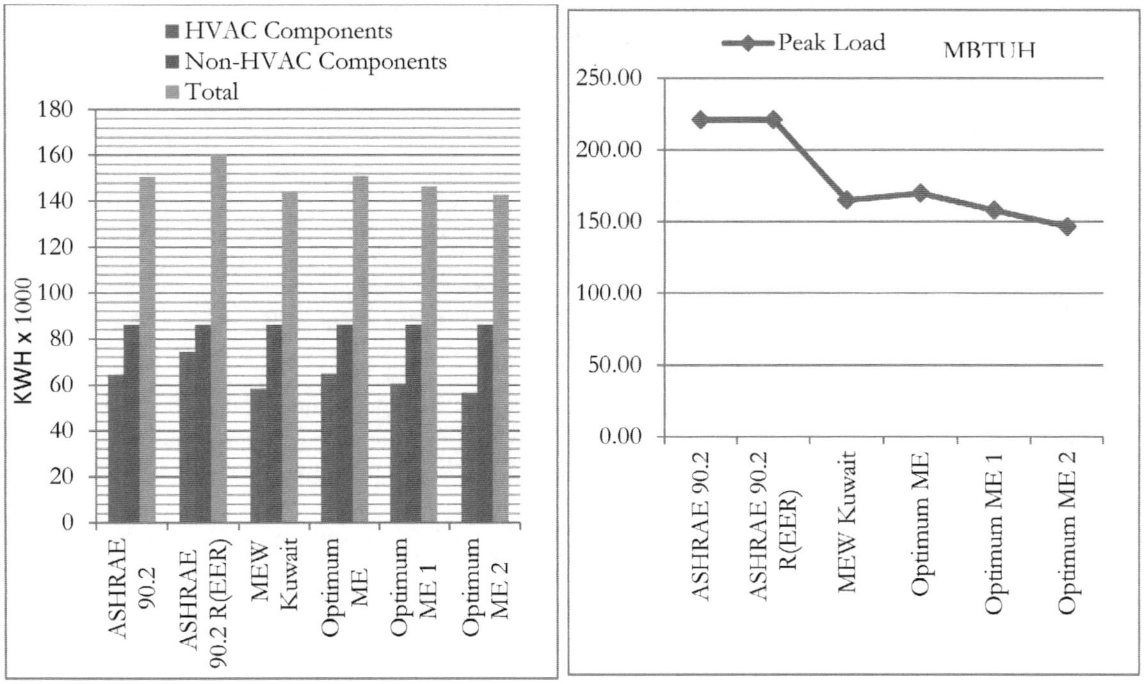

Figure 1: Annual Energy Consumption and Peak Load Bar Chart for Kuwait City, Kuwait

CASE STUDY – 2: DUBAI, UAE

Energy simulation and peak load results for Case Study 2 are summarized in Table 6, which indicates the annual energy consumption for HVAC, non-HVAC components and Total components and Peak cooling loads with various cases, and the associated percentage savings. Percentage savings are compared to ASHRAE 90.2 R(EER) for energy consumption and ASHRAE 90.2-2010 for peak cooling loads.

Table 6. Tabulated Summary of Annual Energy Consumption and Peak Load Results for Case-Study 2 - (Dubai, UAE)

Component	ASHRAE 90.2	ASHRAE 90.2 R(EER)	DM UAE	Savings (%)	Optimum ME	Savings (%)	Optimum ME 1	Savings (%)	Optimum ME 2	Savings (%)
HVAC	67,479	78,333	67,649	13.6%	70,854	9.5%	65,984	15.8%	61,615	21.3%
Non-HVAC	86,087	86,087	86,087	0.0%	86,087	0.0%	86,087	0.0%	86,087	0.0%
Total	153,566	164,420	153,736	6.5%	156,941	4.5%	152,070	7.5%	147,701	10.2%
Peak Cooling Load	163.00		120.20	26.3%	127.90	21.5%	119.20	26.9%	112.20	31.2%

Note: Energy Consumption is in KWH (Kilo-watt hour) and Peak Cooling Load in MBTUH (Million Btu/hr)

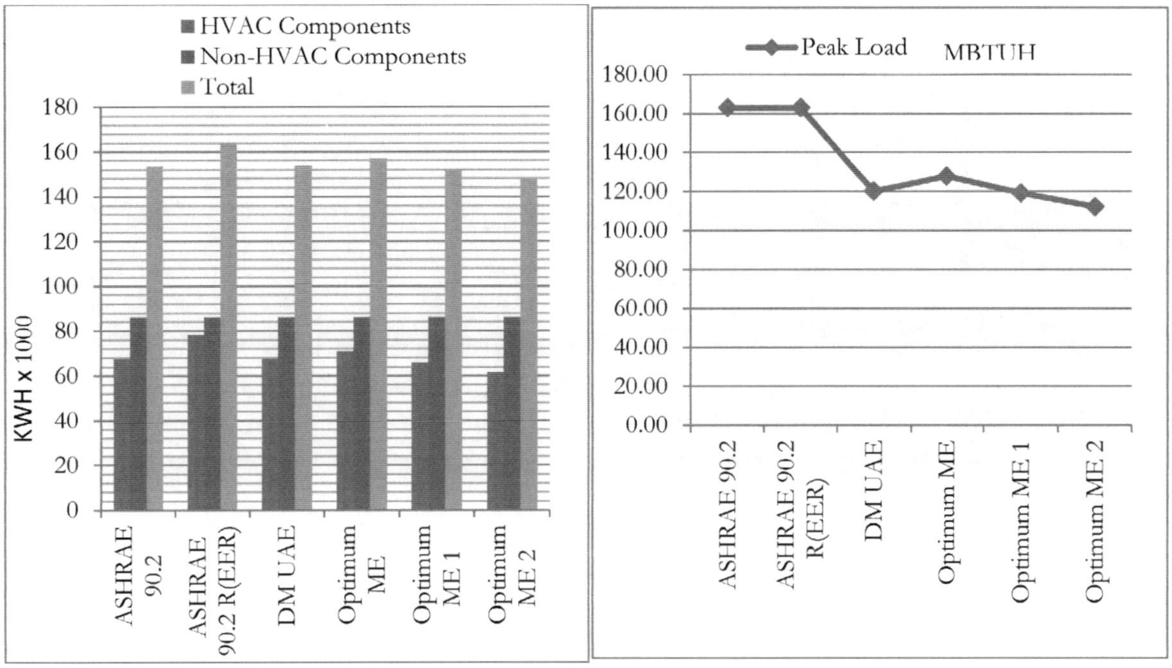

Figure 2: Annual Energy Consumption and Peak Load Bar Chart for Dubai, UAE.

CASE STUDY – 3: RIYADH, KSA

Energy simulation and peak load results for Case Study 3 are summarized in Table 7, which indicates the annual energy consumption for HVAC, non-HVAC components and Total components and Peak cooling loads with various cases, and the associated percentage savings. Percentage savings are compared to ASHRAE 90.2 R(EER) for energy consumption and ASHRAE 90.2-2010 for peak cooling loads.

Table 7. Tabulated Summary of Annual Energy Consumption and Peak Load Results for Case-Study 3 - (Riyadh, KSA)

Component	ASHRAE 90.2	ASHRAE 90.2 R(EER)	SBC 601 KSA	Savings (%)	Optimum ME	Savings (%)	Optimum ME 1	Savings (%)	Optimum ME 2	Savings (%)
HVAC	76,320	88,661	82,773	6.6%	75,316	15.1%	69,806	21.3%	64,840	26.9%
Non-HVAC	86,087	86,087	86,087	0.0%	86,087	0.0%	86,087	0.0%	86,087	0.0%
Total	162,407	174,747	168,860	3.4%	161,403	7.6%	155,893	10.8%	150,926	13.6%
Peak Cooling Load	164.40	-	127.20	22.6%	126.80	22.9%	118.50	27.9%	109.30	33.5%

Note: Energy Consumption is in KWH (Kilo-watt hour) and Peak Cooling Load in MBTUH (Million Btu/hr)

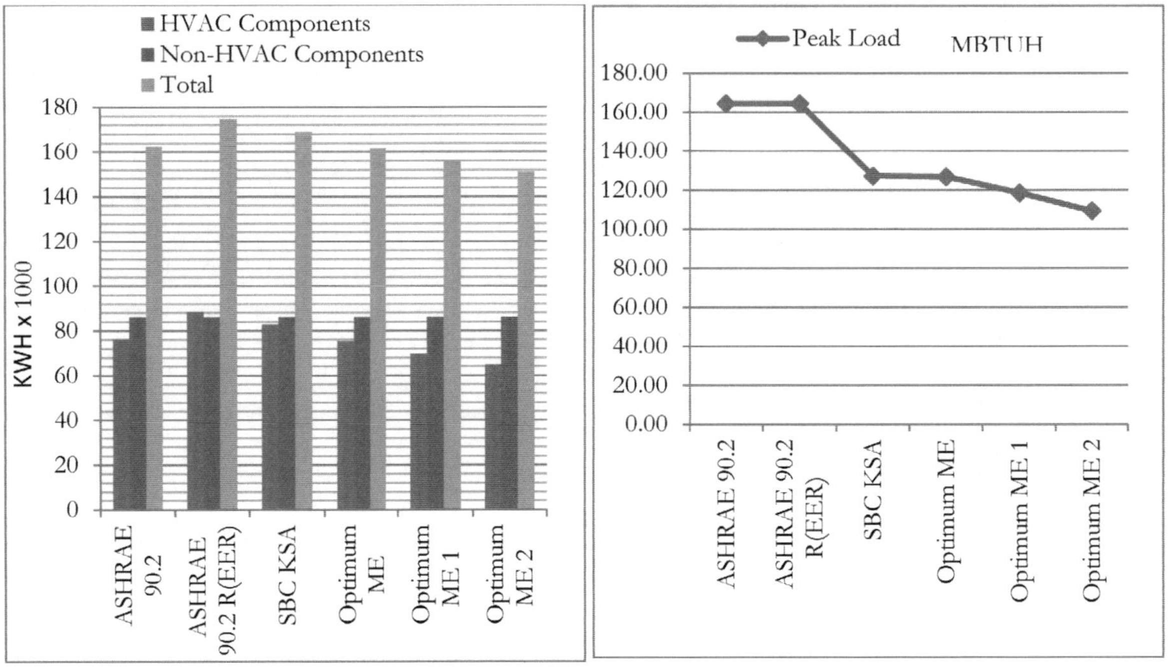

Figure 3: Annual Energy Consumption and Peak Load Bar Chart for Riyadh, KSA.

CASE STUDY – 4: DOHA, QATAR

Energy simulation and peak load results for Case Study 4 are summarized in Table 8, which indicates the annual energy consumption for HVAC, non-HVAC components and Total components and Peak cooling loads with various cases, and the associated percentage savings. Percentage savings are compared to ASHRAE 90.2 R(EER) for energy consumption and ASHRAE 90.2-2010 for peak cooling loads.

Table 8. Tabulated Summary of Annual Energy Consumption and Peak Load Results for Case-Study 4 - (Doha, Qatar)

Component	ASHRAE 90.2	ASHRAE 90.2 R(EER)	Kahramaa Qatar	Savings (%)	Optimum ME	Savings (%)	Optimum ME 1	Savings (%)	Optimum ME 2	Savings (%)
HVAC	76,485	88,904	89,246	-0.4%	75,964	14.6%	70,795	20.4%	66,242	25.5%
Non-HVAC	86,087	86,087	86,087	0.0%	86,087	0.0%	86,087	0.0%	86,087	0.0%
Total	162,572	174,991	175,333	-0.2%	162,050	7.4%	156,881	10.3%	152,329	13.0%
Peak Cooling Load	165.80	-	133.20	19.7%	128.00	22.8%	119.70	27.8%	111.10	33.0%

Note: Energy Consumption is in KWH (Kilo-watt hour) and Peak Cooling Load in MBTUH (Million Btu/hr)

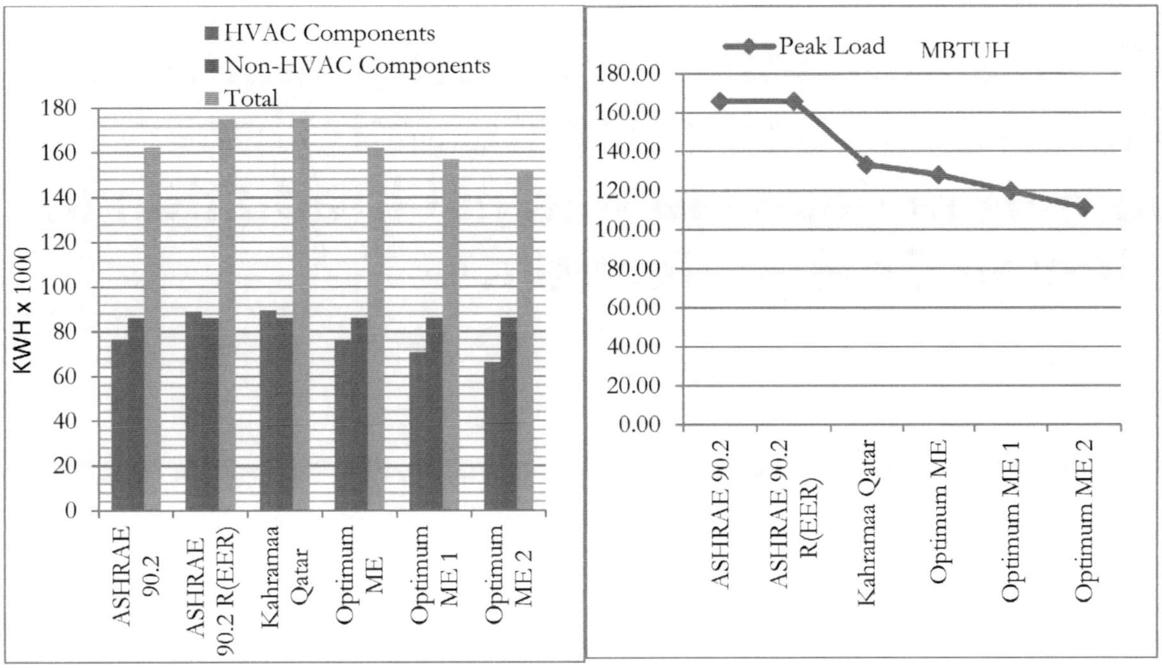

Figure 4: Annual Energy Consumption and Peak Load Bar Chart for Doha, Qatar.

CONCLUSION

The results from the four case studies shows that there is significant opportunity to improve the building envelope performance in the Middle Eastern region, and that energy efficiency ratio (EER) of HVAC systems is an important factor of achieving overall building efficiency in low rise residential buildings.

In most regions of the world, heating and cooling loads represent the largest building-sector energy end-use. The building envelope – the boundary between the conditioned interior of the building and the outdoors – can be significantly improved to reduce the energy needed to heat and cool buildings. Actually, with innovative technologies such as advanced facades, highly insulating windows, high levels of insulation, well-sealed structures, and cool roofs in hot climates, the need for interior conditioning can be avoided in many parts of the world, including some of the fastest-growing regions in hot climates.

Furthermore, while research and development (R&D) will offer improved performance and greater economic viability, there are many products and technologies that are cost-effective and ready for deployment today. Much more can be done to aggressively pursue systems level policies such as effective building codes and improved system efficiencies.

REFERENCES

ASHRAE/IES Standard 90.2-2010, Energy Efficient Design of Low-Rise Residential Buildings.

Energy Conservation Program, Code of Practice MEW/R-6/2010, Ministry of Electricity and Water, Kuwait.

Saudi Building Code Energy Conservation Requirements (SBC 601), Kingdom of Saudi Arabia.

Green building Regulations & Specifications, Dubai Municipality Regulations, Dubai, United Arab Emirates.

Kahramaa Best Practices for Conservation of Electricity and Water, Qatar.

Technology Roadmap, Energy Efficient building Envelopes, International Energy Agency, France.

Energy on demand: the future of GCC Energy Efficiency Middle East Energy and Resources Managing Scarcity for the future, Deloitte & Touche (M.E.)

CFD Investigation of the Performance of Localized Air-Conditioning with Upper-Room Ultraviolet Germicidal Irradiation in Reducing Cross-Infection

Mohamad Kanaan | **Nesreen Ghaddar, PhD** *Member ASHRAE* | **Kamel Ghali, PhD** *Member ASHRAE* | **Georges Araj, MD**

ABSTRACT

Localized airflow has been proven able to provide both temperature and pollutant concentration segregations with limited air mixing between adjacent occupied zones. However, it is not clear whether this type of air distribution can prevent cross-infection within the locally conditioned zone. The aim of this paper is to investigate using computational fluid dynamics (CFD) the transmission of *Staphylococcus aureus*, a common infectious type of airborne bacteria, and the applicability of upper-room ultraviolet germicidal irradiation (UR-UVGI) in minimizing cross-infection in spaces with localized air-conditioning systems. Lagrangian tracking model is used to predict the distribution of pathogens, and the microbiological air quality is assessed based on the pathogen concentration at the breathing level. Results reveal that UR-UVGI is needed to protect occupants in the localized environmental zone from airborne cross-infection in extreme cases of indoor bacteria generation, that delivering 54 W (184.25 BTU/h) UVC to the space may achieve 70% decrease in *S. aureus* concentration at the breathing level.

INTRODUCTION

The heating, ventilation, and air conditioning (HVAC) industry is moving more towards localizing the air conditioning needs of occupants in buildings to minimize energy cost. The concept of localized zonal conditioning is attractive in open large office spaces where zoning is possible by using multiple angled supply jet diffusers with central return for each zone. Lo and Novoselac (2010) used localized air flow distribution system to divide an open cubicle office into thermally and pollutant independent virtual zones without solid partitions. They reported 12% savings on total cooling energy with better air quality.

An important criterion of indoor air quality is the concentration of bioaerosols such as bacteria and viruses. Bioaerosols may be introduced into indoor environments from human expiratory activities such as normal breathing, talking, coughing, and sneezing. Previous studies reported bacteria concentrations up to 13,000 CFU/m^3 (368.12 CFU/ft^3) in air exhaled by infected people depending on the type of disease and expiratory activity (Fennelly et al., 2004; Wainwright

Mohamad Kanaan is a PhD student. Nesreen Ghaddar and Kamel Ghali are professors in the Department of Mechanical Engineering, American University of Beirut, Beirut, Lebanon. Georges Araj is a professor and director of clinical microbiology in the department of pathology and laboratory medicine, American University of Beirut Medical Center, Beirut, Lebanon et al., 2009; Xu et al., 2012). Another indoor source of bioaerosols is biologically contaminated *HVAC* systems whose internal components are considered as ideal environments for bacteria and fungi to multiply and colonize (Kemp et al. 2001, Schmidt et al. 2012, Bródka et al. 2012). Researchers reported that the household air conditioning units can discharge up to 2,500 CFU/m^3 (70.79 CFU/ft^3) of bacteria above ambient levels on initial startup (Jo and Lee, 2008). Bioaerosols have been related since long time to health issues of human beings (Cole and Cook, 1998; Kearns et al., 2000). Therefore, the concentration is an important criterion of indoor air quality. The World Health Organization (WHO, 1988) recommended not more than 100 colony forming units (*CFU*) of bacteria/m^3 of air for hospital environments and a maximum of 500 *CFU*/m^3 (14.16 CFU/ft^3) in office spaces (Ismail et al., 2010; Ross et al., 2004).

Even though localized air flow distribution system has been proven effective in protecting the localized zone from external contaminant transmission, it is not clear whether it can prevent cross-contamination within one environmental zone. In this paper *CFD* modeling is used to address this questionable issue and study the applicability and performance of *UR-UVGI* in improving the biological air quality in a space conditioned using localized air flow distribution system. The *UR-UVGI* has shown efficacy with conventional air distribution systems (Wells et al., 1942; Miller and Macher, 2000; Noakes et al., 2006; Zhu et al., 2013), but to our knowledge no study has investigated the combination of localized zonal air distribution at lower mixing with *UCVI*. The study considers a locally conditioned zone with steady emissions of *S. aureus* from the contaminated cooling unit and infected occupants.

METHOTOLOGY

The localized zonal air distribution system that will be studied is shown in Figure 1. The air distribution system includes a peripheral angled four-sided diffuser which forms a canopy for localizing air flow around the targeted occupied zone and a central exhaust diffuser which aims to withdraw air from the center of the conditioned area.

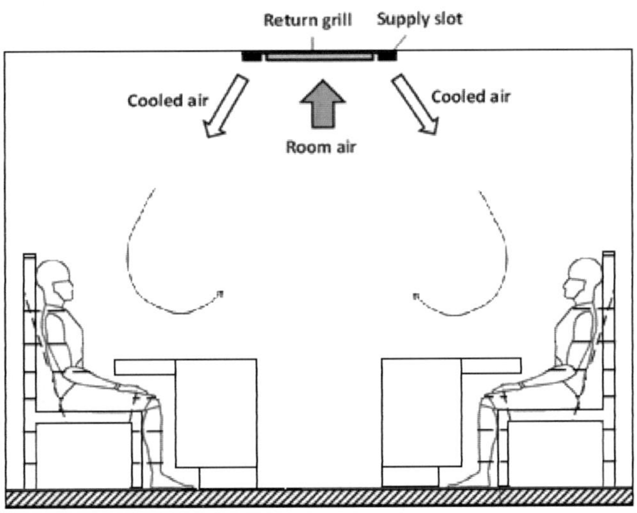

Figure 1 Schematic of the zonal airflow distribution.

This study uses CFD modeling for airflow, thermal field, and bioaerosol transport and inactivation using UV. The simulations will predict the average bacterial concentration at the breathing level, with and without the use of UR-UVGI, and compare obtained values with the WHO standard limit to assess the potential cross-infection in the localized zone.

Modeled Space

A large office space with dimensions of 6 m (19.68 ft) (L) × 6 m (19.68 ft) (W) × 2.7 m (8.86 ft) (H) equipped with a localized zonal air distribution (Figure 2) is simulated using ANSYS 14.5, a well-known *CFD* software. The considered environmental zone contains 4 occupants simulated by heated cylinders of height 1.2 m (3.94 ft) and diameter 0.3 m (0.98 ft) each of sensible load 75W (255.9 BTU/h). Two of the occupants (Cylinders 1 and 2) are infected and assumed to emit each 1,400 CFU/min of *S. aureus*. The bacteria concentration in the supply air is assumed to be 2,500 CFU/m^3 (70.79 CFU/ft^3). These emission rates represent an extreme case to study. Three louvered 18 W (64.42 BTU/h) UV lamps are installed at a height of 2.3 m (7.54 ft).

A flow rate of 0.014 m^3/s (29.66 ft^3/min) is used for a single side supply jet making angle 45° with the vertical (total flow rate is 0.056 m^3/s (118.64 ft^3/min)). Air is supplied at a temperature 16°C (60.80 °F). These supply conditions provide thermal comfort with mean air velocity 0.11 m/s (21.65 ft/min) and temperature 24.1°C (75.38 °F) in the microclimates of occupants, which results in a percentage of dissatisfied people less than 6% (ASHRAE standards 55).

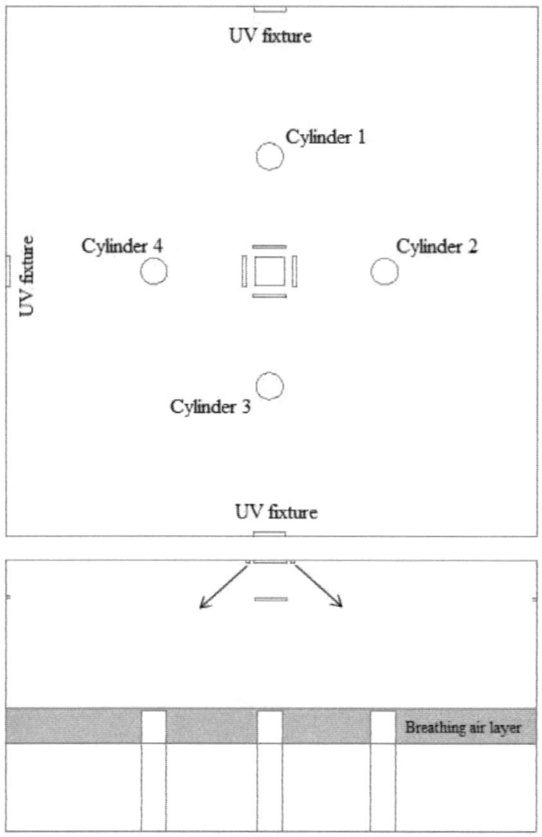

Figure 2 Plots of the climatic room (a) top view and (b) side view.

Airflow and thermal field modeling

The Reynolds averaged Navier-Stokes equations along with standard k-ε are used to simulate turbulent indoor airflow. The domain of the modeled space was discretized into approximately 1,600,000 cells and refined at the supply diffusers and species sources (Figure 3). The option of "standard wall function" was activated requiring the y^+ to be within 30 and 300 in the first cell. An inflation boundary layer was generated at the walls and heated cylinders with a first layer thickness of 1.5 mm (0.059 in) , a growth rate of 1.2 and total of 4 layers. All variables, except the pressure, are discretized with the second-order upwind scheme. STANDARD scheme was used for the pressure term and the SIMPLE scheme was used for coupling pressure and velocity. Numerical convergence is obtained based on many criteria: scaled residuals less than 10^{-5}, stabilized species concentration at the room exhaust, and net imbalance of mass flow less than 1%.

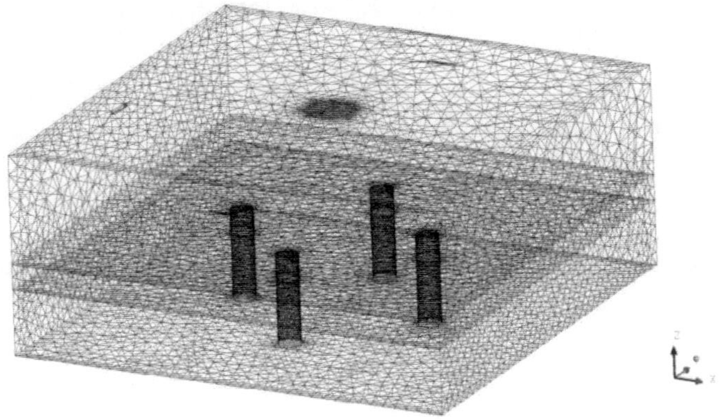

Figure 3 Grid of the simulated office space

Bioaerosol transport modeling

Lagrangian Particle tracking with stochastic discrete random walk (*DRW*) model was used to represent the Eddy interactions of the discrete phase. Throughout the literature review, this model can well predict the transport and dispersion of particles and bioaerosols in indoor environments (Hathway et al. 2011; Heidarinejad and Srebric, 2013).

The expiratory droplets carrying the *S. aureus* cells are simulated as inert spherical particles, 1µm in diameter, with density of 1000 kg/m³ (62.4 lb/ft³) (Hathway et al. 2011). In this method, a total of 500,000 particles are released from the cooling unit supply slots and the infected occupants, and tracked in the simulations.

Particle trajectories are calculated by a fifth order Runge-Kutta method by considering the change in particle velocity u_p due to drag force, inertia $(u-u_p)$, lift force F_L and Brownian motion $n(t)$ thus:

$$\frac{du_p}{dt} = \frac{1}{\tau}\frac{c_d \operatorname{Re}_p}{24}(u-u_p) + g + F_L + n(t) \tag{1}$$

where τ is the particle relaxation time given by:

$$\tau = \frac{\delta d^2 c_s}{18\nu} \tag{2}$$

where, δ is the particle-fluid density ratio, d the particle diameter, ν the fluid kinematic viscosity and c_s is the Cunningham-Stokes slip correction factor given by:

$$c_s = 1 + \frac{2\lambda}{d}\left(1.257 + 0.4\exp\left(-\frac{1.1d}{2\lambda}\right)\right) \tag{3}$$

where λ is the gas molecular mean free path

The deposition of bacteria was simulated using “trap” boundary condition on the walls and other surfaces in the domain.

UR-UVGI effectiveness

The effectiveness of the *UV* device in inactivating a particular microorganism is described using the killing rate (disinfection rate) given by:

$$KR = 1 - S \quad (4)$$

Where S is the microorganism survival rate, S, defined by:

$$S = \frac{C_{UV}}{C_0} = e^{-ZD_{UV}} \quad (5)$$

C_0 and C_{UV} are respectively the concentrations of viable microorganisms before and after *UV* exposure and Z is the susceptibility of the microorganism to *UV* (m^2/J, ft^2/BTU). D_{UV} is the *UV* dose received by the microorganism and it is defined as:

$$D_{UV} = I \times t \quad (6)$$

Where I is the *UV* irradiance intensity, in W/m^2 ($BTU/h.ft^2$), and t is the exposure time is seconds.

UV irradiance modeling

The predictive *UV* model of Wu et al. (2011) is used to integrate the *UV* irradiance field into *CFD* simulations. Local *UV* intensities are implemented into *CFD* control volumes during the post-processing stage. Then, based on the residence time of each particle provided by Lagrangian tracking and the *UV* susceptibility of the microorganism, the received *UV* dose received by each particle can be computed using Equation 6. A particle simulating one bacterium is deleted when the integrated UV dose that it has received can achieve bacterial disinfection rate of 90%. This local *UV* irradiance intensity model for bacteria inactivation is performed by developing a supplementary program (user-defined function) that is compiled by the *CFD* software.

RESULTS AND DISCUSSION

Steady-state *CFD* simulations of the modeled space are carried out for the given conditions. *CFD* plots are presented on a vertical sampling plane in the middle of the space passing through the centers of Cylinders 2 and 4 (Figure 2).

Figure 4 shows the air flow providing a localized environmental zone surrounded by downward air curtains from the peripheral supply jets. This configuration helps to have thermal and flow isolated spaces, then occupancy-based control strategies can also be realized in this setting leading to energy savings and better indoor air quality.

Figure 4 Localized air flow in the conditioned space

In Figure 5 showing *S. aureus* distribution in the conditioned space, the *CFD* result visually displays the containment effect of the localizing air flow, where the majority of exhaled bacteria are confined in the source zone. The indoor air quality is assessed based on the average bacterial concentration in the breathing zone defined as being the horizontal air layer between z = 0.88 m (2.89 ft) and z = 1.22 m (4.00 ft). This concentration was found to be 658 CFU/m^3 (18.63 CFU/ft^3), exceeding the WHO standard limit, when the UV device was OFF, and dropped to the acceptable value 196 CFU/m^3 (5.55 *CFU/ft*^{3}) when 54W (184.25 BTU/h) output was delivered to the space. The UV susceptibility value used for S. aureus is Z = 0.3476 m^2/J (3947.59 ft^2/BTU) (Sharp, 1940).

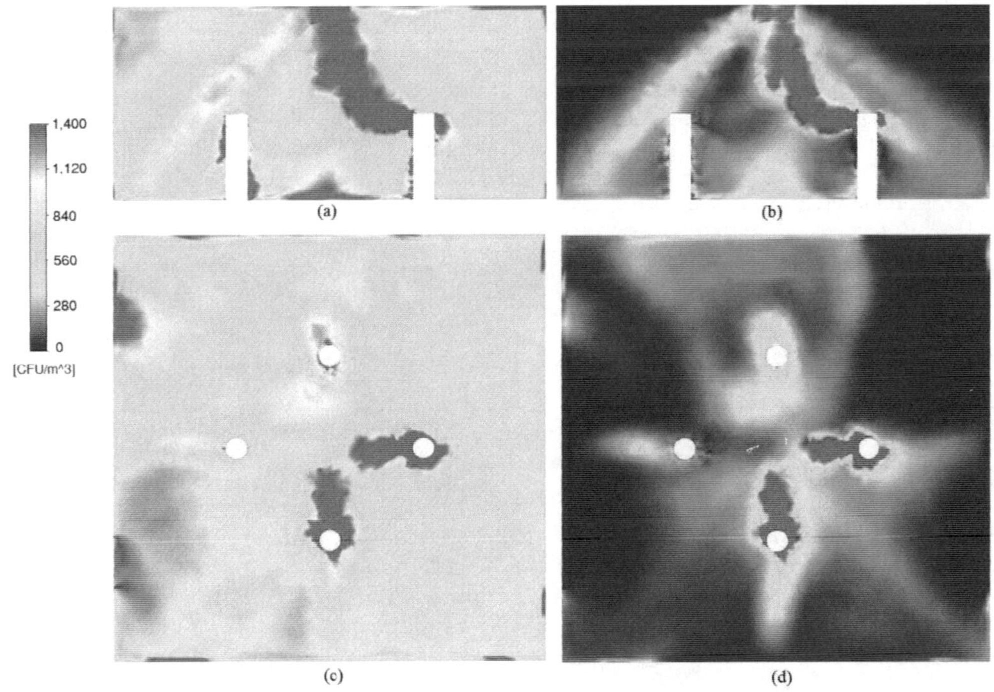

Figure 5 S. aureus distribution in the space on the vertical cut plane y = 3 m (9.84 ft) when (a) UV is OFF and (b) UV is ON and on the horizontal cut plane z = 1.05 m (3.44 ft) when (c) UV is OFF and (d) UV is ON.

CONCLUSION

The *CFD* simulations showed that localized air flow can provide containment of exhaled agents in one localized environmental zone. However, it may not prevent cross-infection within that zone, especially in extreme cases such as the case of high density of infected occupants and high bacterial concentrations in the supply air due to return air recirculation or bacteria growth inside the biologically contaminated *HVAC* system. Results revealed that this issue may be resolved using *UR-UVGI* that can significantly reduce the bacterial concentration at the breathing level in localized air-conditioned spaces.

NOMENCLATURE

n = Brownian motion term
u = velocity
F_L = lift force
t = time
C = bacteria concentration
KR = *UV* killing rate
S = bacteria survival rate
c_d = drag coefficient
c_s = Cunningham-Stokes slip correction factor
d = particle diameter
D = dose
Z = *UV* susceptibility
I = *UV* irradiance intensity
δ = particle-fluid density ratio
λ = gas molecular mean free path
ν = kinematic viscosity
τ = particle relaxation time

Subscripts

p = particle

REFERENCES

ASHRAE. 2004. ASHRAE Standard 55, Thermal Environmental Conditions for Human Occupancy, Atlanta: American Society of Heating, Refrigeration and Air-Conditioning Engineers, Inc.

Brodka, K., Sowiak, M., Kozajda, A., Cyprowski, M., and Irena, S. S. 2012. Biological contamination in office buildings related to ventilation/air conditioning system. Med.Pr., 63(3): 303-315.

Cole, E.C., and Cook, C.E. 1998. Characterization of infectious aerosols in health care facilities: an aid to effective engineering controls and preventive strategies. American Journal of Infection Control, 26: 453-464.

Fennelly, K. P., Martyny, J. W., Fulton, K. E., Orme, I. M., Cave, D. M., and Heifets, L. B. 2004. Cough-generated aerosols of Mycobacterium tuberculosis: a new method to study infectiousness. American Journal of Respiratory and Critical Care Medicine, 169(5): 604-609.

Hathway, E., Noakes, C., Sleigh, P., and Fletcher, L. 2011. CFD simulation of airborne pathogen transport due to human activities. Build.Environ., 46(12): 2500-2511.

Heidarinejad, M., and Srebric, J. 2013. Computational fluid dynamics modelling of UR–UVGI lamp effectiveness to promote disinfection of airborne microorganisms. World Review of Science, Technology and Sustainable Development, 10(1): 78-95.

Ismail, S. H., and Leman, A. 2010. Indoor air quality issues for non-industrial work place. International Journal of Research & Reviews in Applied Sciences, 5(3): 235-244.

Jo, W., and Lee, J. 2008. Airborne fungal and bacterial levels associated with the use of automobile air conditioners or heaters, room air conditioners, and humidifiers. Archives of Environmental & Occupational Health, 63(3):101-107.

Kearns, A. M., Barrett, A., Marshall, C., Freeman, R., Magee, J. G., Bourke, S. J., and Steward, M. 2000. Epidemiology and molecular typing of an outbreak of tuberculosis in a hostel for homeless men. J.Clin.Pathol., 53(2):122-124.

Kemp, P., Neumeister-Kemp, H., Lysek, G., and Murray, F. 2001. Survival and growth of micro-organisms on air filtration media during initial loading. Atmos.Environ., 35(28): 4739-4749.

Lo, L. J., and Novoselac, A. 2010. Localized air-conditioning with occupancy control in an open office. Energy Build., 42(7): 1120-1128.

Miller, S. L., and Macher, J. M. 2000. Evaluation of a methodology for quantifying the effect of room air ultraviolet germicidal irradiation on airborne bacteria. Aerosol.Sci.Technol., 33(3): 274-295.

Noakes, C., Sleigh, P., Fletcher, L., and Beggs, C. 2006. Use of CFD modeling to optimize the design of upper-room UVGI disinfection systems for ventilated rooms." Indoor and Built Environment, 15 (4): 347-356

Ross, C., Menezes, J. R. d., Svidzinski, T. I. E., Albino, U., and Andrade, G. 2004. Studies on fungal and bacterial population of air-conditioned environments. Brazilian Archives of Biology and Technology, 47(5): 827-835.

Schmidt, M. G., Attaway, H. H., Terzieva, S., Marshall, A., Steed, L. L., Salzberg, D., Hamoodi, H. A., Khan, J. A., Feigley, C. E., and Michels, H. T. 2012. Characterization and control of the microbial community affiliated with copper or aluminum heat exchangers of HVAC systems. Curr.Microbiol., 65(2): 141-149.

Sharp DG. 1940. The effects of ultraviolet light on bacteria suspended in air. J. Bactriol, 39:535-54.

Wainwright, C.E., France, M.W., O'Rourke, P., Anuj, S., Kidd, T.J., Nissen, M.D., Sloots, T.P., Coulter, C., Ristovski, Z., Hargreaves, M., Rose, B.R., Harbour, C., Bell, S.C., and Fennelly, K.P. 2009. Cough-generated aerosols of Pseudomonas aeruginosa and other Gram-negative bacteria from patients with cystic fibrosis. Thorax, 64(11):926-931.

Wells, W.F., Wells, M.W., and Wilder, T.S. 1942. The environmental control of epidemic contagion I. An epidemiologic study of radiant disinfection of air in day schools. Am.J.Epidemiol., 35(1): 97-121.

WHO. 1988. Indoor air quality: biological contaminants. World Health Organization, European Series, n. 31, Copenhagen, Denmark.

Wu, C., Yang, Y., Wong, S., and Lai, A. 2011. A new mathematical model for irradiance field prediction of upper-room ultraviolet germicidal systems. J.Hazard.Mater., 189(1): 173-185.

Xu, Z., Shen, F., Li, X., Wu, Y., Chen, Q., Jie, X., and Yao, M. 2012. Molecular and microscopic analysis of bacteria and viruses in exhaled breath collected using a simple impaction and condensing method. PloS One, 7(7):e41137.

Zhu, S., Srebric J., Rudnick S.N., Vincent R.L., Nardell E.A. 2013. Numerical investigation of upper-room UVGI disinfection efficacy in an environmental chamber with a ceiling fan. Photochemistry and Photobiology, 89: 782–791

Fire Management of Underground Car Park Using Computational Fluid Dynamics

Charbel Habchi, PhD **Mahmoud Khaled, PhD**

Mohamad Ramadan, PhD **Georges Harika, Eng**

ABSTRACT

Fire management consists on the design of adequate mechanical ventilation system to allow the smoke extraction within a short time in order to insure the safety of the personals. The zonal calculation method based on global estimations is widely applied to car parks. This method does not permit to test the fire extraction system and to see what happens locally. In this context, the present study report sample results obtained by computational fluid dynamics (CFD) for the ventilation and fire dynamics in a car park basement using the NFPA (National Fire Protection Association) standards. The tenability criteria are the temperature and soot visibility. The fire dynamic simulator (FDS) is used in the present study and which is based on the large eddy simulation (LES) method. Two cases are studied: car park with and without jet fans. It is shown that the jet fans enhance the smoke extraction by reducing the dead spots in the car park.

INTRODUCTION

Underground car parks are confined spaces which are relatively dangerous due to the high risk of vehicle fire. For example, let's consider a car park with 1000 m^2 area which fits about 100 vehicles. Assuming each vehicle is filled with an average of 20 L of gasoline with a heat of combustion of about 40 MJ/kg. Hence, this confined space includes around 56 GJ of flammable energy which corresponds to 13.4 tons of TNT. Thus, the design of adequate ventilation system to allow the smoke extraction within a short time is fundamental for the safety of the personals.

Three methods can be distinguished for fire management systems. The first is the experimental method involving costly measurements and it is also physically risky due to the human presence. The second method is the zonal calculation based on global estimations. The third method is the computational fluid dynamics (CFD) study which is based generally on the finite volume method to discretize the Navier-Stokes and energy equations in every location of the car park (Chow et al., 2006).

Many CFD codes can be found in the literature and which may be used for fire and smoke dynamics simulations. The fire dynamics simulator (FDS) (McGrattan et al., 2010) is a one of the most used CFD software for fire-driven fluid flows (Gissi, 2009, Chow et al., 2006). FDS uses the large eddy simulation method to compute the Navier-Stokes equations appropriate for low-speed, thermally-driven flow with an emphasis on smoke and heat transport. FDS has been widely used to study the smoke and heat control in large atria, car parks and tunnels (Deckers et al., 2013, Weng et al., 2014, Viegas, 2010, Zhang et al., 2007, Enright, 2014) due to its ability to acutely predict the smoke dynamics in confined spaces if it is properly performed (Smardz and Novozhilov, 2006). (Zhang et al., 2007) used FDS to study the effect of ventilation on fire and smoke spread in an underground car park. It is found that the smoke layer reaches the top of the vehicle after

Charbel Habchi, Mahmoud Khaled and Mohamad Ramadan are Assistant Professors in the School of Engineering at the Lebanese International University LIU, Beirut, Lebanon
Georges Harika is Mechanical Engineer at WTA – MEP, Sin El Fil, Lebanon.

about 15 minutes from the fire beginning and the high temperature flows in a random manner accelerating thus the fire spread. (Viegas, 2010) studied the effect of impulse jet fans on the smoke and fire dispersion in underground car parks. They show that for a longitudinal distance of 30 m between 50N thrust jet fans two rows are enough. Moreover, it is shown that the dilution of the smoke near the ceiling decreases the ceiling jet temperature reducing thus the buoyancy. The impact of jet fan ventilation systems on sprinkler activation is studied numerically by (Enright, 2014). It is shown that the increased air flow due to the jet fans can disrupt the heat detection by disrupting the fire plume and ceiling jet. This can cause in delaying sprinkler activation. (Weng et al., 2014) studied experimentally and numerically the smoke movement in metro tunnel and showed a good agreement revealing the accuracy of FDS as an accurate tool. The present study reports the results obtained by CFD for the smoke and fire dynamics in a car park basement using the NFPA and Qatar Civil Defense standards (QCD, 2010, NFPA, 2012a).

NUMERICAL PROCEDURE

FDS uses the LES method to compute the Navier-Stokes and energy equations. In LES, the large flow eddies are explicitly solved while small eddies are implicitly modeled by subgrid-scale (SGS) models (M. Lesieur et al., 2014). In Cartesian coordinates, the filtered governing equations are as follows:

Continuity equation:

$$\frac{\partial \bar{\rho}}{\partial t} + \frac{\partial \bar{\rho}\tilde{u}_j}{\partial x_j} = 0 \tag{1}$$

Momentum equation:

$$\frac{\partial \bar{\rho}\tilde{u}_i}{\partial t} + \frac{\partial \bar{\rho}\tilde{u}_j\tilde{u}_i}{\partial x_j} = \frac{\partial(\bar{p}-\bar{p}_0)}{\partial x_i} + \frac{\partial}{\partial x_j}\left[\mu\left(\frac{\partial \tilde{u}_i}{\partial x_j} - \frac{\partial \tilde{u}_j}{\partial x_i} - \frac{2}{3}\frac{\partial \tilde{u}_k}{\partial x_k}\delta_{ij}\right)\right] + \bar{\rho}g_i\delta_{i3} + \frac{\partial \tau_{ii}}{\partial x_j} \tag{2}$$

Energy equation:

$$\frac{\partial \bar{\rho}c_p\tilde{T}}{\partial t} + \frac{\partial \bar{\rho}c_p\tilde{u}_j\tilde{T}}{\partial x_j} = \frac{\partial \bar{p}_0}{\partial t} + \frac{\partial}{\partial x_j}\left(k\frac{\partial \tilde{T}}{\partial x_j}\right) + \frac{\partial h_j}{\partial x_j} + \tilde{q}_c + \tilde{q}_r \tag{3}$$

Equation of state:

$$\bar{p}_0 = \bar{\rho}R\tilde{T}\sum\frac{\tilde{Y}_s}{M_s} \tag{4}$$

Where τ_{ij} and h_j are the subgrid Reynolds stresses and turbulent heat fluxes respectively and which are modeled using the Smagorinsky model (Smagorinsky, 1963). The signs $\tilde{}$ and $\bar{}$ denote Favre and normal filtering respectively. For more details about the LES method used in FDS, the readers may refer to (Zhang et al., 2007, McGrattan et al., 2010).

OPERATING CONDITIONS

The car park considered in the present study consists of 1125 m^2 surface area and 3 m ceiling height. It is equipped with supply and exhaust air ducts as shown in **Figure 1**. The green arrow in the upper right corner corresponds to the emergency exit. The car park includes 5 impulse jet fans adequately distributed. The mesh used for the present study consists of 153,600 structured cells (160x80x12). Thus the resolution of the computational domain is 0.31x0.26x0.33 m^3.

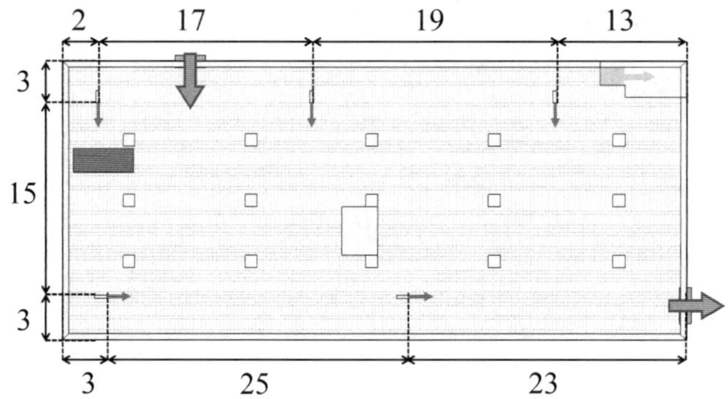

Figure 1 Cark park geometry and dimensions (in meters) showing the location of the supply, exhaust and jet fans as well as the source of fire.

The total volume of the car park is about 3375 m^3. Thus considering 10 air changes per hour (ACH) (ASHRAE, 1989), the flow rates in the supply and exhaust fans are determined as shown in Table 1. The impulse jet fans have 42 N thrust each and the discharge velocity is about 21 m/s. It is assumed that the ventilation system run on full capacity after 2 minutes from fire ignition.

Table 1. Operating conditions of the mechanical ventilation system

Type	Number	Dimensions (cm)	Flow rate per fan (m^3/s)
Supply fan	1	240x100	9.40
Exhaust fan	2	120x100	4.70
Impulse jet fans	5	30x30	1.50

As per the Qatar Civil Defense Standards and Minimum Requirements FSS 7.2 (QCD, 2010), the minimum design fire size for sprinkler protected enclosed parking that permit only passenger cars is 2x5 m^2 area and 4 MW heat release rate (HRR) based on polyurethane properties. With respect to (NFPA, 2012b) rule, this leads to use a fast growing *t*-squared HRR curve which is commonly used to estimate transient fire growth for fire protection design purposes:

$$\dot{Q} = \alpha t^2 \tag{5}$$

where $\dot{Q}$ is the HRR in kW, α the fire growth coefficient which is in the case of fast growing fire equal to 0.0469 kW/s^2 and t the time in seconds. According to the (NFPA, 2012b) 204 rule, the radiative fraction is taken to be 0.35 which means that the HRR is reduced by a fraction of 0.65.

Using equation (5), the fire reaches its maximum HRR of 4 MW at 5 minutes. After this time we consider the fire to stay constant for 100 seconds as shown in **Figure 2**. After that, the fire is considered to be distinguished permitting thus to study the time required to extract the smoke from the car park. Moreover, the fire source must be in the most onerous location, which is the most remote location from the exhaust grills (QCD, 2010).

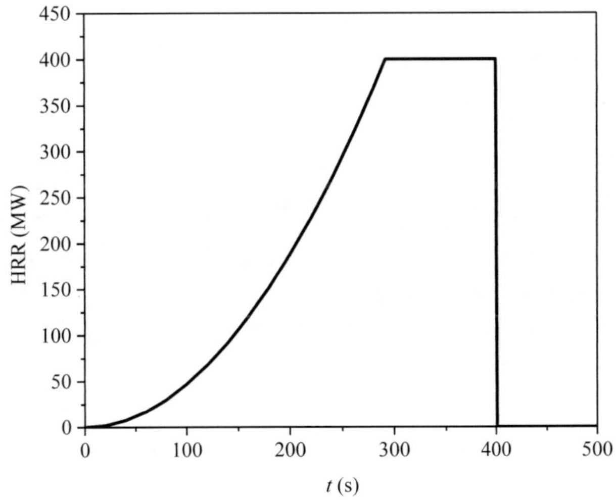

Figure 2 Heat release rate variation with respect to time.

RESULTS AND DISCUSSIONS

The evaluation of the smoke management system is based on standard criteria for evaluating tenable environments as proposed by (NFPA, 2012a) and (QCD, 2010). The following acceptance criteria at 1.8 m above the floor level must be attained within 10 m radius of the design fire:

1. visibility ≥ 10 m
2. smoke temperature ≤ 60°C

These parameters are discussed in this section.

Velocity Field

Figure 3 shows the velocity contours at z = 1.8 m above floor level for t = 30 s. It is observed that high velocity is almost maintained in all the car park area which plays an important role in driving the smoke towards the exhaust grill. The jet fan locations are highlighted by high speed values. Moreover, no dead spots are observed.

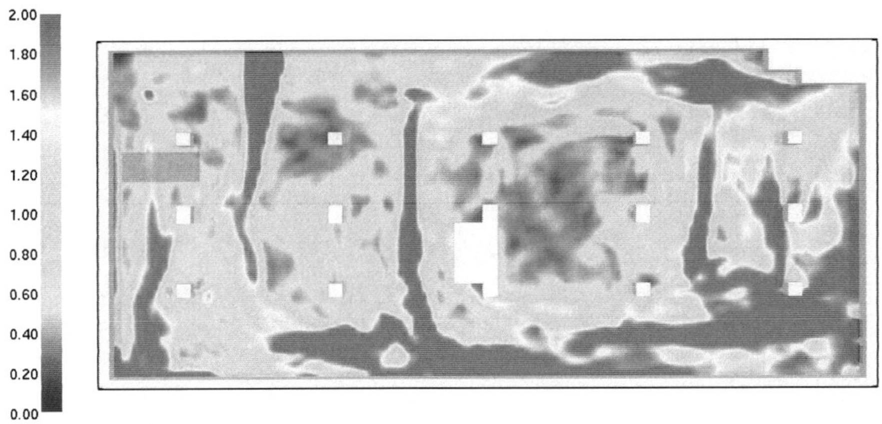

Figure 3 Velocity (m/s) field at z = 1.8 m above floor level for t = 30 s.

Soot visibility

The visibility through smoke is obtained from the following equation:

$$S = \frac{C}{K} \tag{6}$$

where C is the visibility factor characterizing the type of object being viewed through the smoke, for instance light-reflecting signs with $C = 3$. The light extinction coefficient, K, is a product of the density of smoke particles, ρY_s, and a mass specific extinction coefficient, K_m, that is fuel dependent:

$$K = K_m \rho Y_s \tag{7}$$

A value $K_m = 8700\ \text{m}^2/\text{kg}$ is commonly used (Mulholland and Croarkin, 2000).

Figure 4 shows the time evolution of the soot visibility contours at z = 1.8 m above floor level. It is observed for t = 400 s that the visibility stays above the tenability criteria at 10 m radius from the fire source. The jet fans drive the smoke towards the exhaust grills enhancing thus the soot extraction. After the fire is distinguished, the soot visibility increases quickly as the exhaust fans extract the smoke from the car park as shown for t = 500 s.

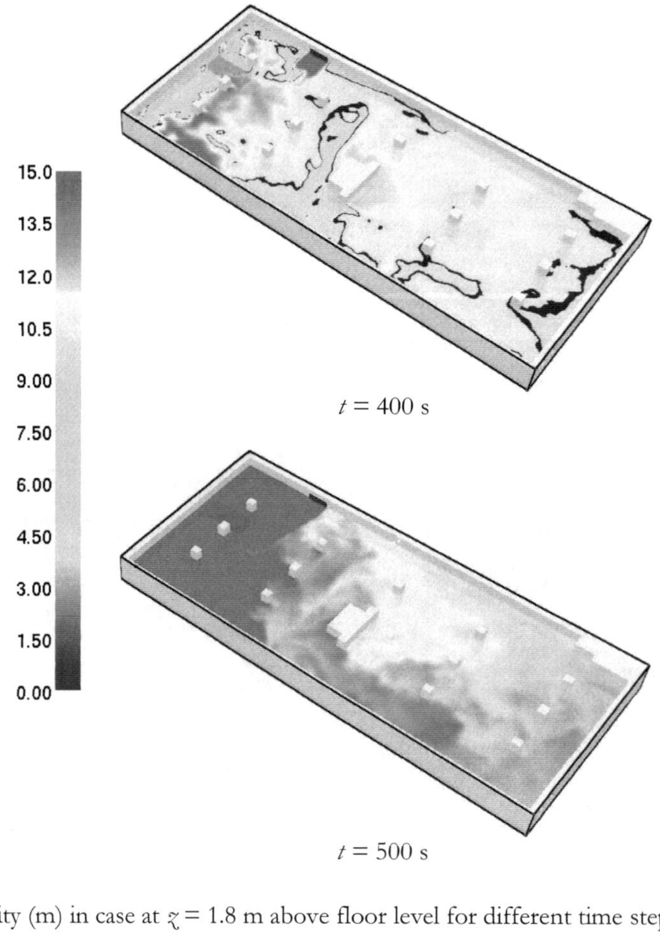

Figure 4 Soot visibility (m) in case at z = 1.8 m above floor level for different time steps.

Figure 5 shows the quantitative time evolution of the minimum soot visibility calculated on a plane at z = 1.8 m above floor level. From this curve it is shown that the minimum visibility reaches around 8.6 m which is slightly below the tenability criterion in case of jet fans while it reaches around 6 m for the case without jet fans. The visibility increases then after the fire is distinguished.

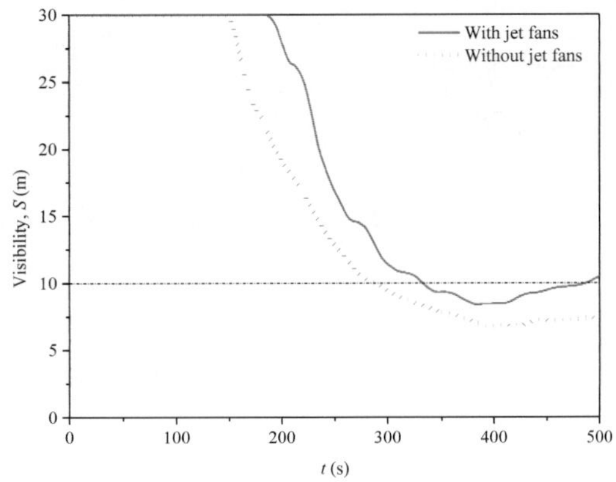

Figure 5 Soot visibility variation with respect to time for both cases calculated at z = 1.8 m above floor level.

Temperature

Figure 6 shows the contours of the smoke temperature for different time steps. It is observed that the temperature increases in time and reaches the tenability criterion after about 320 s. However, near the emergency exit, the maximum temperature reached does not exceed 50°C letting the time for evacuation. **Figure 7** shows the average temperature calculated on a plane at z = 1.8 m above ground level. Once the fire is distinguished the temperature decrease quickly below 60°C as shown in **Figure 7**. In case of jet fans, the temperature stays smaller than in case without jet fans as shown in this figure by about 20%.

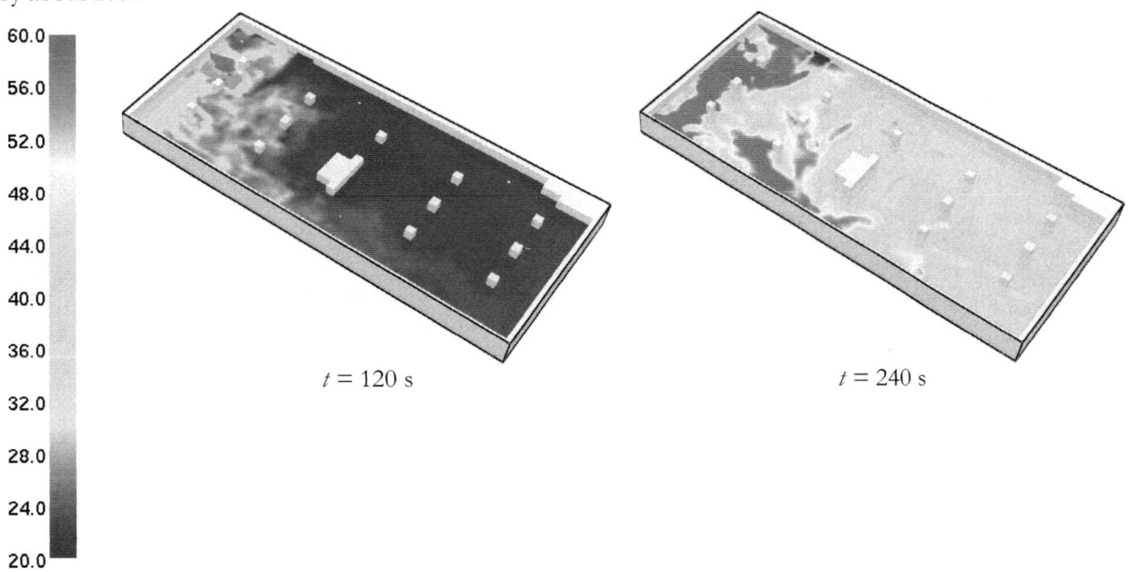

t = 120 s

t = 240 s

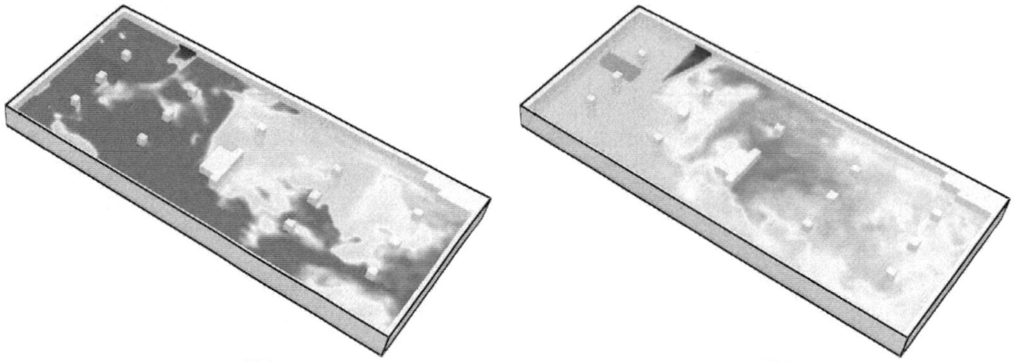

t = 320 s *t* = 480 s

Figure 6 Smoke temperature (°C) at z = 1.8 m above floor level for different time steps

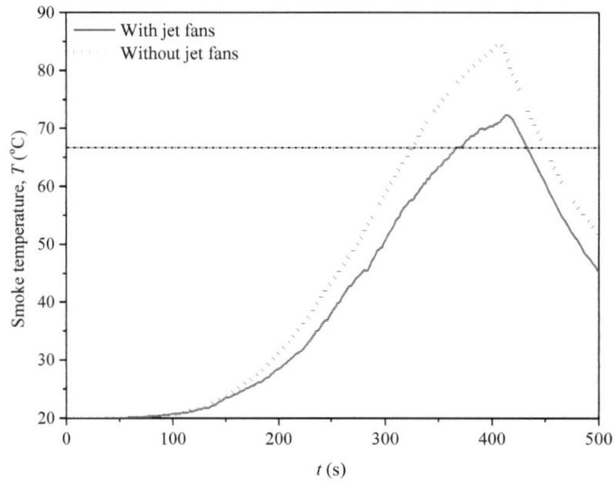

Figure 7 Temperature variation with respect to time calculated at z = 1.8 m above floor level.

CONCLUSIONS

The FDS code is used to compute the Navier-Stokes and energy equations in a car park consisting of 1125 m^2 surface area and 3 m ceiling height. The car park is equipped with supply and exhaust air. Five impulse jet fans are also used to reduce the dead zones. 10 air changes per hour (ACH) is used for the supply and exhaust air flow rates. The fire source corresponds to a passenger car with 2 x 5 m^2 area and 4 MW heat release rate (HRR) with a radiative fraction equal to 0.35.

It is shown that the jet fans play an important role in driving the smoke towards the exhaust and in reducing dead spots. Soot visibility stays above the tenability criteria in all the car park. The temperature exceeds slightly the tenability criterion after 320 s. However, near the emergency exit, the maximum temperature reached does not exceed 50°C letting the time for evacuation.

This study highlights the importance of using CFD simulations to access the local behavior of the smoke dispersion allowing better design of fire extraction systems, emergency exists and smoke detector localization. Future study would deal with alternate ventilation solutions such as under floor soffit and fire control/smoke exhaust at floor level.

NOMENCLATURE

h = turbulent heat fluxes (W/m^2)

p = pressure (Pa)

Q = heat flux (W/m^2)
S = soot visibility (m)
t = time (s)
T = temperature (°C)
u = velocity (m/s)
x = Cartesian coordinates (m)
τ = Reynolds shear stress (Pa)
ρ = density (kg/m^3)

REFERENCES

ASHRAE 1989. ANSI/ASHRAE Standard 62-1989, Ventilation for acceptable indoor air quality. Atlanta: American Society of Heating, Refrigerating and Air-conditioning Engineers, Inc.

CHOW, T. T., LIN, Z. & TSANG, C. F. 2006. Applying large eddy simulation in the study of fire and smoke spread at underground car park. *International Journal on Architectural Science,* 7, 35-46.

DECKERS, X., HAGA, S., TILLEY, N. & MERCI, B. 2013. Smoke control in case of fire in a large car park: CFD simulations of full-scale configurations. *Fire Safety Journal,* 57, 22-34.

ENRIGHT, P. A. 2014. Impact of jet fan ventilation systems on sprinkler activation. *Case Studies in Fire Safety,* 1, 1-7.

GISSI, E. 2009. An introduction to Fire Simulation with FDS and Smokeview. http://www.corbezzolo.org/.

M. LESIEUR, O. METAIS & COMTE, P. 2014. *Large-Eddy Simulations of Turbulence,* Cambridge Cambridge University Press.

MCGRATTAN, K., MCDERMOTT, R., HOSTIKKA, S. & FLOYD, J. 2010. Fire dynamics simulator-user's guide (Version 5). *NIST Special Publication 1019-5.*

MULHOLLAND, G. W. & CROARKIN, C. 2000. Specific Extinction Coefficient of Flame Generated Smoke. *Fire and Materials,* 163, 227–230.

NFPA 2012a. National Fire Protection Association: Life Safety Code. NFPA 101.

NFPA 2012b. National Fire Protection Association: Standard for Smoke and Heat Venting. NFPA 204.

QCD 2010. Qatar Civil Defense Standards and Minimum Requirements: Enclosed car park smoke control ventilation systems performance requirements. FSS 7.2.

SMAGORINSKY, J. 1963. General circulation experiments with primitive equations-I: the basic experiment. *Monthly Weather Review,* 91, 99–105.

SMARDZ, P. & NOVOZHILOV, V. 2006. *Validation of Fire Dynamics Simulator (FDS) for forced and natural convection flows.* Master of Science in Fire Safety Engineering, University of Ulster, UK.

VIEGAS, J. C. 2010. The use of impulse ventilation for smoke control in underground car parks. *Tunnelling and Underground Space Technology,* 25, 42-53.

WENG, M. C., YU, L. X., LIU, F. & NIELSEN, P. V. 2014. Full-scale experiment and CFD simulation on smoke movement and smoke control in a metro tunnel with one opening portal. *Tunnelling and Underground Space Technology,* 42, 96-104.

ZHANG, X. G., GUO, Y. C., CHAN, C. K. & LIN, W. Y. 2007. Numerical simulations on fire spread and smoke movement in an underground car park. *Building and Environment,* 42, 3466-3475.

Cooling PV Arrays Using the Return Air Flow of Air Conditioning Systems

Wassim Salameh, PhD **Ali Assi, PhD**

Charbel Habchi, PhD **Mahmoud Khaled, PhD**

ABSTRACT

Shading and increased temperature decrease the efficiency of photovoltaic (PV) systems. Dust accumulation is one type of shading that PV systems face during operation, while the increased temperature can be attributed to environmental conditions at places where the PV system is installed. Both issues affecting the efficiency of PV systems can be avoided by using a well-designed ventilation system that helps in reducing dust accumulation and the temperature of PV modules. The present work suggests the use of forced air flow of the return air of an air conditioning system to reduce the PV modules' temperature of operation in addition to reducing the dust accumulation on their surface. An analytical study is presented considering the problem as a forced convection between fluid flow and plane surface. The upstream effects on the Heating, Ventilating and Air Conditioning (HVAC) system are investigated and evaluated.

INTRODUCTION

The use of fossil fuel-based and nuclear energy constitutes today the bulk of total energy needed worldwide. The photovoltaic (PV) systems with other renewable energy forms came to alleviate the problem of using traditional forms. However, being basically a silicon P-N junction, the efficiency of solar cell decreases when the temperature of its junction increases [1, 2]. At high temperatures, a considerable percentage of solar energy is converted into thermal energy instead of being converted to electricity. This fact requires innovative techniques to cool down the junction temperature of solar cells. PV/T is a photovoltaic system based on the circulation of a fluid of a lower temperature than the PV in order to extract heat from the panel and decrease its temperature.

Several studies have been done on PV/Ts to cool down PV module using forced air flow [3, 4, 5, 6]. Tonui et al. suggested two modifications for the reference system (REF) placed by forced convection behind the PV module [7]. REF is a system attached behind the PV module providing a single pass air channel. The first REF modification is the thin flat metallic sheet (TMS). The sheet in the TMS creates a double-pass configuration in order to extract the heat surface. FINS is the second modification in which fins of rectangular profile are attached to the back wall of air duct and oriented parallel to the flow direction for good thermal contact. Both modules increase heat transfer from the channel walls to the circulating air but their negative effect is the increasing pressure drop through the system. Table I summarizes the maximum thermal efficiency through the three systems using the following equation:

$$\eta_{th} = \frac{\dot{m}C_p\left(T_{out} - T_{in}\right)}{A_a G} \tag{1}$$

The authors are faculty members in the School of Engineering at the Lebanese International University LIU, Beirut, Lebanon

Where $\dot{m}$ is the air mass flow rate, C_p is the air specific heat capacity, T_{out} is the air outlet temperature, T_{in} is the inlet air temperature, A_a is the aperture area of the PV module and G is the solar irradiance.

As the outlet air temperature increases, the transfer of heat increases resulting in the increase of the thermal efficiency. This table shows that Fins type system gives the higher thermal output than the TMS system with better electrical performance due to the achieved PV cooling. Moreover, an iron glass was inserted to each of REF, TMS and FINS techniques where heat extraction was increased by 275 K (35.6 °F), 277 K (39.2 °F) and 283 K (50 °F) respectively [8].

Table 1. Summary of results obtained in [7]

	REF	TMS	Fins
Efficiency improvement	25 %	28 %	30 %

A numerical study of two case studies on a typical single pass hybrid PV/T air collector has been investigated in [9]. This work revealed that the PV/T is more efficient in winter days; the electrical, thermal and overall collector efficiencies were 12.3%, 19.4% and 53.6% respectively in winter while that for the summer were 9%, 22.8% and 47.8% [9]. Another work [10] showed that the PV/T cells were being affected independently during the cooling process. The electrical efficiency was increased by 2% [10].

A fan is well-known as traditional source of air convection that helps in cooling down PV/Ts. When using fans in PV/Ts, the air flows internally. However, in this paper, a new design using the air return of HVAC system is presented. In this system, no additional modifications are added to the PV system and the air flows externally through surfaces of the PV modules.

THE PROPOSED PV COOLING METHOD

The return air from air conditioning systems is mainly used to reduce the surface temperature of PV modules. The proposed method is illustrated in Figure 1.

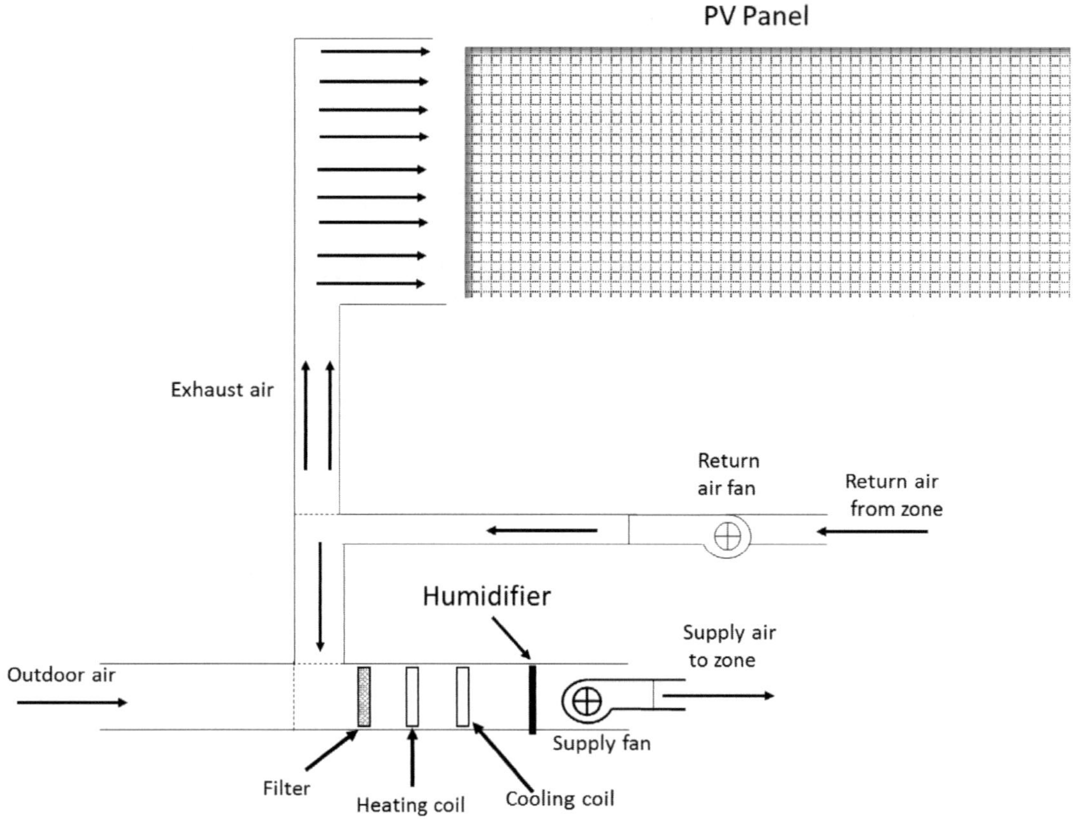

Figure 1 The proposed PV cooling system.

Referring to figure 1, a part of the return air will be used to cool down the surface temperature of the PV array (the exhaust air) while the other part will be re-used in the HVAC system. In [11], these proportions are 50-50 %. The exhausted air has two functions: it reduces the surface temperature in addition to removing the dust so that the PV array functions efficiently.

Therefore, the return air from HVAC systems reduces the high temperature of the PV module surface. The mass flow rate of the return line can be calculated using the first law of thermodynamics [12]:

$$\dot{m} = \frac{\dot{Q}}{i_{room} - i_{supply}} \tag{2}$$

Where $\dot{m}$ is the mass flow rate, $\dot{Q}$ is the total flow rate of heating or cooling, i_{room} is the enthalpy of the air inside the zone and i_{supply} is the enthalpy of the air exit from the cooling or heating coil. Note that enthalpies can be found using the Psychometric chart [14].

The base of the PV module is considered insulated and the heat transfer occurs only through its upper surface. The heat transfer from the solid surface to the adjacent fluid layer is done via pure convection:

$$Q_{panel} = Q_{conv} = h \times (T_{surf} - T_{fluid}) \tag{3}$$

Therefore, the surface temperature can be expressed as follows:

$$T_{surf} = \frac{Q_{panel}}{h} + T_{fluid} \tag{4}$$

where Q_{panel}, Q_{conv} are the rates of the solar irradiance and the rate of convection heat transfer per unit area respectively, h is the convective heat transfer coefficient, T_{surf} and T_{fluid} are the surface temperatures of the PV module and the fluid temperature of the air stream respectively.

The convection heat transfer coefficient, in general, varies along the flow direction. The mean or average heat transfer coefficient for a surface is determined by averaging the local heat transfer coefficient over the entire surface.

In order to calculate the heat transfer coefficient, the type of flow should be determined by calculating the Reynolds number:

$$\mathrm{Re} = \frac{V \times L}{\upsilon} \tag{5}$$

where V and υ are the velocity and the viscosity of the air stream and L is the length of the module.

The average Nusselt number over the flat plate is [14]:

For laminar flow: $$Nu = \frac{hL}{K} = 0.453\mathrm{Re}^{0.5}\,\mathrm{Pr}^{0.33} \text{ with } \begin{cases} \mathrm{Re} < 5\times10^{5} \\ \mathrm{Pr} > 0.6 \end{cases} \tag{6}$$

For mixed flow: $$Nu = \frac{hL}{K} = (0.037\mathrm{Re}^{0.8} - 871)\,\mathrm{Pr}^{0.33} \text{ with } \begin{cases} \mathrm{Re} < 10^{8} \\ 0.6 \le \mathrm{Pr} \le 60 \end{cases} \tag{7}$$

where Pr is the Prandtl number.

Table 2 gives the solar irradiance in W/m^2 on a normal surface data for Al Ain city – United Arab Emirates [11], which we considered as a case study in this work.

Table 2. Solar irradiance in W/m² on a normal surface for Al Ain city

Month	January	February	March	April	May	June	July	August	September	October	November	December
Q_{panel}	662	606	711	628	719	663	587	660	674	747	709	623

Figure 2 shows the monthly average surface temperature of the PV panel with the mass flow rate of the exhaust air calculated using equation (4). The first point represents the surface temperature of the PV without using the return air flow (cooling occurring by natural convection only). On the other hand, the increase in mass flow rate of HVAC system results in lower PV temperatures. The results show that the temperature of PV surface remains constant when the mass flow rate reaches 4 Kg/s (31746.56 Btu/h). The temperature of the PV surface becomes very close to the temperature of the exhaust air. The forced air flow helps the surface of the PV module to reduce its temperature by 293 K (68 °F) (during August).

The heat extracted from the PV surface depends on the mass flow rate of the exhaust air and thus the same arguments hold (figure 3).

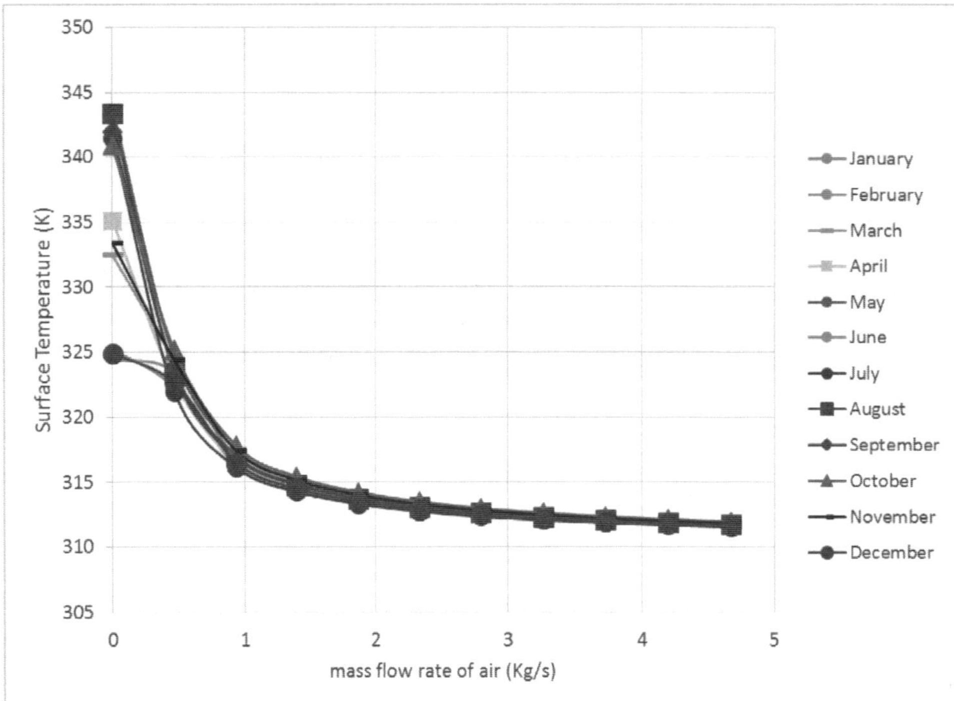

Figure 2. Temperature reduction of PV modules for different mass flow rate of air

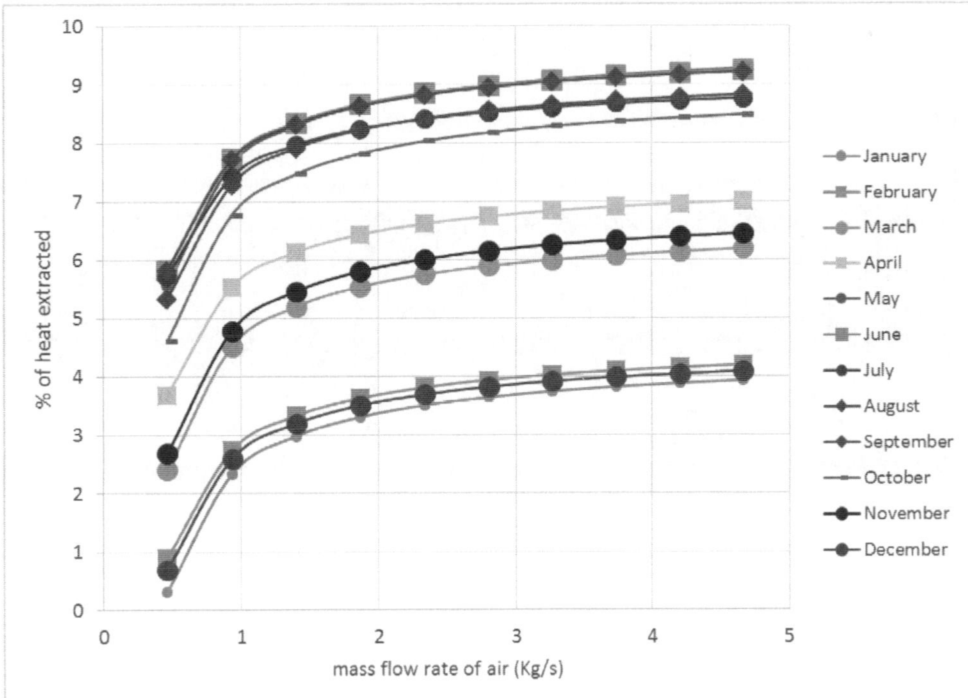

Figure 3. Heat extracted from PV modules for different mass flow rate of air

CONCLUSION

Theoretical and simulation results showed that exhausted air of HVAC systems could be used to reduce the surface temperature of PV modules in PV systems installed within buildings where HVAC is available. Reducing the operating temperature of PV modules will improve the efficiency of the PV system. We are currently planning to set up an experimental model to validate the theoretical and simulation results claimed in this paper.

NOMENCLATURE

$\dot{m}$	= the mass flow rate (kg/s)
$\dot{Q}$	= the total flow rate of heating or cooling (W)
I	= enthalpy (J)
Q_{panel}	= rate of the solar irradiance (W)
Q_{conv}	= rate of convection heat transfer (W)
h	= convective heat transfer coefficient (W/m^2.K)
T	= Temperature (K)
V	= velocity (m/s)
υ	= kinematic viscosity (m^2/s)
L	= Length of the module (m)
Nu	= Nusselt number (-)
Pr	= Prandtl number (-)
Re	= Reynolds number (-)
ϱ	= density (kg/m^3)

K = thermal conductivity coefficient (W/m.K)

C_p = specific heat capacity (J/kg.K)

T_{out} = outlet air temperature (K)

T_{in} = inlet air temperature (K)

A_a = aperture area of the PV module (m^2)

G = solar irradiance (W/m^2)

Subscripts

room = inside the zone

supply = the air exit from the cooling or heating coil

surf = surface

Fluid = fluid

REFERENCES

[1] E. Skoplaki, J.A. Palyvos. 2009. On the temperature dependence of photovoltaic module electrical performance: A review of efficiency/power correlations. Solar Energy, 83, 614–624

[2] V.Jafari Fesharaki, Majid Dehghani, J. Jafari Fesharaki. 2011. The Effect of Temperature on Photovoltaic Cell Efficiency. Proceedings of the 1st International Conference on Emerging Trends in Energy Conservation – ETEC Tehran, Tehran, Iran, November 20-21.

[3] Kelly, N., Strachan, P.A. 2000. Modeling enhanced performance inte-grated PV modules. In: Proceedings of the 16th European PV SolarEnergy Conference, Glasgow, UK, May 1–5, 2025–2028.

[4] Tripanagnostopoulos, Y., Nousia, Th. and Souliotis, M. 2000a. Low costimprovements to building integrated air hybrid PV-Thermal systems.In: Proceedings of the 16th European PV Solar Energy Conference,Glasgow, UK, May 1–5, 1874–1877.

[5] Tripanagnostopoulos, Y., Bazilian, M. and Zoulia, I., Battisti, R. 2000b.Hybrid PV/T system with improved heat extraction modification. In:Proceedings of the PV in Europe – from PV Technology to EnergySolutions Conference, Rome, Italy, October 7–11, 718–721.

[6] Tripanagnostopoulos, Y., Nousia, Th., Souliotis, M., Yianoulis, P., 2002.Hybrid photovoltaic/thermal solar system. Solar Energy 72, 217–234.

[7] J.K. Tonui, Y. Tripanagnostopoulos. 2007. Improved PV/T solar collectors with heat extraction by forced or natural air circulation. Renewable Energy 32, 623–637.

[8] J.K. Tonui, Y. Tripanagnostopoulos. 2008. Performance improvement of PV/T solar collectors with natural air flow operation. Solar Energy. 82, 1–12.

[9] Karima E. Amori, Hussein M. Taqi Al-Najjar. 2012. Analysis of thermal and electrical performance of a hybrid (PV/T) air based solar collector for Iraq. Applied Energy. 98, 384–395.

[10] Anas Al Tarabsheh, Spyros Voutetakisb, Athanasios I. Papadopoulosb, Panos Seferlisb,c, Issa Etiera, Omar Saraereha. 2013. Investigation of Temperature Effects in Efficiency Improvement of Non-Uniformly Cooled Photovoltaic Cells. CHEMIICAL ENGIINEERIING TRANSACTIIONS. 35, 1387–1392.

[11] A. Assi, A. Hassan, M. Al-Shamisi and H. Hejase "Removal of Air Blown Dust from Photovoltaic Arrays Using Forced Air Flow of Return Air from Air Conditioning Systems" The International Conference on Renewable Energies for Developing Countries (REDEC 2012), 28 -29 November 2012, Beirut Lebanon.

[12] Moran J., Shapiro N.M. Fundamentals of engineering thermodynamics. Sixth edition. 2007, Wiley.

[13] McQuiston F., Parker J., Spitler J.. Heating, Ventilation and Air Conditioning - Analysis and Design - Sixth Edition, Willey.

[14] Incopera F., Dewitt D., Bergman T., Lavine A. Fundamentals of Heat and Mass Transfer. Sixth Edition, 2007, Willey.

Efficient Method to Utilize Solar Energy in Buildings Via Combined Solar Thermal Technologies

Armen Gharibyan, PhD
State Engineering University of Armenia

Ken Touryan, PhD
American University of Armenia

Arman Shvagir, Msc
CBS LLC

ABSTRACT

HVAC systems in buildings consume the largest amounts of energy. They also contribute to green house gases significantly, and are responsible for ozone depletion. It has been shown that the best economically viable active solution for reducing the energy consumption from HVAC systems is the utilization of solar energy in buildings. In this study a combined solar space heating, cooling, and solar hot water scheme and methodology is proposed which will allow maximal utilization of solar thermal energy generated from solar thermal collectors, in all the seasons of the year. All calculations and assumptions have been done for the continental climate zones. An optimization objective function has been set up for determination of the optimum area of solar thermal collectors that could yield maximum cost savings, based on monthly energy consumption patterns. It has been shown that the suggested methodology provides full utilization of the available solar thermal energy, year round.

INTRODUCTION

Solar water heating for domestic hot water supply is the most straightforward way to use the solar thermal collectors in an efficient way. Solar domestic hot water heating is already being widely used in number of countries. However, worldwide it contributes to only 0.4% of the overall energy demand for domestic water heating. Space heating using solar collectors is also gaining widespread use, frequently combined with heating of domestic water, in relatively cold but sunny climates. In the case of combined solar water heating and solar space heating for commercial and residential buildings, sizing of solar collectors is mostly based on the space heating demand, which in turn leads to the installation of much larger size collectors then those needed for solar water heating. This kind of sizing strategy leads to a significant surplus of generated hot water in warmer seasons, which in turn makes the combined solution neither economically feasible, nor efficient when utilizing solar energy. This problem is more prominent in temperate or continental climates subject to colder winters and hotter summers. The better way to utilize extra hot water is to direct it to the hot water driven absorption chiller and to utilize it for refrigeration. The purpose of this study is to suggest an efficient way of utilizing solar energy in buildings by combining solar space heating, solar water heating with solar cooling. All the studies and calculations for this research have been carried out for the continental climate zones, using in particular data available for the republic of Armenia.

METHODOLOGY

There are a number of ways of using combined solar space heating and cooling systems, where solar thermal space heating is accomplished using under floor heating in winter and hot water driven absorption chillers in summer. In continental climate regions, during warmer months, there is always surplus energy when installing relatively large solar thermal systems. In the case of combined water heating and space heating scenario a significant surplus of energy is generated during the summer months and in case of combined space heating and cooling scenario an energy surplus is generated during relatively mild weather months (mostly in autumn and spring). To make sure that the thermal energy generated by solar thermal plant is used efficiently in continental climates, an efficient scheme and methodology has been developed for commercial and residential buildings, presented in this work. The schematic of proposed system is shown in the Figure 1.

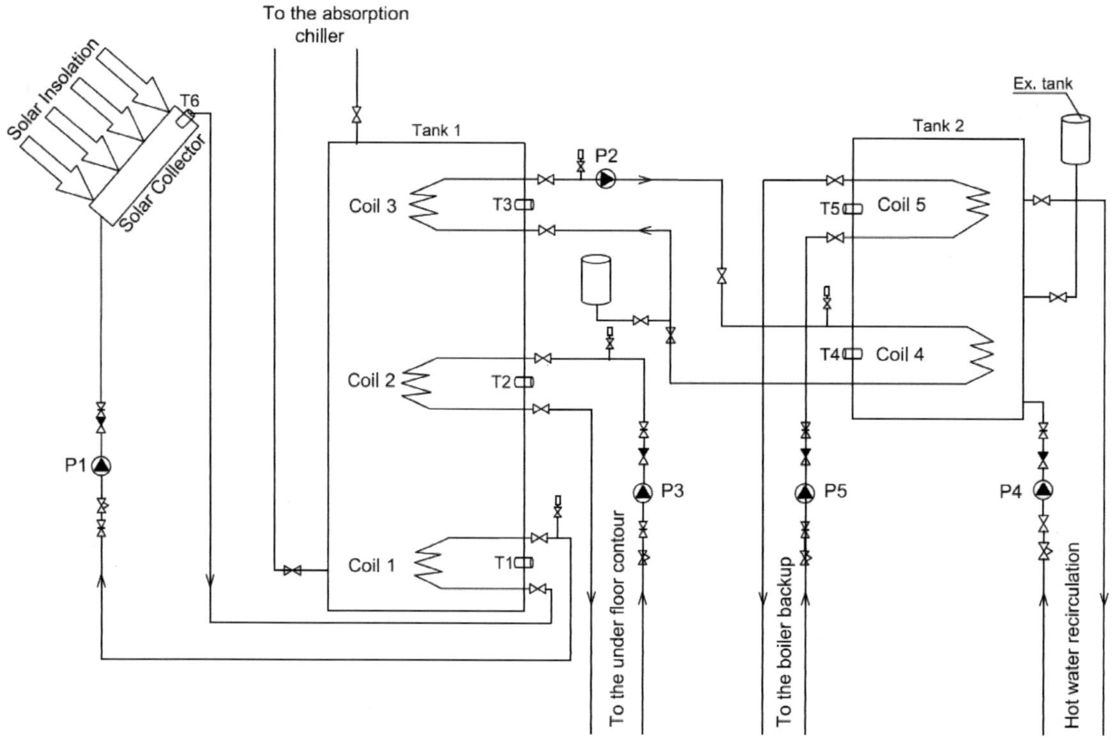

Figure 1 Schematic of combined solar space heating, solar cooling and solar water heating system

The sequence of operation of the proposed system is as the follows: If $T_6+8>T_1$ then start P_1. Otherwise stop P_1. If $T_5<65^0C$ (149 °F) & $T_3<45^0C$ (1113 °F) then start the boiler & P_5 or else stop P_5. If $T_3>T_4$ & $T_5<65^0C$ (149 °F) start the P_2 or else stop P_2. During winter season if $T_2<40^0C$ (104 °F) then start the P_2 to transfer energy from Tank 2 from Tank 1. As far as the hydraulics is concerned, the proposed scheme will work perfectly, because all the contours are separated by utilizing tank-integrated heat exchangers. Separation of all the coils also prevents the mixing of the liquids inside different loops thus increasing the operational safety and the lifetime high efficiency of the heat transfer in the coils. In case solar capacities are greater than 200kW (682kBtu/hr), tank integrated heat exchangers become quite expensive and the coils 1,3,4,5 are recommended to be external ones, with additional pumps for the circulation of the first contour. In conventional schemes suggested by other authors, few three way and two way valves are used for switching between heating and cooling contours. In the scheme suggested in this paper, reliability of the combined system is increased up to a maximum level by having simple thermostatic controls which eliminates the mixing of liquids among different contours. One purpose in the development of the above scheme was to keep it simple and to make the operation and maintenance as easy as possible. 85/75 supply return temperature scheme is recommended to consider for the contour of absorption chiller.

The solar assisted air-conditioning systems use the heat from the solar radiation to drive a thermally-driven chiller such as absorption chiller. Solar assisted air conditioning system produces cooling with considerably less electricity demand than conventional air-conditioning systems. With current solar collector technology like evacuated tube solar collector which able to produce high temperature approximately 88°C, it has made heat source from solar energy viable to drive the absorption chiller to produce chilled water for indoor cooling purposes. In addition, the working fluids like Lithium Bromide used in the absorption chiller does not contribute to global warming, contrary to most working fluids in conventional compression chiller

Capacities of coils 1,3 &4 should be the same and they can be determined from the installed number of solar water heaters. As heat pipe evacuated tube solar water heaters have been considered in this study, correlation of power capacities of coils 1, 3 &4 have been determined as a function of the number of vacuum tubes, expressed below.

$$P_{coil} = N_{tube} * 0.085 * Eff_{summer} * E_{peak}/1000 \,, \tag{1}$$

Where P_{coil} is the coil capacity in kW (kBtu/hr), N_{tube} is the number of installed vacuum tubes, Eff_{summer} is the highest efficiency of vacuum tubes recorded during summer, E_{peak} is the maximum insolation in kWh (Btu) per square meter (10,8 square feet) on a horizontal surface recorded during summer, 0.085 is the absorbing surface of each 1.8m (5 feet) length 58mm (0.3 feet) diameter vacuum tube in square meters recommended by the manufacturers. For most of the continental climate countries where around 900-1000W/m^2 (285-317 Btu/(h*ft2)) solar irradiation is recorded during the summer season, equation 1 can be simplified and written in the form:

$$P_{coil} \approx 0,07 * N_{tube} \tag{2}$$

UTILIZATION OF SOLAR ENERGY

Solar collectors

There are different types of solar water heaters available in the market of which flat plat collectors and evacuated tube collectors are the most efficient ones and are also most popular. In the continental climatic zones evacuated tube collectors demonstrate higher generated annual energy outcome than flat plate collectors because of higher vacuum based insulation and ability to absorb diffused solar infrared irradiation during the winter season. There are two types of evacuated tube collectors available; water inside and heat pipe. The most important difference between water inside evacuated tube solar collectors and heat pipe evacuated tube solar collectors is that the heat carrying fluid inside the copper heat pipe is not connected to the solar loop. It should be noted that in case of providing space heating, hot water heating and cooling via solar energy, it is reasonable to make all the loops closed and pressurized. That is why heat pipe evacuated tube solar water heaters are proposed for such systems. Heat pipes are widely used for heat recovery and energy saving in various ranges of applications because of their simple structure, flexibility, high efficiency, compactness and excellent reversibility. The heat pipe vacuum tube collects heat from the sun at high efficiency. It is important for the heat pipe solar collectors to be installed with a right tilt depending on the application to generate the maximum amount of energy on annual basis. They operate like a thermal diode where the flow of heat is in one direction only. This type of collector is commonly filled with alcohol or water in vacuum and operates in two versions, one with a dry and one with a wet connection.

Optimum solar collector area

It is obvious, that if more solar collectors are installed, more energy savings will be achieved. However, up-front costs can be higher and the project can become uneconomical. Optimal quantity of collectors depends on a few key parameters. These are: the climate data where the project is located, tilt of the collectors, solar insolation data, efficiency of the collectors and the price of the fossil fuel used. Average solar insolation data and fuel prices for the republic of Armenia have been considered in the optimization process, to represent the continental climate zone. Monthly solar insolation data in kWh/m^2 (Btu/ft^2) and corresponding monthly energy generated from each square meter of active absorbing area are presented in the Figure 2.

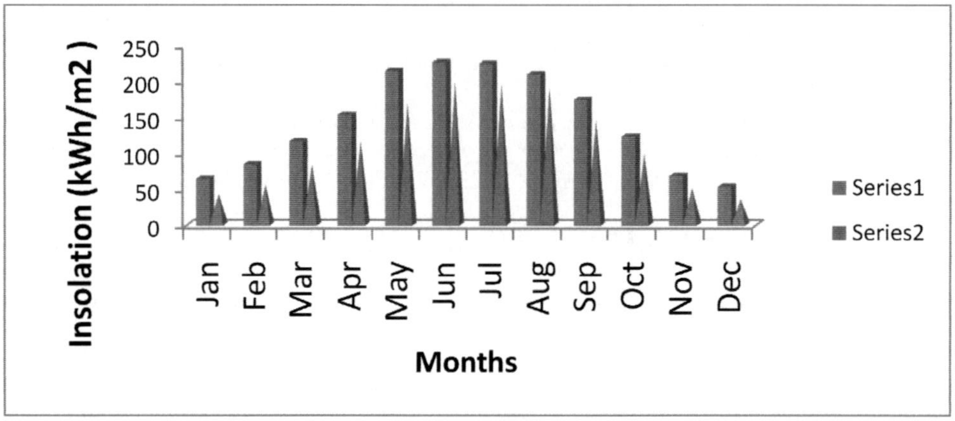

Figure 2 Typical monthly solar insolation data per meter square (Series 1) and thermal energy generated from one meter square absorbing area for continental climate zones (Series 2)

From the Figure 2 it is seen that efficiency of heat pipe evacuated tube solar collector has a seasonal variation, depending on the outdoor temperature. To find the optimal collector size, first the monthly energy consumption demand should be determined. Monthly consumption of typical 10,000 m^2 ($10{,}76*10^4$ ft^2), medium insulation building located in a continental climatic zone is presented in the Figure 3.

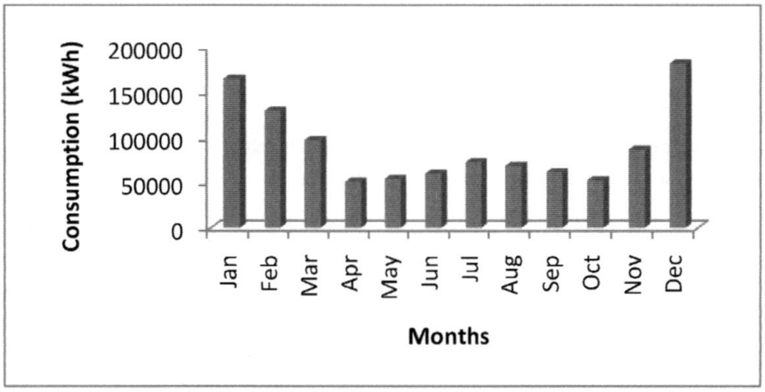

Figure 3. Monthly consumption of typical 10,000 m^2 (10,76*104 ft^2)commercial building with medium thermal insulation, located in a continental climate zone

Typical commercial buildings in Armenia are usually heated by natural gas and cooled by electricity driven, air cooled heat pumps. That is why the volume of consumed gas is always converted to kWhrs (Btu), considering the heat generation calorific value of the local gas. Because gas and electricity prices are different, the cost of thermal energy generated during the heating season is different from the cost of thermal energy generated during the cooling season. Calculating the monthly energy demand of a building in kWh (Btu), and the energy generated by one square meter solar collector area in kWh (Btu), one can convert both data to equivalent amount of money, and thus can utilize a cost minimization function. As natural gas is consumed during the heating season and electricity is consumed during the cooling season, the cost savings of 1kWh (Btu) thermal energy generated by solar collector will be different during different seasons. To find the maximum surface area, or number of tubes of solar collectors, an objective function has been developed, denoted by Z, which represents a summation over 12 months, using the absolute values of the difference between the monthly consumption of the building and cost savings generated by the solar collector:

$$Z = \sum_{m=1}^{12} \left| C_{dem}(m) - C_{gen}(m) \right| \quad (3)$$

Where $C_{dem}(m)$ is the energy consumption (or energy demand) of the building, $C_{gen}(m)$ is the monthly cost generated by the installed solar collector capacity. The monthly values of $C_{dem}(m)$ should be determined either by simulation or from actual energy consumption data of the existing building. As $C_{gen}(m)$ depends on the monthly efficiency of the solar collectors and monthly solar insolation, it can be presented by one of the following formulas depending on the type of the fuel used during the month under consideration.

$$C_{EL\_gen} = 11N_{tube} * C_{tube} * Eff_m * T_{el} * E_m \quad (4)$$

$$C_{Gas\_gen} = 11N_{tube} * C_{tube} * Eff_m * E_m * T_{gas}/Q_{Gas} \quad (5)$$

Where N_{tube} is the quantity of evacuated heat pipe tubes which is also the main optimization parameter, C_{tube} is cost of a single tube in the country, Eff_m is the monthly efficiency of the collector, E_m is the monthly solar insolation expressed in kWh/m^2 (Btu/ft^2), T_{el} is the tariff for 1kWh (3412Btu) electricity, T_{gas} is the tariff for 1 cubic meter natural gas and Q_{Gas} is the energy contained in 1 cubic meter gas. Number 11 in the formulas above reflects the number of vacuum tubes in $1m^2$ absorbing area. An optimization model has been created in the "Mathematica" software package. Parasitic sources of energy consumption like pumps, valves and fans have been excluded from the monthly energy demand, because only primary thermal energy sources are of interest. An average cost of $35 has been considered per installed tube. Typical 10,000 m^2 (10,76*104 ft^2) building thermal energy demand for cooling, heating and water heating is depicted in the Figure 4. The figure represents the monthly cost savings generated by the solar collectors during heating and cooling seasons.

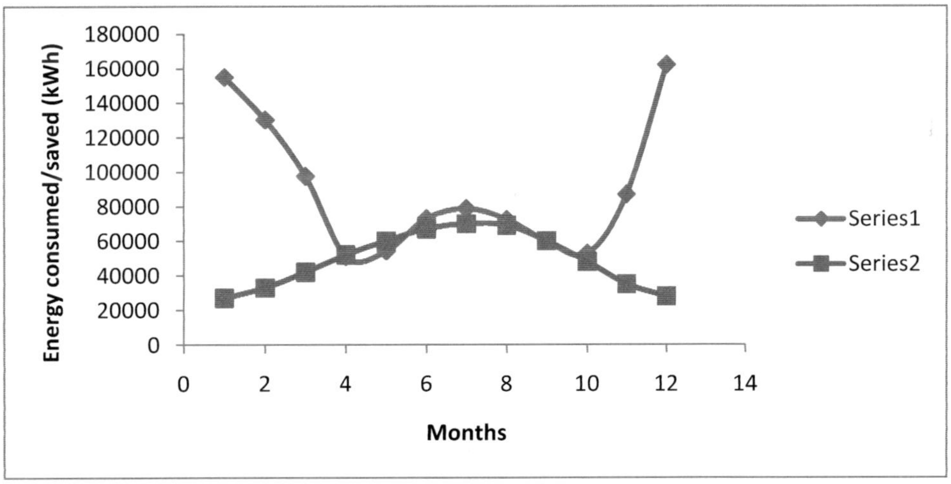

Figure 4 Monthly consumption of typical 10,000 m^2 (10,76*104 ft^2) commercial building (Series 1) and cost savings generated by the optimal quantity of vacuum tube solar water heaters (Series 2)

Fig. 4 can be divided in three zones. First zone is located between months 1-4, 6-8 and 10-12 where the deficit in solar energy is covered by the boilers. The second zone starts from month 4 and lasts around 1.5 months where a little surplus is observed because of free cooling through ventilation, using the temperature advantage gained by the outdoor air. The third zone is located in the region of months 8-10 where the energy cost generated by the solar collectors coincides with the thermal energy demand of the building. Solar fraction is a very important quantity in solar engineering which indicates the part of the load covered by the energy generated by sun. Solar fraction is usually determined by the ratio of generated monthly solar energy of the system and monthly actual consumption of the building. In this case it is reasonable to use the solar saving fraction (SSF) which is determined by the ratio of monthly cost savings generated by the suggested combined system and monthly energy demand of the considered building. SSF monthly values for the proposed system derived from the ratio of the curves in Figure 4 are depicted in the Figure 5.

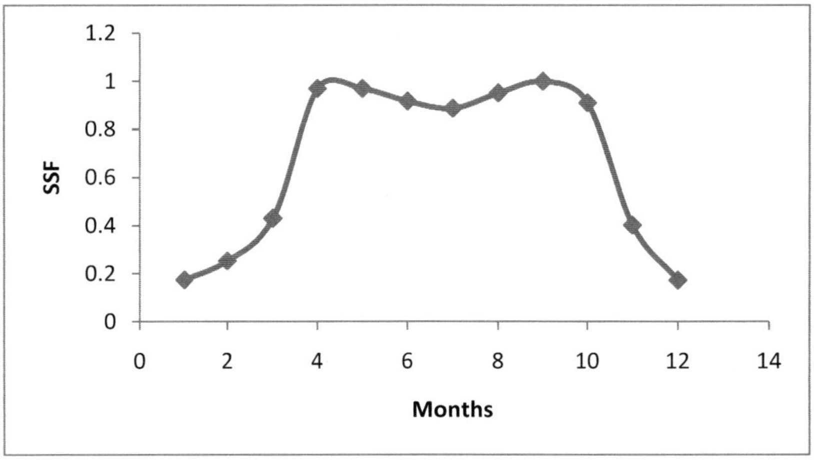

Figure 5 Fraction of solar savings

CONCLUSION

A new scheme for combined solar space heating, cooling and hot water heating has been proposed, along with the required sequence of operations. The suggested scheme allows for the full utilization of solar energy in continental climate zones. A calculation method has been proposed for determining the optimal collector area, such as one can take advantage of the maximum quantity of solar energy with minimum expense.

ACKNOWLEDGMENTS

Thanks for the support of the top board members and donors of Avedisian High School project located in the republic of Armenia as the proposed idea has been committed first time to finance and implement in the new constructed school pursuing a LEED silver certification. The research for this paper was financially supported by the Civilian Research and Development Foundation, grant No ECSP-12-01.

REFERENCES

A. Ataei, M. Khalaji Assadi, R. Parand, N. Sharee, M. Raoufinia, A. Kani, Solar combi-systems a new solution for space heating in buildings, Journal of Applied Sciences 9 (2009) 1458-1465.

Falahatkar A, Khalaji M. 2011. Analysis of solar lithium bromide-water absorption cooling system with heat pipe solar collector. World Renewable Energy Congress. 3889-3896

F. Assilzadeh, SA. Kalogirou, Y. Ali, K. Sopian, Simulation and optimization of a Libr solar absorption cooling system with evacuated tube collectors, Renewable Energy 30 (2005) 1143-1159.

H. Henning, Solar Assisted Air Conditioning in Buildings – A Handbook for Planners, Springer, 2003.

ASHRAE Handbook-HVAC Applications. 2011. Atlanta: American Society of Heating Refrigeration and Air Conditioning Engineers, Inc.

Hybrid PV Battery System for a Typical House in Lebanon Inland Region

Raghid Farhat **Ghaith Tibi** **Nesreen Ghaddar** *Member ASHRAE* **Kamel Ghali** *Member ASHRAE*

ABSTRACT

Power shortages are forcing residential sector to rely on backup diesel generators (DG). Using photovoltaic (PV) technology can be one of the clean solutions to this problem; however its adoption is challenged by its relative high capital cost. To overcome this challenge, we propose reducing the initial construction cost through the use of local low-embodied energy construction materials and use the savings resulting from envelope replacement costs to offset the PV cost. A typical house in the dry desert climate of inland Lebanon is thermally modeled using commercial software for the base case and when using local low-embodied energy construction materials, mainly hemp and straw. The savings resulting from reduced electrical demand and from the initial cost change between the base case model and the local material model were found to be \$41/m² (\$3.81/ft²). The savings are used to assess the investment in installing a hybrid PV battery system while reducing operating cost of the diesel generator. HOMER *software is used in the assessment of hybrid PV battery system while considering two scenarios. In scenario 1, we used the savings from the initial cost of envelope material and downsized DG which resulted in net energy savings up to 47% with payback period of 3 years. In scenario two, the DG replacement by PV is based on life cycle assessment in which further investment in the PV system is evaluated and optimized. This resulted in a system able to operate without need for the DG, to decrease the electricity purchased form the grid, and to increase the electricity sold to the grid with net energy savings reaching 130% with a payback period of 6 years.*

INTRODUCTION

Challenges to the electricity sector faced in several countries in the Middle East are increasing day after day due to the increase in energy consumption, lack of new projects to cover this increase, and the rapid increase of fossil fuel prices. Diesel generators are being used to cover the shortage and serve as a backup during cut-off hours. But diesel generators not only impact the environment negatively but also pose a high financial burden on the individuals. Since the energy sector has a significant role in the overall development of countries, we should look for other alternatives to cover and decrease the energy consumption. Efficient passive building design has proven to be the first step in decreasing the energy consumption (Da Graca et al, 2012). In addition, renewable energy is another possible solution where investments in renewable energy systems will not only help in covering the electricity supply shortage and decreasing the CO_2 emissions, but will also help in reducing the dependence on fossil fuels which are all imported. This can result in some security to the country in addition to the economic growth and the overall sustainable development.

Since the residential sector accounts for a high percentage of the energy consumption, for instance it reaches 47% in Lebanon, any possible consumption savings in this sector would result in improvement in the environmental conditions

Raghid Farhat and Ghaith Tibi are students in the Department of Mechanical Engineering, American University of Beirut. Nesreen Ghaddar and Kamel Ghali are professors in the Department of Mechanical Engineering, American University of Beirut.

(Ghaddar and Bsat, 1998). In this study we aim to replace conventional envelope material of a typical Lebanese passive house by local construction material to assess reduction in the initial cost of the building and investigate the soundness of investing the savings in a renewable energy system such as photovoltaic (PV) for power generation and solar water heating (SWH), thus decreasing the energy consumption while maintaining initial cost at its conventional level. Further investment in PV systems is also evaluated as potential replacement of the backup diesel generator and the associated payback period is determined.

METHODOLOGY

In this study, cheap local construction materials are used as a method to decrease the initial construction cost of a typical passive house. The typical house will serve as a base case model to which the integration of PV system and SWH will be examined. As presented in Figure 1 the PV system will be grid connected but with the use of battery storage. The three main components that home owners are concerned about include the decrease in the initial capital cost, the increase in the lifetime benefits, and minimizing of the payback period.

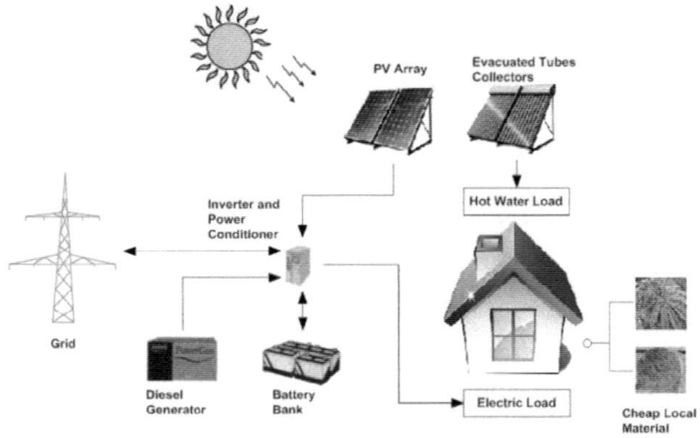

Figure 1: System Schematic

The typical passive house is selected using the work of Tibi (Tibi, 2012) to which local material will be applied to replace conventional material and reduce initial construction cost. Having this house model characteristics, eQUEST (2006) simulation is done to get the average energy consumption, daily load profile for all months, peak demand, and CO_2 emissions. The SWH size is calculated based on the water consumption of a typical family. Applicable different PV sizes with the needed number of batteries and inverter size will be estimated based on the chosen type of PV modules to be installed, the percentage of roof area available for installation of the PV modules, and the maximum electric load. Solar data, results of eQUEST, PV modules sizes, number of batteries, and inverter size are then served as input to HOMER software (2012) to study and optimize the best system based on life cycle assessment. Energy savings, initial cost, payback period, and CO_2 emissions reduction are the main indicators to evaluate the studied system.

This methodology is followed to evaluate two scenarios; the first is designing a system constrained by investing only the savings attained from decreasing the initial cost of envelope construction material and the savings from decreasing the required backup diesel generator. The second scenario designs a PV system to replace the backup diesel generator based on life cycle assessment and not just initial cost. Both scenarios have to cover the needed electric energy demand.

Typical Passive House Description

The above-mentioned methodology is applied to the typical single-family duplex passive house located in the Lebanese inland region, Bekaa area. The house has typical rectangular floor plans, as shown in Figure 2, with a footprint of

120 m² (1292 ft²), and a total built area of 240 m² (2583 ft²). The number of family members is 7. The local materials with low embodied energy and relatively high thermal storage capacity selected for the study include rammed earth, straw, and natural stone. Conventional building materials used in Lebanon include hollow blocks, polystyrene insulation, steel reinforced concrete, low conductance double skin envelope, and double glazing window. The house model is shown in Figure 2. The envelope description for the base case is as follows; the roof and slabs are built with reinforced concrete and has a U-value of 2.96 W/m².K (0.52 Btu/h.ft².°F), the walls are built using hollow blocks and plaster on both wall faces and have a U-value of 3.22 W/m².K (0.567 Btu/h.ft².°F), and the windows are single plane with aluminum frames having a U-value of 5.41 W/m².K (0.953 Btu/h.ft².°F).

In the envelope replacement scenario, the roof and slabs are made up of steel reinforced concrete of 12 cm (4.72 in) thickness that is insulated using straw boards of 5 cm (1.97 in) thickness decreasing the U-value to 0.82 W/m².K (0.144 Btu/h.ft².°F), The walls are made of hemp-reinforced rammed earth blocks of 20 cm (7.87 in) thickness on the interior, insulated using straw boards of 5 cm (1.97 in) thickness sealed with cement board on the exterior making the U-value reach 0.51 W/m².K (0.09 Btu/h.ft².°F). The rammed earth construction is a method of building walls using a balanced mixture of clay, sand, and aggregate. This method has a promising potential for future construction for it is abundant, recyclable and has a minimal environmental impact (Nowamooz and Chazallon, 2011); also, the use of natural fibers such as hemp to enhance the tensile strength of rammed earth construction proved to be effective (Li et al, 2006). The windows are double glazed with a low-e film and aluminum frame decreasing the U-value to 2.41 W/m².K (0.424 Btu/h.ft².°F). The windows facing south have operable horizontal shading elements to be used during summer. Keeping the internal loads, occupancy schedules, and infiltration rate matching the base case model, the house model was simulated for the base case and envelope replacement case.

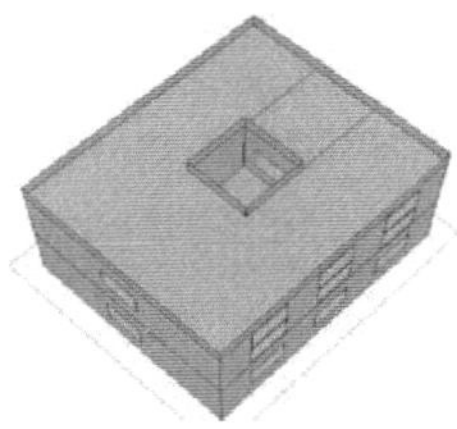

Figure 2: House Model Geometry

The evaluation was based on four parameters; namely the Construction Materials Cost, the Initial Embodied Energy, the Annual Heating & Cooling Electric Energy Consumption, and the Total Annual Electric Energy Consumption. The local material envelope case was compared to the base case accordingly. The annual embodied energy is found to be 62.1 kWh/m² ($1.9x10^4$ Btu/ft²) and 13.8 kWh/m² ($4x10^4$ Btu/ft²) for the base case with and without insulation respectively but it decreased to 12.4 kWh/m² ($3x10^3$ Btu/ft²) for the case with local insulating material. The annual electric energy consumption is 177.4 kWh/m² ($5.6x10^4$ Btu/ft²) without insulation and 125.3 kWh/m² ($3.97x10^4$ Btu/ft²) with insulation which decreased to 66.5 kWh/m² ($2.1x10^4$ Btu/ft²) for local material case model. According to the local market rates, the initial construction cost decreased by $41/m² ($3.81/ft²) when local material is used for insulating the house for the best case. These savings will be used to invest in SWH and PV. A total of $9840 savings will be the first constraint for the first scenario; the sizing of the system will depend mainly on this constraint.

The eQUEST software (2006) was used to get the hourly energy consumption of the typical house. All data was entered for the building envelope, HVAC systems, lighting, and equipment. By using ASHRAE IWEC2 Weather Files for International Locations and EnergyPlus Energy Simulation Software weather data, the weather files for Damascus were used since the weather for Bekaa area is the same as that for Damascus (ASHRAE, 2012). After simulating the house, the total energy consumption per month as well as the average daily profile through the year is predicted. The peak demand reached 6.6 kW ($2.25x10^4$ Btu/hr) during summer season due to the cooling load. The daily average consumption reached 32 kWh (10^5 Btu), and the total energy consumption for one year was found to be equal to 11,826 kWh ($4x10^7$ Btu).

Another data that was extracted from eQUEST was the annual hot water load. The domestic hot water system used in this study is considered to use gas. It was found that around 4500 kWh (1.5×10^7 Btu) is consumed yearly to heat the water. Based on 40 L/day/occupant (10.5 gal/day/occupant) (Kalogirou, 2013), the average daily usage of hot water was estimated to be around 280 L (74 gal). This value will be used to size the solar hot water system to replace the gas heaters whenever possible.

Designing the Alternative system

Domestic Hot water Sizing. The SWH sizing depends on the solar radiation available at the site, the type and efficiency of the collectors, the house hot water demand or the number of occupants, and the desired hot water temperature (DeKay, 1998). Usually the hot water use is estimated according to ASHRAE application standards (RETScreen, 2004). RETScreen (2012) is one of the softwares used to do the sizing of SHW. By using the number of occupant of 7 with an average of 65% occupancy rate, RETsreen estimated the average daily use of around 273 L (72 gal) which in accordance of the estimated value in eQUEST with an error of only 2.5%. The base case was modeled as per the results of eQUEST, and the proposed case consisted of two SWH collectors of 2.8 kW (9.56×10^3 Btu/hr) capacity with a total storage capacity of 400 L (106 gal) at a cost of $2000 as per current market prices. The area required to install these solar collectors is around 7 m^2 (75.35 ft^2). The proposed system resulted in savings of 80% and the financial analysis estimated around 6.5 years to payback the invested amount in the SWH system. This was based on the current market rate of $1.2/Kg ($0.55/lbs) of propane gas (Nzeraan, 2014) which is usually used in gas.

PV Sizing. To predict the maximum size of PV modules that can be used to cover the available roof area, the following equation is used;

$$P_{dc} = A * \eta_{PV\ cells} * \frac{1kW}{m^2} \text{ of insolation} \qquad (1)$$

Where P_{dc} is the DC rating power of the PV modules, A is the total area of these modules, and 1kW/m^2 (3.4×10^3 Btu/hr) is the maximum solar insolation that can be reached at the peak sun hours.

If considering one third of the roof area to be partially filled with cooling systems or electrical and mechanical rooms, SWHs, and the remaining to be the space lost between the PV modules, the area left that can be filled with PV modules is equal to 80 m^2 (861 ft^2) . Since the efficiency of the PV cells available in the market ranges between 7 to 18% as mentioned earlier, a 15% efficient module under standard test conditions (STC) is considered for this study. Due to the fact that PV efficiency drops by 0.5% for each one degree Celsius increase above STC temperature of 25°C (77°F) (Bergene and Lovvik, 1995), this considered 15% efficiency will slightly decrease during summer time, in July and August specifically, where it might decrease to around 14.5% if considering an average temperature of 30°C (86°F) during this period of the year. Thus, the maximum possible size of the PV modules to be considered is 12 kW (4.1×10^4 Btu/hr). However, to reach the best optimized system, a range of PV sizes from 0 kW (0 Btu/hr) to 12 kW (4.1×10^4 Btu/hr) is considered. The selected size will depend on the initial construction cost savings in scenario 1 and on DG elimination for Scenario 2. For the different scenarios studied, the total energy produced for the best optimized cases are presented later in the results.

Batteries and inverter are also needed for the PV system. The batteries are sized in a way to cover the intermittent nature of solar energy. For this study, batteries of 12V rated capacity with a maximum storage capacity that ranges between 200 and 500 Ah will be considered. To get rid of the diesel generator, the maximum capacity required to cover the demand needs is predicted to be around 24 batteries. However, to have the best optimized system for both scenarios, a range of batteries number from 0 to 28 is considered. To size the inverter, the total Watt of all appliances operating at the same times should be considered. For safety, the inverter should be considered at least of 30% bigger size (Sharma at al, 1995). Since our peak demand reaches 6.6 kW (2.25×10^4 Btu/hr), the maximum inverter size needed is predicted to be around 8.5

kW (2.9 x 10^4 Btu/hr) if the diesel generator will be eliminated. However, to have the best optimized system for both scenarios, a range of inverter sizes from 0 (0 Btu/hr) to 10 kW (3.4 x 10^4 Btu/hr) is considered.

Size of the Generator. In a typical house in Lebanon, people are relying either on personal small home diesel generators or on monthly subscription to private companies that sell electricity being generated from large diesel generators. Based on the highest demand of 6.6 kW (2.25 x 10^4 Btu/hr), 8 kVA (6.8 kW) (2.32 x 10^4 Btu/hr) rated generator is assumed to be used to cover the utility cut off hours. So, the base case study is a 6.8 kW (2.32 x 10^4 Btu/hr) diesel generator used with the utility grid to cover the electricity demand. A range of diesel generators with lower capacities are then considered to get the best size, if still needed, to assist the PV panels to cover the electricity shortage based on cost optimization results.

Hybrid PV System Sizing and Selection Model

To study the feasibility of installing a hybrid PV-battery system in a typical house, HOMER (2012) software was used. Its sensitivity and optimization analysis algorithms help in evaluating the economic and technical feasibility of a different technology options while accounting for variations in technology costs and energy resource availability. Since it was validated by Bekle et al (2010), Bludszuweit et al (2006), Jamil et al (2012), Shaahid et al (2007), and many others as a good tool to access and analyze the importance of using different renewable energy sources, HOMER was used in this study to access the importance of PV.

By using the hourly solar radiation data for the whole year, hourly load profile as resulted in eQUEST, PV, diesel generator (DG), batteries, converter, and grid, the model for the hybrid PV-grid system for the house is as shown in Figure. 3. The optimization parameters are the PV size, the generator size, the number of batteries, and the converter size. 0 to 12 kW (0 to 4.1 x 10^4 Btu/hr) PV sizes, 4 to 6.8 kW (1.3x10^4 to 2.32x10^4 Btu/hr) DG, 0 to 24 batteries of 12V 200Ah, 0 to 10 kW (0 to 3.4x10^4 Btu/hr) converter are the variable ranges used for each of these parameters. The schedule of operation for the grid has 4 consecutive cutoff hours at night and 4 consecutive cutoff hours during the day randomly alternating through months. All component costs used in the model are based on current market prices. PV module prices are dramatically going down as per Swanson (2006) who predicted that the module price will reach \$1.44/W (\$0.42/ Btu/hr) in 2013 which is in accordance with what we see currently in the market. So the cost of 1kW (3.4x10^3 Btu/hr) of PV is assumed \$1500, and the cost of 6.8 kW (2.32x10^4 Btu/hr), 5kW (1.7x10^4 Btu/hr), & 4 kW (1.36x10^4 Btu/hr) DG are assumed \$4000, \$3200, and \$3000 respectively. The cost of each 1kWh (3.4x10^3 Btu) being purchased from the grid and sold to the grid is equal to 0.1\$.

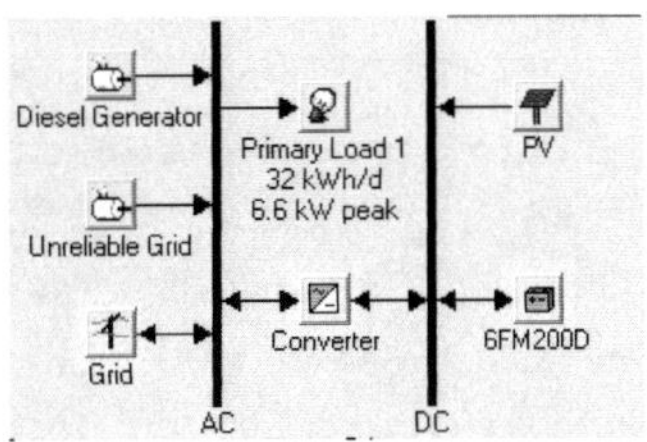

Figure 3: HOMER Model

HOMER simulated these different configurations and scaled them according to their total lifetime installation and operational cost which is also known as net present cost (NPC). First it estimated if the proposed system can serve the electrical load, then it estimated the system's NPC. The model was done over one operational year based on an hourly time-step simulations. The full year hourly load data obtained from eQUEST served as the input to HOMER. One of the main calculations that are hourly performed is the PV-battery supply which was compared to the required electrical load. If this supply satisfies the load, the excess will be sold to the grid. If the supply does not satisfy the load, the shortage will be covered by either the grid or the diesel generator.

Scenario 1. Choosing a system by only using the savings attained from decreasing the initial cost of envelope construction material plus the savings of decreasing the required backup diesel generator is the first scenario studied and optimized. The cost function is represented by:

$$C_{PV,capital} + C_{b,capital} + C_{I,capital} = Sav_E - C_{SWH,capital} + Sav_{DG,capital} \quad (2)$$

Where $C_{PV,capital}$, $C_{b,capital}$, $C_{I,capital}$, $C_{SWH,capital}$, $C_{DG,capital}$ are the PV, batteries, invertor, SWH, and diesel generator capital cost respectively. Sav_E is the total savings in envelope material, $Sav_{DG,capital}$ is the savings in decreasing the generator capital cost.

So the most feasible combination between the backup generator, size of the PV, batteries, and converter for scenario 1 is constrained by the initial cost. The best optimized case should have a capital cost not exceeding the cost of the DG plus the savings of building material ($9840) minus the cost of the SWH system ($2000) plus the savings due to decreasing the generator size. This capital cost should not be much less than the constraint to benefit as much as possible from the savings attained. So, the initial capital cost should be less than or equal to the remaining savings ($7840) plus the DG cost ($4000) as in the base case but not less than or equal to $11400. After meeting the initial cost constraint, the best optimized case should also cover the needed electric consumption at the lower possible NPC. This would be based on life cycle assessment; that is decreasing the total life cycle cost of the system components.

Scenario 2. The second scenario will evaluate and optimize further investment in the PV system to replace the backup diesel generator based on life cycle assessment; that is its main objective is to decrease the total life cycle cost of the DG free system. So the most feasible combination between the backup generator, size of the PV, batteries, and converter for scenario 2 is constrained by the DG elimination. The best optimized case should have the lowest NPC, cover the needed electric load, and be DG free.

RESULTS AND DISCUSSION

Scenario 1

For scenario 1, the best optimized case had an initial capital of $11,400 and consisted of a 2 kW (6.8×10^3 Btu/hr) PV, a 4 kW (1.36×10^4 Btu/hr) DG, 12 batteries, and a 2 kW (6.8×10^3 Btu/hr) converter in addition to the grid. The yearly energy production of this system exceeded the needed electric load of 11826 kWh (4×10^7 Btu) and thus could have some grid sales of 5% of the total consumption. The PV system energy production reached yearly 3,842 kWh (1.3×10^7 Btu) covering 3,491 kWh (1.2×10^7 Btu) of the total electric load as shown in figure 4. So, the net energy savings reached 35%.

Comparison with Base Case. To check how profitable investing the envelope savings in the PV system, the comparison with the base case is done. The base case is the 6.8 kW (2.32×10^4 Btu/hr) DG with the grid. The return on investment is around 37% and the simple payback period is around three years. The total savings from covering a 30% of the total electric load and from selling to the grid reached around $2,556 for each year.

System Performance. To study the system performance, four months are chosen to represent the four seasons. During January and July, and due to heating and cooling load respectively, the dependence on the DG during the cutoff hours is still high. However, during April and October, the PV output with the batteries storage almost eliminated the need to use the DG. The percentage of the daily load covered by the PV and battery storage reached 11% during January, 57% during April, 27% during July, and 58% during October. The maximum attained PV power reached 1.8 kW (6.15×10^3 Btu/hr) during October but with so close values for April and July. The daily PV output profile for the four seasons does not differ much, thus the reliability of the PV system can be considered similar all year round; it is not seasonal based. However, the penetration of the PV and the ability to cover the required load in all months depend on the required load itself. Here rises the need to have larger PV system to accommodate for the peak load and thus get rid of the DG.

Scenario 2

For scenario 2, the best optimized case is the DG free case with the lower NPC. In addition to the grid, the best case components consisted of a 12 kW ($4.1x10^4$ Btu/hr) PV, 28 batteries, and a 9 kW ($3.07x10^4$ Btu/hr) converter. The initial cost of this system is $30,600. The PV system energy production reached yearly 23,052 kWh ($7.87x10^7$ Btu) covering 6,100 kWh ($2.08x10^7$ Btu) of the total electric load (as shown in figure 4) and selling 11,117 kWh ($3.79x10^7$ Btu) to the grid thus allowing for net energy savings of 140%.

Comparison with Base Case. Same as in Scenario 1, the comparison with the base case is done to convey the cost effectiveness of this system. A payback period of around 6 years is an important indicator that can encourage home owners to cover their roofs with PV panels. The total savings reached around $4,095 for each year.

System Performance. The percentage of the daily load covered by the PV and battery storage reached 45% during January, 66% during April, 62% during July, and 64% during October. Most of the PV energy output is sold to the grid where these sales reached 82% of the PV output during the peak hours due to the fact that during this time the house load is minimal.

The grid sales are not only important to home owners but also to the grid itself where it will be able to cover more of the needed load for other houses or non-residential sector. For instance; the grid is also supplying companies that have their peak load mainly between 10 a.m. and 4 p.m., so the excess in energy generated by the house with the PV can be used to assist the grid in covering the peak load for this company. If more houses invest is this system, they will make the grid stronger and increase its ability to cover more demand especially during the day. This will affect the Lebanese energy sector by having more reliable electric supply.

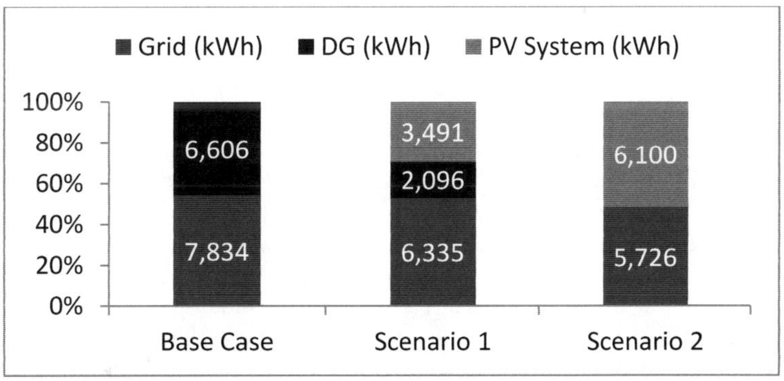

Figure 4: Electric Load Percentages covered by the Grid, DG, and PV system for the Base case, Scenario 1, and Scenario 2

Table 1: Summary of Important indicators for investing in the PV system and SWH for both Scenario 1 and Scenario 2

PV system & SWH	Scenario 1	Scenario 2
Net annual Energy Savings	47%	135%
Initial Cost	$9,400	$32,600
Payback Period (years)	3	6

	Annual Savings	$2,933	$4,472

CONCLUSION

In the view of the above, the crucial role the solar energy can have in solving the electricity problem in Lebanon and countries with similar climate is proved. Integrating a hybrid PV system and a SWH heater in houses will generate not only direct profit to house owners but also to the whole society and energy sector. By using savings attained from integrating cheap local material for a passive house design of $41/m^2 ($3.81/ft^2) to invest in these renewable systems will result in further savings that can each up $2,933 each year from both covering part of the electric load, hot water, and selling to the grid as shown in Table 1. However, if increasing the size of this system to eliminate the DG, the total savings can reach around $4,472 for each year.

That being so, the crucial role the solar energy can have in solving the electricity problem in Lebanon is proved. Integrating a hybrid PV system and a SWH heater in houses will generate not only direct profit to house owners but also to the whole society and energy sector. The being so, the government should stress more in this area and provide further incentives to both the end users and suppliers. The end users will be driven by low interest loans, low prices for PV and SWH, experienced suppliers and installers, and high prices of each 1 kWh (3.4×10^3 Btu) being sold by them to the grid.

NOMENCLATURE

A	=	Area	[m^2]
P	=	Power	[kW]
U-value	=	Thermal Transmittance	[W/m^2.K]
η	=	Efficiency	[%]

Subscripts

dc	=	Direct Current

REFERENCES

ASHRAE, ASHRAE IWEC2 Weather Files for International Locations . Moraga, CA : White Box Technologies, 2012

Bekele, G., Palm, B., 2010. Feasibility study for a standalone solar–wind-based hybrid energy system for application in Ethiopia. Applied Energy, 87(2), 487-495.

Bergene, T., Løvvik, O.M., 1995. Model calculations on a flat-plate solar heat collector with integrated solar cells. Solar Energy, 55(6), 453-462.

Bludszuweit, H., Domínguez, J. A., Bernal, J.L., 2006. Pre-feasibility study of a grid connected wind-PV hybrid system with energy storage and power prediction. Renewable Energy & Power Quality Journal, 4, 400-401.

Da Graça, G., Augusto, A., Lerer, M. 2012. Solar powered net zero energy houses for southern Europe: Feasibility study. Solar Energy, 86(1), 634-646.

DeKay, M., 1998. Translating ASHRAE for architects: Sizing solar hot water systems and spacing rooftop collectors in preliminary design. In proceedings of the 23rd National Passive Solar Conference, 14-17 June 1998, Albuquerque, New Mexico. Boulder, CO: American Solar Energy Society, 23, 271-277.

eQUEST: the QUick Energy Simulation Tool. Lab, Lawrence Berkeley National, 2006.

Ghaddar, N., Bsat, A., 1998. Energy conservation of residential buildings in Beirut. International Journal of Energy Research, 22, 523–546.

HOMER Software, Laboratory National Renewable Energy. Colorado, USA, 2012.

Jamil, M., Kirmani, S., Rizwan, M., 2012. Techno-Economic Feasibility Analysis of Solar Photovoltaic Power Generation: A Review. Smart Grid & Renewable Energy, 3(4), pp

Kalogirou, S., 2013. A. Solar Energy Engineering: Processes and Systems, 2nd Edition. Academic Press, New York (2013).

Li, Z., Wang, X., Wang, L., 2006. Properties of hemp fibre reinforced concrete composites. Composites Part A: Applied Science and Manufacturing, 37(3), 497-505.

Nowamooz, H., Chazallon, C., 2011. Finite element modelling of a rammed earth wall. Construction and Building Materials, 25(4), 2112-2121.

Nzeraan, E., Announcement Number 88: Determining the selling price of liquid fuels. s.l., Lebanon : Lebanese Ministry of Energy and Oil, Directorate General of Oil, (2014).

RETSCREEN, Minister of Natural Resources. Canada, (2012).

RETScreen®, International. Solar Water Heating Project Analysis (2004).

Shaahid, S.M., Elhadidy, M.A., 2007. Technical and economic assessment of grid-independent hybrid photovoltaic–diesel–battery power systems for commercial loads in desert environments. Renewable and Sustainable Energy Reviews, 11(8), 1794-1810.

Sharma, V.K., Colangelo, A., Spagna. G., 1995. Photovoltaic technology: basic concepts, sizing of a stand-alone photovoltaic system for domestic applications and preliminary economic analysis, Energy Conversion and Management, 36(3), 161-174.

Swanson, R., 2006. A Vision for Crystalline Silicon Photovoltaics. Progress in Photovoltaics: Research and Applications, 14(5), 443–453.

Tibi, G. 2012. Sustainable design guidelines for detached housing in the Lebanese inland region. International Journal of Sustainable Built Environment, 1(2), 177–193.

Princess Nora Bint Abdul Rahman University (PNU) Solar System

Houssam Chebaro, PhD **Wissam Jarmak**

ABSTRACT

The Princess Nora Bint Abdul Rahman University (PNU) is designed with the goal of employing LEED strategies focusing on utilizing sustainable designs to reduce and improve the efficiencies, of energy consumptions, and reducing CO2 emissions (Green design). One such measure towards achieving this goal is the use of solar energy as an obvious measure particularly in the KSA where solar energy is abundantly available over most of the year. Using the sun as a heating source for space heating and domestic hot water is a perfectly logical sustainable design solution. The principle of using a solar water heating is simple and the technology is both well-known and reliable.

The PNU is designed with a central district water heating system that serves space heating and domestic hot water. At the heart of the central heating system is a boiler plant having a heating capacity to cover all the space heating and domestic water, i.e. to cater for 100% of the demand, especially for cloudy winter days and at winter night periods when the heating demand is at maximum and where there will not be any available solar energy. However, during winter non-cloudy days, and as well as the remaining days of the year, there will be abundant solar energy to harvest and use for heating, and for this reason, the PNU is provided with a central solar heating system .

This solar heating system relies on highly reliable large flat plat collector technology installed on the roof of the logistic center located nearby the boiler plant. The system has a peak power yield of 17.3 MW (59,030 MBtu/h), and an annual yield of 26,011 MWH (88,753,216 MBtu)

This paper will focus on the design of PNU solar system particulars (solar panels, circulating pumps, piping, buffer tanks, heat exchangers, control system, etc..) and the system engineering that took place and evolved in selecting the flat plate collector technology over other technologies such as compound parabolic collectors and parabolic trough collectors. Further, the paper will focus on the design and engineer measures that are implemented to protect the PNU solar system from overheating, damage or failure.

Houssam Chebaro is a Senior Associate at Dar Al-Handasah Consultants (Shair and Partners)
Wissam Jarmak is a Mechnical Engineer at Dar Al-Handasah Consultants (Shair and Partners)

INTRODUCTION

Several technologies to harvest the power of the sun and transform it to a useful form of energy exist. The greatest advantage of solar energy is that it is clean and doesn't cause any environmental pollution. In this paper we are interested in Solar Thermal Systems. The major part of all solar systems is the solar collector. It's a heat exchanger that absorbs solar radiation and converts it to heat. The following types of solar collectors are considered: Flat plate collectors (FPC), Compound Parabolic Collectors (CPC) and Parabolic Trough Collectors (PTC).

PROJECT CHARACTERISTICS

Princess Nora Bint Abdul Rahman University is one of the largest universities in the world with a capacity of 40,000 students. The project aimed to implement LEED strategies by using renewable energies and improve energy consumptions and efficiencies. It was obvious to use solar energy since it's available abundantly in Saudi Arabia. Particularly the use of Centralized Solar Water heating system was adopted in the domestic hot water and space heating design.

This paper study focuses on the criteria and analyses in ultimately adopting the large flat plat collector technology at PNU. During the course of developing the PNU solar design, several solar collection technologies were considered and studied.

The paper presents a comparison of the energy that can be harvested by the three technologies and in addition to the technical pros and cons for each of the technology, and the main reasons behind selecting the flat plate technology.

Lastly, the paper addresses the means employed to protect the solar collector system against overheating during low hot water demand (low energy usage).

DESIGN SPECIFICATION

The system is designed to maximize the solar winter energy yield for the university's space heating and domestic hot water requirements, in accordance with the following specifications:

1. Peak system power output of 17.3 MW (59,030 MBtu/h) to meet domestic hot water demand.
2. Preheat the return water from the boiler under normal operating conditions for space heating.
3. Buffer tanks included to improve system performance in relation to peak capacity and to address load demand variation throughout the day, and to provide heating during night period.
4. Overheat protection features and redundancies included.

In the months where space heating will not be required, the solar system would be able to provide energy required for domestic hot water supply subject to having enough heat storage to address night time demand, where in these months the boiler will be shut down.

SOLAR WATER HEATING CALCULATIONS

Peak Power and Energy Yield

Solar radiation varies through the day, from none before sunrise to peak at noon then to none after sunset, Solar Collectors modules output varies during the day as well. When assessing the power output of a Solar Collector Plant the term Mega Watt Peak (in Watts or Btu/hour) is used, it is assessed in specific test conditions and represents a name plate capacity only. The term of interest is the annual energy yield (in Watt-Hours or Btu), which represents the cumulative energy that the plant produces and supplies during the year, regardless of the instantaneous power delivered at a certain moment.

Types of radiation

Sunlight on the earth's surface has two portions, one that is direct from the sun through sun rays and the other is diffusion which is the portion of radiation is scattered and reflected from different particles in the atmosphere. Global Radiation is the sum of the direct radiation and diffuse radiation on a horizontal surface.

Solar Altitude and Azimuth

To be able to track the sun's path two angles are used which are the azimuth angle, and the elevation angle.

The azimuth defines the sun's position with respect to the earth on the horizontal plane, while the elevation defines the angle that the sun makes with the horizontal. A high angle is when the sun seems to be high in the sky, a low angle means that the sun is close to the horizon.

The shortest route through the atmosphere that a sun ray takes to reach the earth is when the solar altitude (elevation) is perpendicular to the earth surface. For flatter angles (sun is next to the horizon) the route becomes significantly larger, thus requiring the rays to go through a longer route which will increase scattering and absorption. The result is less radiation intensity.

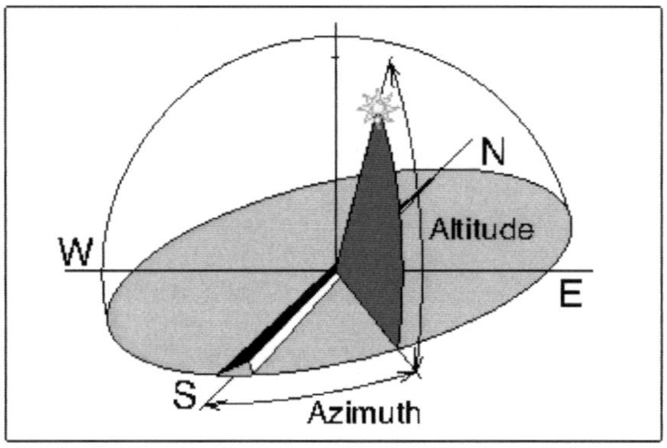

Figure 1 - ***Solar Azimuth and Altitude [5]***

Estimated Monthly Average Daily Solar Radiation on Collectors Tilted Surface

Solar radiation data in Riyadh can be collected by many ways like using a software, database, measured data or can be calculated. ASHRAE IWEC2 Weather files [4] were used in order to be more accurate in the calculations, shown in Table 1.

The monthly average daily solar radiation on collector's tilted surface can be estimated with equation (A) from that collected on horizontal surface according to Isotropic diffuse model assumption, by assuming that diffuse and ground reflected radiation are each assumed to be isotropic. Then the radiation on the tilted surface was considered to include three components: beam, isotropic diffuse and solar radiation diffusely reflected from the ground.

$$H_T = H_b R_b + H_d\left(\frac{1+\cos(\beta)}{2}\right) + H\rho_g\left(\frac{1-\cos(\beta)}{2}\right) \quad \text{(A)}$$ "Solar engineering of thermal Processes, Second Edition"

Given the fact that Parabolic Trough Collector tracks the sun, the monthly average radiation on its surface is equivalent to the average radiation on a surface having the hourly solar altitude as an inclination. Having the collected data and using equation (A) the monthly average radiation on Parabolic Trough Collector's surface was calculated. Results of solar radiation on different configuration of collectors are shown in Table 1.

Table 1. Monthly Average Daily Radiation

Month	Horizontal surface (kWh/m^2 or Btu/ft^2)	On tilted surface (45°) (kWh/m^2 or Btu/ft^2)	On Parabolic Trough (variable) (kWh/m^2 or Btu/ft^2)
Jan	4.28/1,357	6.23/1,975	5.9/1,870
Feb	5.06/1,604	6.3/1,997	5.87/1,861
Mar	5.77/1,829	6.02/1,908	5.83/1,848
Apr	6.41/2,032	5.56/1,763	5.89/1,867
May	7.34/2,327	5.48/1,737	7.01/2,222
Jun	8.03/2,545	5.5/1,743	8.53/2,704
July	7.82/2,479	5.55/1,759	7.91/2,507
Aug	7.43/2,355	6.02/1,908	7.63/2,419
Sep	6.89/2,184	6.74/2,137	7.52/2,384
Oct	5.99/1,899	7.23/2,292	7.7/2,441
Nov	4.71/1,493	6.71/2,127	5.87/1,861
Dec	3.60/1,141	5.27/1,671	4.79/1,518
Average	**6.11/1,937**	**6.05/1,918**	**6.71/2,127**

Collector Field Layout

The space used for the solar plant is a warehouse's roof having an area of 60,000 m^2 (645,834 ft^2). Taking into consideration a space for corridors used for access and maintenance the remaining area for solar collectors is 44,000 m^2 (473,612 ft^2).

Figure 2 *-Solar Collectors Plant*

To ensure optimal selection of collectors' technology that meet peak and yield performance requirements, an in depth analysis of FPC, CPC and PTC collectors was conducted (table 2). Each of the selected collectors presents various challenges and opportunities in delivering on the required peak and winter yield conditions. Each collector was analyzed and its optimum configuration selected to meet the specification requirement.

A tilt angle of 45 degrees to gain the best performance in winter when the sun elevation is minimal, and spacing between collector rows, to ensure minimum shading losses, of 0.8 meters (2.62 feet) for CPC and 1 meters (3.28 feet) for Flat Plates and Parabolic Trough Collectors aligned to the axis of the building were considered. Manufacturer's requirements were taken into account in these layouts with regards to the maximum number of collectors that can be mounted in series.

Table 2. Different Plant Configuration

Type and Model	Large Flat Plate GOT GK3100[6]	Compound Parabolic Collectors CPC 1518 Paradigma[7]	Parabolic Trough Collectors PolyTrough 1200[8]
Module Efficiency	50 %	51 %	72 %
Perfomance Ratio	0.8	0.8	0.8
Length (mm or ft)	2064/6.77	1640/5.38	26019/85.36
Width (mm or ft)	4867/15.96	2080/6.82	1211/3.97
Collectors aperture area (m^2 or ft^2)	9.43/101.5	3.00/32.29	28.8/310
Area of plant (m^2 or ft^2)	44,000/473,612	44,000/473,612	44,000/473,612
Spacing between collector rows (m or ft)	1/3.28	0.8/2.62	1/3.28
Tilt angle	45°	45°	Variable
Number of panels used	3,675	10,794	753
Average Shading Effect	85%	81%	86%
Total collector aperture area (m^2 or ft^2)	34,655/373,023	32,382/348,557	22,003/236,838

Calculation Method

In order to calculate the total energy output of the system the following formula was used:

Monthly Yield (kWh or Btu) = Total Collector Aperture Area (m^2 or ft^2) x Monthly Average Daily radiation on tilted surface (kWh/m^2 or Btu/ft^2) x Number of days in each month x Collector's Efficiency (%) x Performance Ratio (%) x Shading Effect (%) (B)

The peak power output of the system with maximum power of Wp (Btup/h) is given by:

Peak power (Wp or Btup/h) = Total Collector Aperture Area (m^2 or ft^2) x Solar panel Efficiency (%) x 1000 (W/m^2 or Btu/h.ft^2) (C)

Table 3 summarizes three configurations of collectors fitted on the warehouse's roof.

Table 3. Calculations Results

Type and Model	Large Flat Plate GOT GK3100[6]	Compound Parabolic Collectors 1518 Paradigma[7]	Parabolic Trough Collectors PolyTrough 1200[8]
Month	**Yield (MWh or MBtu)**	**Yield (MWh or MBtu)**	**Yield (MWh or MBtu)**
January	1,882/6,422,678	1,669/5,695,752	1,729/5,901,009
February	2,100/7,167,183	1,838/6,270,418	1,816/6,197,456
March	2,296/7,835,113	2,105/7,181,439	2,065/7,046,032
April	2,055/7,011,951	1,912/6,525,561	2,016/6,879,539
May	2,148/7,328,250	1,996/6,810,764	2,543/8,676,974
June	2,101/7,169,043	1,952/6,660,767	3,018/10,296,291
July	2,195/7,489,003	2,041/6,962,871	2,898/9,886,865
August	2,322/7,922,942	2,160/7,369,445	2,729/9,310,114
September	2,535/8,650,891	2,350/8,019,382	2,643/9,016,789
October	2,613/8,916,537	2,370/8,088,158	2,614/8,919,819

November	2,204/7,518,927	1,912/6,524,506	1,850/6,310,756
December	1,560/5,322,173	1,379/4,704,077	1,366/4,660,504
Annual Yield (MWh or MBtu)	**26,011/88,753,216**	**23,684/80,813,162**	**27,286/93,103,696**
Peak Power (MWp or MBtu/h)	**17.3/59,030**	**16.5/56,300**	**15.9/54,253**
Yield/Area (Kwh/m² or MBtu/ft²)	591/187	538/170	625/196

- Parabolic trough collectors have the highest Yield per square meter and results show that they can harvest the highest amout of energy yearly. However trough collectors are still very expensive and require a complicated control and sun tracking mechanism. So they are mainly used in industrial applications and not recommended in domestic hot water heating applications.
- Flat plates have lower energy yield per m^2 (ft^2) than parabolic trough collectors have. However this difference can be reduced by being able to employ more collectors for the same area thus harvesting a similar amout of energy yearly. In addition to their high energy yield, flat plates don't require a complex maintenance and cleaning procedure and they are heavy duty collectors.
- Compound parabolic collectors have a considerable energy yield, however due to their complex shape they are difficult to maintain particulary difficult to clean.

All these technical and qualitative assessments results show that large flat plates are more appealing in term of energy yield and maintenance simplicity; therefore encourage their adoption for the PNU solar hot water system.

OVERHEATING AND HEAT DISSIPATION

A key challenge in applying a fixed collector solar field is the ability to dissipate heat when there is no load demand. Heat dissipation measures are taken in design and operation of the project to ensure there is no overheating which would lead to having the flat plates reaching their stagnation temperature and the occurance of steam in the system, which require a tedious process of emptying and refilling the system to remove the steam.

The following heat dissipation protection measures were taken:

1. Cloth covers for the seasonal change-over and in anticipation of prolonged holidays where the solar collector field will be covered to shut off.
2. Pressure relief valves are open when the pressure rises in the system caused by overheating.
3. Installation of insulated buffer tanks. It is used to store heat for night time and non-sunny hours use.
4. Dry air coolers to reduce temperature for the whole system capacity, at 17.3 MWp (59,030 MBtu/h) requirements. The coolers operate automatically, fixed in line with the system. This measure is for emergency protection – as a measure of last resort – and when there is not sufficient time for taking the first measure, covering the collectors.

NOMENCLARURE

H_T = Estimated Monthly Average Daily Solar Radiation on Collectors Tilted Surface
H = The monthly average daily solar radiation on a horizontal surface
H_b = The monthly average daily beam radiation
H_d = The monthly average daily diffuse radiation
R_b = Ratio of the monthly average daily beam radiation on the tilted surface to that on a horizontal surface
β = Collectors Tilt Angle

ϱ_g = Ground reflectivity (albedo), which is equal to 0.2 when the monthly average temperature above 0°C (273°K) (natural ground) and 0.7 when the monthly average temperature below -5°C (268°K) (Snow ground)

REFERENCES

[1] ASHRAE 2011, Solar energy use, ASHRAE Application
[2] Soteris A. Kalogirou 2004, Solar thermal collectors and applications, Elsevier
[3] John A. Duffie and William A. Beckman 1991, Solar engineering of thermal Processes, Second Edition
[4] White Box Technologies, Moraga, California, Y. Joe Huang, Principal Investigator
[5] www.new-learn.info/packages/daymedia/axel/sky/sunpath_calc.html
[6] www.solfex.co.uk/Product/120-gk-3000-series/
[7] www.linuo-ritter-international.com/products/evacuated-tube-collectors/pressurized/
[8] www.nep-solar.com/products/polytrough-1200/